ENGINEERS OF HAPPY LAND

EDITORS

Sherry B. Ortner, Nicholas B. Dirks, Geoff Eley

PRINCETON STUDIES IN

CULTURE / POWER / HISTORY

ENGINEERS

OF

HAPPY LAND

TECHNOLOGY AND NATIONALISM
IN A COLONY

Rudolf Mrázek

PRINCETON UNIVERSITY PRESS

PRINCETON AND OXFORD

LIBRARY OF CONGRESS CATALOGING-IN-PUBLICATION DATA

MRÁZEK, RUDOLF.
ENGINEERS OF HAPPY LAND : TECHNOLOGY AND NATIONALISM IN
A COLONY / RUDOLF MRÁZEK.
P. CM. — (PRINCETON STUDIES IN CULTURE/POWER/HISTORY)
INCLUDES BIBLIOGRAPHICAL REFERENCES AND INDEX.
ISBN 0-691-09161-7 (ALK. PAPER) — ISBN 0-691-09162-5 (PBK. : ALK. PAPER)
1. INDONESIA—CIVILIZATION—20TH CENTURY. 2. TECHNOLOGY—SOCIAL
ASPECTS—INDONESIA. 3. NATIONALISM AND TECHNOLOGY—
INDONESIA. 4. NATIONALISM—INDONESIA—HISTORY—20TH
CENTURY. I. TITLE. II. SERIES.

DS643 .M73 2002
959.803–DC21 2001036860

BRITISH LIBRARY CATALOGING-IN-PUBLICATION DATA IS AVAILABLE
THIS BOOK HAS BEEN COMPOSED IN ACASLON
PRINTED ON ACID-FREE PAPER. ∞
WWW.PUPRESS.PRINCETON.EDU
PRINTED IN THE UNITED STATES OF AMERICA
1 3 5 7 9 10 8 6 4 2

For Jana

CONTENTS

ILLUSTRATIONS

ACKNOWLEDGMENTS

THROUGHOUT THE WRITING of this book, I was supported by the environment at the University of Michigan in Ann Arbor. The Fulbright-Hayes and Social Science Research Council Fellowships allowed me to spend a year of research in the Netherlands and in Indonesia in 1993–1994. The Michigan Humanities Award gave me the most significant six-month time for writing in 1996–1997.

In Mary Murrell, I met an inspiring editor. A community of colleagues and friends read the manuscript, arranged for me to speak at their seminars, or encouraged me otherwise: Oliver Wolters at Cornell, Pramoedya Ananta Toer, Goenawan Mohamad, and Taufik Abdullah in Jakarta, Geoffrey Eley, Raymond Grew, and Alton Becker at Michigan, Joseph Errington and James Scott at Yale, Mary Steedly at Harvard, Paul Israel at MIT, Anne Booth and John Sidel in London, Bernard Dahm in Passau, Kees van Dijk in Leiden, Wim Wertheim and Henk Schulte-Nordholt in Amsterdam, John Legge at Monash, and Jan Mrázek wherever I caught up with him.

Three friends, however—very long before the book was conceived and through the gestation—have been most special to me. Without George Kahin, Benedict Anderson, and James Siegel I would not have made it. Because of their scholarship and their immediate presence, I never (too seriously) regretted that I had decided to study Indonesia and to stay in America, of all places.

About half of an earlier version of chapter 3 was published in Vicente Rafael, ed., *Figures of Criminality in Indonesia, the Philippines, and Colonial Vietnam* (Ithaca: Cornell Studies in Southeast Asia, 1999). An early version of chapter 4 was published in Henk Schulte Nordholt, ed., *Outward Appearance: Dressing, State, and Society in Indonesia* (Leiden: Royal Institute for Linguistics and Ethnography, 1997). An earlier version of chapter 5 appeared in *Comparative Studies of Society and History*, vol. 39, no. 1 (1997). An earlier version of the epilogue can be found in *Indonesia* (Ithaca: Cornell Southeast Asia Program), vol. 61 (April 1996).

ARS in New York for the estate of Alexander Calder, MAGNUM PHOTOS in Paris for the estate of Henri Cartier-Bresson, Niels Douwes-Dekker archives in Ithaca, New York, the Historical Documentation Center of the Royal Institute of Linguistics and Anthropology (KITLV) in Leiden, the Tillema Archives at the Royal Museum of Ethnography (RMV) in Leiden, and the Photo-Archives of the Royal Tropical Institute (KIT) in Amsterdam, graciously gave me permission to use images from their collections.

PREFACE

And so it is with our own past. It is a labor in vain to attempt
to recapture it: all the efforts of our intellect must prove futile.
The past is hidden somewhere outside the realm, beyond the
reach of intellect, in some material object (in the sensation
which that material object will give us) of which we have no
inkling. And it depends on chance whether or not we come
upon this object before we ourselves must die.
—Proust, *Swann's Way*[1]

THIS BOOK suggests an alternative way to study twentieth-century culture, identity, and nation.

The *place* is the Indonesian archipelago, the home of multiple civilizations, and, through a greater part of the modern era, the principal colony of the Netherlands. The rich carpet of Indonesian ways of life, created by the incessant culture invasions and innumerable combinations of resistance, is one of my reasons for choosing this place for study. The complexity of the invasions of the Dutch way of life into the Indies is the other. In modern memory, there have been few events as momentous as the decline of the superpower Dutch culture of the seventeenth century to the small-country ways of the nineteenth and twentieth centuries. This decline included the clinging to the Indies possession as the last relic of the past. In the early twentieth century, this epic "loss of identity" inspired one of the boldest and most profound modernist movements in Europe, the Dutch art, literary, philosophical, architectural, and political avant-garde. Few places in the Netherlands were as receptive to it as the Dutch colony in Asia. The Netherlands East Indies is such an alluring place to study, because it appears to exist simultaneously at the farthest reaches of modernity and close to its most dynamic center.

The *time* is late-colonial: the last three-quarters of a century of Dutch rule. I prefer to think of the time period less as a chronology than as, let us say, shifting sands. Only the era's end—the date the Dutch capitulated to the invading Japanese armies, March 8, 1942—is sharply inscribed in this book. The rest appears as an accumulation, shifting, and running out of culture, identity, and sense of nation. Once we loosen time, I believe, we can better feel and convey the surreal significance of late-colonial duration, and the time of the Indies as a premonition. Often, the sound of hands clapping in Europe—clapping for fascism, clapping for the avant-garde—was heard in the Indies before the hands actually clapped. In the sense of time as well as place, the Netherlands East Indies was both at the periphery and next to the center of the empire and the world. When we loosen time just a bit, late-colonial culture, identity, and sense of nation emerge as the most revealing reflection of the global.

Technology in the book's title refers to a method more than to a topic. I have chosen technology to cast late-colonial culture, identity, and nation in an unusual light, and to agitate the picture in a less predictable way. People in the Indies, both the Indonesians and the Dutch, felt awkward about new technologies, as we all do, but in a specific late-colonial way. Encountering the "unseemly" technologies, people in the Indies often began to move, speak, and write in a way that broke through—or at least scratched—the otherwise smooth surface of their behavior and language. While the people handled, or were handled by, the new technologies, their time, space, culture, identity, and nation came to feel awry. These moments give us a chance to recognize the culture differently, more nakedly, and, hopefully, more deeply.

Indonesia is extraordinarily rich in extant sources. The archives and libraries in the Netherlands—Leiden, the Hague, or Amsterdam—are meticulously ordered and easy to use. The archives and libraries in Indonesia—Jakarta, Bogor, Bandung, or Medan—are rather messy, unwelcoming, and thus, even more exciting. I read the abundant sources with almost exclusive concern for small, even trivial, technological, details, looking for a "material object," in Marcel Proust's words, "the sensation which that material object will give us." In there, Proust believed, and I now do as well, "the past is hidden." The texts of public and private statements of the period—political speeches, letters and diaries, essays on cultural themes, works of art, novels, poetry, photographs, paintings, song lyrics, daily newspaper reports—"giggled," and sometimes opened up, as I searched them for the unseemly technology trivia. In order to keep on "tickling the sources," and to maintain a symmetry, I next tested the contentedly dry technical texts of the period—demographic, communication, architectural, and urban-planning reports, drawings, maps, graphs, and statistics—for their poetry.

The book is divided into six chapters, each dealing with a different type of technology. The introductory chapter looks at the late-colonial sense of touching the ground with one's feet or wheels on trains or cars, as well as roads, velocity, and the conceptualization of moving. The second chapter studies the technology of architecture and urban planning, and the sense of dwelling, sheltering, and hiding. The third inquires into optics and glass, and into looking, seeing, and watching. The fourth considers fashion and the body—machined, standing up, and posing. In the fifth chapter, I turn to telephone, radio, and other communication technologies, and to the culture of listening, making voice, and keeping silent. Finally, as a kind of epilogue, I make explicit the idea of continuity that pervades the whole book, the insistent sense of late-colonial culture overgrowing into the post-colonial period and the present. It is a chapter on the technology of exile, and on the possibility—or impossibility—of disconnecting oneself.

Some words that appear often in the book need to be explained.

Engineers, to use Karl Marx's words, are "a superior class of workers."[2] They believe in their language as we all believe in ours. More than the rest of us, however, engineers believe that their language and everything else can be taken apart and reassembled (and taken apart again) for the language's and everything's benefit. Engineers dream and plan as often and intensely as the rest of us. More than the rest of us, however, they believe that there is a calculated sameness between the planning and the dreaming. Engineers, in their essence, remain unchanged when they carry their beliefs to the limits. They merely become more impressive to the rest of us, more tragic, or more dangerous. As they reach the edge, some of them, in the words of one of the most tragic among them, may even declare themselves "engineers of human souls." Of course, there is an engineer in each of us.

Kromoblanda, a more specifically Indies word, is a neologism invented by Hendrik Freerk Tillema—a Dutch pharmacist, wholesale dealer in pure water and whiskey, motor-car enthusiast, city councilor, film pioneer, and, without a formal degree, in his inner self, most of all an engineer—who lived in the colony between 1896 and 1914. *Kromo* was how the Dutch called the natives,[3] mostly the Javanese. *Blanda* was how the Javanese and other Indonesians called the Dutch. *Kromoblanda*, as Tillema presented it in a six-volume opus published between 1915 and 1927,[4] was a dream and a plan—in its sweep comparable to Le Corbusier's *Ville contemporaine*—of the *kromo* and the *blanda* living in a future, well-equipped, modern, happy, and efficient Indies, together.

Nationalism is a word that does not appear often in the book, but that expresses the book's undercurrent. In both its main variants—Dutch and Indonesian—nationalism is the missing hyphen in *Kromo(-)blanda*.

ENGINEERS OF HAPPY LAND

Ferry over the river Progo, Central Java, ca. 1900. On the ferry there
are three Europeans with their automobile (KITLV)

ONE

LANGUAGE AS ASPHALT

> The need of a constantly expanding market for its products
> chases the bourgeoisie over the whole surface of the globe.
> It must nestle everywhere, settle everywhere, establish
> connections everywhere.
> —Marx–Engels, *Manifesto of the Communist Party*[1]

Bare Feet

IT MAY help, just for a moment, to think, with Bergson, about a land-scape in the beginning, as if it were "the surface of a soul that is thoroughly calm and unruffled."[2]

On February 13, 1891, an expedition embarked on a route out of Padang Pandjang, a small town in West Sumatra, a point at the edge, a station in the late-colonial Netherlands East Indies beyond which trains could not go. The aim of the expedition was to survey the area yonder for a new railway to Siak, across the island, to its east coast. At the head of the Siak expedition was Dr. Jan Willem IJzerman, the main engineer of the Dutch state railways. He was forty at the time and a renaissance man of sorts. Besides being, possibly, the most influential technician behind building the new Indies railways, he was a well-known amateur archeologist and a member of the archeological society in the princely town of Yogyakarta.[3] Toward the end of his life, in 1924, IJzerman, too, had been recognized as the initiator of the first and only technical college in the late-colonial Indies. His bronze bust had been placed in IJzerman Park, near the school campus in Bandung.[4]

On February 13, 1891, at six o'clock in the morning, the IJzerman Siak expedition started up "with a little word *madjoe*," which meant "forward" in Malay, the lingua franca of the colony. The plan was to walk "from six till four," every day, from sunrise to two hours before sunset.[5] In addition to IJzerman and three other Dutchmen, there were about a dozen Javanese servants, twenty Javanese railway workers, and about 120 helpers recruited locally from among the Sumatrans. The expedition carried "trunks, field beds, chairs, imitation-leather sheets for shelters, mats, ammunition, wire, ropes, nails, paraffin, wicks, a photographic camera, and food."[6]

Dr. IJzerman, as they were on the way, never parted with his pocket revolver. He was as happy, or so it seems, as a man in his forties might be: "Here," he wrote, "the saying of Mr. Potter from his famous American novel rings true: '*What is a man without a revolver in Texas?*'"[7] The men of the IJzerman expedition killed fish for dinner with dynamite cartridges.[8] Their camp in the jungle was attacked by "bandits."[9] The sense of the Wild West was heightened by an image of a pioneer's grave. This was, actually, how IJzerman opened his story in the expedition memorial book: "Since the discovery of the Oembilin coal fields by the genius of a mining engineer, W. H. de Greve, in 1868, people started to speak and to write about a desirability to reach the big rivers that flowed as far as into the Straits of Malacca."[10] A large photograph of the engineer Greve's grave in the jungle, "merely a little pile of earth in the shade of a big *tjoebadak* tree," took up the entire title page of the memorial book.[11]

IJzerman's men, as they walked, came repeatedly upon fresh *sporen*, "tracks" or "footprints," of "elephants, rhinoceros, tapirs, tigers, boars, deer." The footprints were "sharply engraved in the damp clay."[12] Most of the time, it was just footprints the men saw; rarely did they spot an animal itself.[13] It was, largely, just sensing, prints of animal feet on the thoroughly calm and unruffled surface of earth. Yet in Dutch—the language of the expedition—*sporen* meant both the footprints of the jungle animals the men sensed and the railway tracks they came to build.

In their own account, the Dutchmen in the expedition appeared exposed and their senses seemed bared. The untamed nature through which they passed, too, appeared as if opening itself up, and—if a man could take it—in a sense, friendly:

> In the forest, far from inhabited world, deep silence reigned most of the time. There were no monkey colonies to raise echoes with their merry shouting, no flock of songbirds to start their crystal melodies, all the large animals as if were extinct. . . . No mosquito disturbed our rest at night, we did not even have to unpack our mosquito nets. No poisonous snake, no centipede, and no scorpion made our sleeping in the open, under the trees, on the mounds of molding leaves dangerous, no rat gnawed at our rice reserves, cans with our food could be left around opened. . . . The plants were even more innocent, no leaf inflamed our skin, no thorn infected our blood. . . . We could even drink forest water safely.[14]

As the men moved, so the memorial book of the expedition describes it, they felt the landscape as kindred and moving with them. The memorial book conveys a sense of fluency between the people and nature. The expedition worked a river, for instance, and the river, thunderously and grandly, streamed past the people as they worked:

> Gigantic tree trunks raced with great speed and broke on the rocks that filled the riverbed, in pieces, crushed in pieces the lumber moved on its way to the sea, or,

ebbed off the stream, gathered on the shallow sands . . . there was a majestic calmness further ahead; noiseless, dark-green woods gently blanketed the river banks.[15]

The men's language was lush with words, as their days were lush with actions and movements, and their senses lush with sounds, shapes, and scents. Or, to put it less innocently, this was a Baroque-like sameness of man and man's surroundings, in which dreams and man's labor as well as conquest might be supposed to come together. One of the Dutchmen on the expedition, Van Bemmelen, captured another moment in this particular sensing of colonial modernity. As they struggled along one of the untamed rivers, "In my elation, I shouted to [IJzerman] and urged him to admire a newly appearing splendor on the right bank. 'I look only to the left side,' he shouted back, 'this is where the railway must go.'"[16]

They worked the river and the jungle so hard, moved with so much exertion, sensed everything so intensely, that there seemed to be little time and motivation left for landscape mapping, animal stuffing, or, for instance, butterfly pinning:

> Initially, it was our intention to check the longitude and latitude of all important spots with the help of chronometer and by astronomical observation, for the remaining assessment patent *boussole* Smalcalder was to be used. But, very soon, it became clear that, given the limited time at our disposal and given our limited forces, we would have to be satisfied with just fleeting measurements made with a simple compass and a tape.

Besides, "The number of containers with alcohol for collecting natural samples, and the quantity of paper for drying and pressing the plant samples, had to be reduced; all other considerations had to be put aside in face of the necessity to carry an amount of rice large enough to feed 300 men for at least 14 days."[17]

All the work was done so that the (railway) wheels, in the future, might turn. Yet, for the expedition to move—and to move the landscape as they moved—meant leaving footprints in the damp clay, like a rhinoceros or a boar. To move on, a particular technology of movement was required: "clambering over fallen trees, balancing on their trunks, skidding down, and stumbling through the muddy holes in between."[18]

The exertion, the working the landscape and the walking, it seems, was powerful enough to create a credible illusion. It even appeared natural, after such a day of prospecting, when a photograph was taken, that it would take time and effort to distinguish IJzerman and the other Dutchmen from the Javanese and Sumatran, as the memorial book formulated it, *reisgenooten,* "traveling companions," and *metgezellen,* "companions."[19] A greater part of the illustrations in the memorial book of the expedition, as usual at the time, were either watercolors or drawings made from photographs. In *An Evening Deliberation*, for instance, a group of men is seen, huddled together. Even the artist who made the drawing did not apparently think it particularly important to retouch the photograph. It is not easy, at all, to tell who is IJzerman, where the other Dutchmen are, and which ones are the Asians.[20]

Softly, even, they seemed to work together. As the work was accomplished, on the first morning after the expedition reached Siak, the final station of the future railway (before they set on their way back onboard a Chinese steamer), Van Bemmelen, as he wrote in the memorial book, woke up, and his thoughts went back over the past weeks. He thought of the days as "days of freedom, camp life without borders," and "adventure." However, it is clear that the Dutchman most eagerly wanted to tell it as a tale of walking. What would stay in his memory most powerfully, he wrote, was "dampness in my half-torn *laarzen* boots, and the gaiters that almost turned brown by the exposure to campfires and mud."[21]

Hard and Clean Roads

In 1840, a new and eager Dutch minister of colonies, Jean Chrétien Baud, ordered forty camels to be shipped to the Dutch Indies. As an afterthought, two hundred donkeys were also to go. A year later, people in the colony were not allowed to slaughter buffaloes, which might be used for work in ports and on roads.[22] This kind of eagerness lasted. As late as 1862, a deputy in the Dutch parliament suggested that a number of llamas and elephants should be sent to the Indies to work as draught animals, especially on the "sugar road" between Semarang and Salatiga in Central Java.[23] However, this deputy already appeared to be behind the times.

Since the early nineteenth century, the Great Daendels Mail Road had been in use in the Indies, cutting through the main island of Java from west to east. It was built, between 1808 and 1811, as a part of the defense against an expected British invasion, and it was "a gigantic road-building project that, with justification, was called 'Napoleonic.'"[24] The Daendels road was a pre-twentieth-century wonder of speed (18 to 20 kilometers per hour in the best places),[25] and, equally so, of order. As native children learned in 1886 from their primary school textbooks, "Along the road, each 16 1/3 minutes, there is one stake [*paal*] to indicate a distance. At each fifth stake, there is a post [*bangsal*] for the government mail to change horses."[26]

For most of the Indies, even most of Java, at the time, the road of course was not very relevant. As late as the mid-nineteenth century, it took as long as three, even five, months for a load of coffee from the interior of Java off the Daendels road to reach the sea port of Semarang.[27] As the traffic got heavier, the highway itself, even for the areas that had access to it, became insufficient for anyone wishing to travel modern: "a European coach, usually overloaded with passengers and their belongings is, frequently too heavy to manage for the little Javanese horses."[28]

In 1842, a year after the Dutch colonial minister Baud suggested the camels, the first ever magazine published in the Dutch Indies, *Kopiïst*, "Copyist," launched a series on inventions, and, as invention no. 11, presented "steamways and steam carriages":

> If we are not mistaken, Asia, until today, does not have a single railway. The invention, in the East, did not extend beyond the northern shores of Egypt. . . . It is a widely known fact that the population of Java, as far as its numbers and potential are concerned, is not evenly distributed. In a country like this, introduction of even a single steam machine, certainly, would free a vast number of hands so urgently needed by our agriculture in other locations. The steam means of transport, generally, has a power to release the population from the unproductive drudgery in which it is bound today.[29]

According to the *Kopiïst*, Asia, and namely, the Dutch Indies, was at least as fit for the wonderful new invention as Europe. The terrain of the Indies, the *Kopiïst* wrote, did not pose any serious obstacle; it laid, waiting for the trains:

> From Soerabaia, to the north of the river Kedirie, towards *dessa* [Malay for village] Menoeng . . . the turf is flat. . . . From Tjirebon to the river of Madioen, there are no great hurdles. . . . Over the river of Madioen, a wooden bridge could easily be built . . . the rugged ground there could easily be leveled and the ravines filled.[30]

Reading the *Kopiïst*, one just could feel how agreeable it would be to build the tracks through Java, from the port city of Surabaya in the east, hundreds of miles to the west, to Batavia, the metropolis of the colony. As the *Kopiïst* saw it, there appeared nothing prohibitive in the costs of the project either:

> the population can do the earth-moving and stone-crushing work for the railway free of charge or, at worst, it can be paid in rice and salt, as this can be made a part of a usual *corvée* service for the government . . . it is evident that the costs will be even more affordable than those in either Europe or America.[31]

To build the main track, the *Kopiïst* calculated, would cost merely 8,704,080 guilders, and the side tracks 3,215,520 guilders, for a total 11,919,600 guilders, 2,000,000 guilders would pay for vehicles and warehouses, 1,000,000 guilders would take care of interest. The grand total might be "21,000,000 guilders only."[32]

In 1842, the same year the *Kopiïst* article was published, in the same vein of engineering optimism, the king of the Netherlands, Willem I, issued the first railway decree for the Indies: "In order to promote the transport of

Opening of a new tramway line in the Indies, 1904 or 1905.
Foto Suit Kan. (KITLV)

products and other goods from Semarang to Kedoe, the Princely Lands in Java, and vice versa, an iron railway will be laid."[33]

It took twenty-five years, in fact, after the first Indies royal railway decree to build the first 25 kilometers of the iron rails in the colony, and it took another decade to build the next 300 kilometers—all of them on Java.[34] By heavy and slow steps, rather than daintily as the *Kopiïst* suggested, the Indies trains came into existence. By 1882, in the words of a special commission of the learned *Indisch Genootschap*, "Indies Society," railways and "the little railways," the tramways, also in the Dutch colony, proved themselves to be "the most useful of the present discoveries, the most admirable victory of man over time and distance, the most powerful incentive to labor, exchange in values, and civilization."[35]

By 1888, eight main railway lines were in operation, all in Java, and the fifteen largest cities of the island had a railway connection.[36] In April 1899, an electric tram was installed in Batavia,[37] and, in 1909, the tramway lines in the city were already 14 kilometers long. That whole year only one fatal accident was reported on the Batavia tramway lines.[38]

Raden Ajeng Kartini—a teenager when Dr. IJzerman went on his expedition, twenty years old when the electric tram started in Batavia—was the daughter of a high-ranking Javanese official in the colonial government, a *boepati*, "regent," of Japara. Japara, where Kartini spent most of her life, was a small place on the north coast of Java. Kartini's friends, mentors, and protectors, almost all Dutch, in the hype of their time, called her a Javanese princess. In post-colonial Indonesia, and until today, almost a century after her death, she has been called the mother of Indonesian nationalism.

"If I were a boy," Kartini wrote in 1900, from Japara to a Dutch friend, "I should not think twice, but would become a sailor at once."[39] Kartini thought of ships often: "We do not want any more to sail on a sinking ship," she also wrote, meaning Javanese society;[40] "courage of the hand at the rudder, and pumping at the leak, could have saved us from destruction."[41]

Japara, where Kartini lived, was very much off all the new and newly emerging modern Indies roads. From Japara, one had to travel on horse or in a horse-driven cart, on a dirt, and often muddy, road to Majong. There, one might board a steam tram to Djoewana or Semarang. Only there was there a "real train."[42] Kartini often warned potential visitors of "the tiring trip."[43] Yet, whenever an opportunity arose, or in her dreams, she quickly got onboard: "we were on the track with the first morning tram . . . long before the tram station of Pemalang appeared, as we steamed on, we were looking out for our darling."[44]

When guests were to come to Japara, Kartini traveled with her aristocratic father to meet them at the station: "how afraid we were that we miss the tram."[45] She was allowed to travel very little, as she was an unmarried Muslim woman. Yet, the moments on the train, carefully counted, were the most intense for her, and, so it sometimes seemed, the only real moments of life.[46] Kartini met people, was touched, received news, and heard rumors on the train or at a railway station: "In the train . . . I pressed my hand on my heart . . . I heard much in the tram."[47] The dreams were most frequent and drew the clearest image: "Now, we fly with a storm over the iron road":[48] "Would I ever be able to forget that divine ride with her to the station? . . . Do not fly so fast on the smooth iron tracks, you sniffling, steaming monster, do not let this beautiful meeting end so quickly. . . . I prayed that the ride would never end. . . . But, alas! the stoker did not hear me."[49]

It was beyond any doubt to the excited Kartini that the modern roads in the Indies had to be made all anew, and hard. The newness, the hardness and cleanness—it was the roads' modernity. Cleanness of the roads, in this logic, was purity of times, democracy even, we might say. There were some people in the Indies, Kartini wrote, who demanded that they be addressed by aristocratic titles; often, these titles did not even belong to them. On the whole, Kartini wrote, "it is a matter of indifference": "but when overseers, railroad engineers (and perhaps tomorrow station masters too), allow themselves to be thus addressed by their servants, it is absurdly funny."[50]

New roads through Java and in the whole colony, to Kartini, were to be fully made of progress, and, as long as they were made of that hard and clean stuff, nothing could stop the wheels. A Javanese girl was run over by a tram not far from Japara, and Kartini reported the accident in a letter. This event proved, Kartini wrote, how a modern system of "the first medical aid" became very important.[51]

Kartini died a year after she was married, at the age of twenty-five, a few days after she gave birth to her first child. She may be written about, perhaps, the way a Viennese author, Robert Musil, at about the same time, wrote about one of his Austrian contemporaries: "She was one of those charmingly purposeful young women of our time who would instantly become bus drivers if some higher purpose called for it."[52]

There were no buses in Japara, and in the entire Netherlands Indies colony, in 1900, however. And also, Kartini, in her liking for newness, cleanness, and hardness, would dream of planes instead of buses. This was, at least, what she wrote, three years before her death, in 1901, to one of her and her father's friends, physician Dr. Anton, in Jena, Germany: "flying machine will have come into use, and on some golden day you will see one of them flutter over Jena's blue horizon bringing a guest from afar. I should indeed have been born a boy."[53]

Struggle for the Roads

Modern roads in the Indies, besides the many wonderful things they did, became from the moment of their inception a battlefield and a space where the Dutch in the colony were clearly uncertain of themselves.

Professor Stokvis, a Dutch liberal and respected colonial expert, in a lecture in 1894, "Man in Tropics in Connection with Colonization," declared that "not a single example is known to me of a European family that has been able to reproduce itself in this tropical land for as much as three generations without regularly traveling to Europe, or without being blood-mixed with one of the local races."[54] This vulnerability to mixing, another respectful Dutch expert wrote at the time, should serve as a warning against "too rosy" an expectation about the future of the Dutch in the Indies.[55]

Many of the honored voices in the colony warned that the Dutch might sink in the Indies. An eminent Dutch engineer and best-selling author, van Sandick, wrote in 1891:

> against every one who might accomplish it, there stand hundreds who will get stuck in the Indies, without a chance to see Europe again. Holland will be reached again only by those who are truly loved by fate; the aggregation of those who, year after year, depart to the tropics will also die there, forgotten by their motherland.[56]

According to official statistics widely publicized at the time, 80 percent of the Dutch population of the colony, by the early twentieth century, had been born in the tropics. An unspecified but large majority of them, according to the same source, had in their veins, indeed, "a drop of native blood or more."[57] In 1900, according to another very seriously heeded estimate, among all the legally Dutch children born in the Indies (i.e., including those with "a drop of native blood or more"), as many as 40 percent could not speak Dutch at all, and 30 percent of them spoke their "native language" with difficulty.[58]

In 1900, about 60,000 Europeans, mostly Dutch, lived in the Indies.[59] In 1930—when the last ever census was taken in the colony—the number had grown to 208,000. Still, it was merely 0.34 percent of the 60 million total population.[60] As one Dutch journalist, Willem Walraven, put it in the 1930s the Dutch lived in and moved over the Indies "like flies upon milk."[61]

Through this landscape of milk the modern roads of the Indies were to push. In 1880, in the most prestigious and scholarly journal of the colony, *Indische Gids*, "Indies Guide," J.F.F. Moet, again, as many before him, argued in favor of trains in the Indies. He emphasized one point in particular. He wrote about how wonderful his experience was in Europe—Kassel, Germany, for instance, or on the steam tramway between the Hague and Scheveningen, in the Netherlands:

> In spite of the fact that the train passes through the busiest roads and streets, it does not cause any difficulty. Horses are not frightened, even in the evenings, as the locomotives move amidst the houses of the towns with their large red lights in front and large white lights on the sides. I saw a machine with two carriages, at [the main square of Kassel], in fact, on the market day; it puffed through the

very middle of the crowd, and the women kept sitting behind their stalls and selling their fruits and vegetables as if nothing at all was happening.[62]

This was, and increasingly so, the late-colonial plan and dream. Horses and crowds would not be frightened in the Indies. It would be, in this aspect, quite like Europe. The time schedule, and the list of stops, depots, and terminals of the new electric tramway in Batavia, in 1910, for instance, read exactly in that programmed and dreamy way. New modern roads would be kept clean and hard, running upon the landscape of milk, calm and orderly, between one and the next point of an undoubted modernity:

> Telephone Office, Photographic Association, Batavia Sporting Club, Military Engineers Workshops, Railway Station Weltevreden, Protestant Church, High School, Racetrack, English Sporting Club, Telephone Office, Officers Barracks, Concordia Club, Waterloo-Square Concert Terrain and Soccer Fields (also for natives) . . .[63]

As late as the mid-1920s, even the rails for the Indies railways were imported from Europe.[64] As late as in the early 1940s, to the very end of the Dutch colonial era, virtually all technical equipment came from the West. Only a few body parts of carriages, and the sleepers, were made of teak and other Asian wood.[65]

Many, and in some parts most, of the skilled railway and road-building workers were Indies Chinese, or natives imported from other islands or other parts of Java.[66] In 1918, the Indies state railways employed 179 Dutch officials born in Europe and 564 Dutch officials born in the Indies.[67] Not a single clerk, station master, or machinist was a non-European. In 1917, a few dozen natives, in the colony of 60 million, had been admitted to a new state-railways training course for the bottom-level clerical positions.[68] The milk should not be stirred. The idea was to install a tradition of just another vocation, inherited, gradually and orderly, from father to son, exactly as in Europe, without frightening the crowds.

Trains in the Indies, however, as everywhere in the world, possessed an amazing power of attraction. Horses, perhaps, might be made not to panic. But, as soon as rails were laid and the first train appeared, people, the whole landscape, turned around and moved to the train.

In 1883, a Dutch official publication about the contemporary colonial Indies noted: "Particularly the native population makes a great use of the existing railways."[69] This became an inevitable appendix of the road optimism. Most often, it was just an awkwardly placed emphasis.

In 1904, a special government investigating committee reported that the number of passengers in the first ("European") class of the Indies trains rose by 4,000 during the previous 3 years; in the second (lower-income "European" and top-level "native") class the number of passengers rose by 33,000. Not very much, in fact. In the third ("native") class, however (or *kambing*, "goat" class, as it was commonly called), the increase was 550,000![70] The committee did not hide its surprise: "The use of trains and trams by *kleine man* [Dutch for "little man," man in the street, native] is rising faster than initially expected."[71]

The same committee hired a group of "four deft native conductors" and sent it to the trains on various lines to engage the passengers in "little kind talks." The data thus collected were also unexpected. The little" men, women, and children, as a railway touched upon their habitat, did not appear to panic, but neither were they in any perceptible awe over the modern technology. In fact, they did not appear to change their traditional ways very much. They just appeared to add a little to the tradition. Pragmatically, en masse, and with an efficient use of the trains, logically, they appeared on the move:

A. Reasons to travel of economic nature; market, search for work: 69.5%
B. Reasons to travel of personal nature: 30.5% . . .
    f. visits to other family members 20.8%
    g. law and order; summons to government office or court of justice 3.6%
    h. faith and tradition; visits to graves and other holy places 3.0%
    i. pleasure travel 3.1%[72]

The natives, as the four conductors found out, were choosing the trains, best of all, that allowed them to carry free baggage of 50 kilograms or more.[73] The little people of the Indies traveled as they always did, with their goats and their hens sometimes, and with bags of clothes and food. The Madurese (of the island off the northern coast of Java), the report noted, had always been known as *echte zwervers*, "real wanderers or drifters." By now, however, as the report put it, "all natives appear eager to use trains and trams for their own goals."[74] The commission discovered, also, that the natives of lower standing, the "simple men," the real masses, traveled more eagerly and often than their higher-ups, native aristocrats, native colonial officials, the Dutch-supported elite, who, in the ideal plan of the empire, should have been transmitting the modern manners in an orderly way to the plebeians of the colony: "the native notables [were found] much more *hokvaster* [home-loving, literally: fond of one's fireside] than the little man. . . . The little man . . . changes his places of sojourn much more frequently than one usually thinks."[75] *Santri*, the Indies Muslim scholars and students—for a long time suspected by the colonial government as a subversive element of the native society, potential leaders of unrest—were, indeed, found by the conductors' report to be one segment of the Indies native population that used trains and trams radically more than the rest.

This was an alarming vision of a breaking down of the ideal plan of empire through an invasion of physicality, crowding, and, most gravely, touching. In Batavia, in 1909, on the electric tram lines, in a single month, 10,404

Funeral hearse of the Aceh tram constructed by the Netherlands Indies Army Engineers, ca. 1890. (*Gedenkboek van het Korps, Marechaussee van Atjeh*)

passengers traveled in first class, 72,623 in second class, and 255,197 in third class.[76] True, there had been, most of the time, a special carriage in every train for each class. But the rhythm of the train, the shaking, and the machine were the same. All the passengers were (traveling) humans, and their uncomfortable sameness could nowhere be seen, felt, and smelled as strongly as in the train.

About the time of the glorious invention of the electric tram in Batavia, Louis Couperus, a famous Dutch writer, published a novel about the Indies, *De Stille Kracht*, "The Hidden Force." It became one of the most widely read books in the colony. Couperus wrote about an undercurrent that he, sooner than most of the others, felt in the colony, gathering against the Dutch colonial presence. As the Couperus story comes to its end, the Dutch woman protagonist of the novel wanders through modern Batavia, and she watches, in particular, the first European class of the new electric tram:

> in the mornings—when she did her errands in the shops on Rijswijk and Mo-
> lenvliet, which, with a few French names, tried to give the impression of a south-
> ern shopping center of European luxury—did Eva see the exodus to the Old
> Town of the white men. . . . The exodus filled the trams with the white burden
> of mortality. Many, already well off, but not yet rich enough for their purposes,
> drove in their mylords and buggies to the Harmonie Club, where they took the
> tram to spare their horses.[77]

It took a whole third of a century from the opening of the first railway line in the Indies for the colonial government, in 1908, to decree a single time for its railway, post, and telegraph service throughout the island of Java.[78] Even after this, and until the 1920s, in fact, "Java time remained inconvenient . . . local time was maintained for all other purposes, though there was some tendency for this so-called *midden Javatijd* [Central Java Time] to be used more widely. Local times continued to be calculated for places outside Java and Madura."[79] The modern Indies roads, until the end of the colonial rule, were strips, largely, with a special time. An economic nightmare, certainly, but also a comforting exclusivity against the vision of panicking horses and people.

Special time, indeed, went hand in hand with special security. As Henri Van Kol, well-known Dutch engineer and socialist expert on colonial issues (also Kartini's mentor at one time), rode on the Atjeh tram in North Sumatra in 1903—passing through the region only a few months before the scene of the fighting between natives and the Dutch, the most bloody colonial Aceh war (1874–1903)—Van Kol was thrilled, excited, and truly a pure-idea model colonial passenger: "Such a short time ago, so many fallen. . . . [And now, there] by tram, from Lho Seumawé to. . . . Maneh . . . All through, and safely."[80]

Mas Marco Kartodikromo was a contemporary of all of them—IJzerman,

Kartini, Van Kol. He was a most extraordinary journalist of the fledgling Indonesian nationalism in the early twentieth century. As far as the Indies modern roads and trains were concerned, he had the acuteness of a guerrilla fighter. Even at a time when there seemed to be a perfect peace in the colony (no uprising, no political unrest), Mas Marco saw the Indies "little man"— *boemipoetra*, "son of the soil," *saudara*, his "brother," as he called him[81]— when about to take a modern road, or board a train, as approaching a thin yet well embanked battle line. One of Mas Marco's several journals was called *Doenia Bergerak*, "World on the Move." When one happened to be a son of the soil, *Doenia Bergerak* wrote, in 1914, for instance, one would do well to approach a railway station in a state of alert:

Disappearance of platform tickets

As every one of our readers knows, platform tickets used to be available at all the big stations, in Koetoardjo, Djocja, Bandoeng etc. . . . As we read in the newspapers, however, Malay as well as Javanese ones, at the station of Koetoardjo, for example, platform tickets recently disappeared. Somebody may write to us, hopefully, and explain it. This does not seem to be done in order to please the public, and to make it possible for the people to see off and welcome friends at the train without having to pay an additional charge. . . . Now, a Javanese dad, let us say, cannot go inside the station gallery, and has to wait in the open, at a safe distance from the station. . . . However!!! when someone happens to belong to the Dutch or Chinese stock, well, please, do enter and sit on the bench inside the station. Hmm!!! . . . The Javanese dad is made into one-fourth of a man.[82]

When a son of the soil, a native, penetrated as close to the battle line as the ticket office:

Tramway tickets in Tjepoe railways station are sold out

. . . When we asked for the green (3rd class) tickets to Solo, we were told by the ticket-office man: *We hebben niet meer in voorraad* (Dutch for "we have no more in stock"). Thus, we were forced to pay for our tickets from Tjepoe to Goendih (2nd class price) 1.80 guilder, and, in Goendih, from Goendih to Solo, again another 0.90 guilder.[83]

When a son (or a daughter) of the soil got on board:

Various News

On July 19, 1914, several women boarded a train of the state railways to *Kalisoso*, where they were going to visit some holy places together. The Dutch conductor on the train (the train leaving Solo at 5.40), in manners not too polite, chose to amuse himself with the women on board. So the story goes . . . the conductor pinched the cheeks of the women, for instance, and, throughout, he expressed himself in words that rather should not be heard in public. As he was doing this, the women became pale, ashamed, and afraid. If this report is true, then, why do these women themselves, their husbands, or others in their families not complain to the railway management? What is there to be afraid of? Don't be scared! We stand ready to help as much as needed! Our cause is just![84]

Some sons of the soil, natives, made it eventually as close to the line as to become themselves native road or railway personnel. Here is a letter full of technology from one of these as printed in the same Mas Marco paper:

Stationery and Pencil Inventory

. . . Beginning with 1913 . . . we were asked to use pencils instead of pens. We do need *inktpotlood* (indelible pencils) of course, and very strong ones, because we write, as a rule, through one or more sheets of *doordrukpapier* (carbon paper). Ink pencils came from the inventory of the state railways. . . .

Beginning January 1914, I was transferred to the present place in N. Since then, and until April 1914, I used the pencils I still had from my previous station—three stubs of Koh-I-Noor pencils. Then the pencils were finished, and, in April 1914, I asked the chief of the station in N. for another pencil. He gave me a red ink pencil model "Johanfaber's No. 400." After I finish this, I will not get another one, since there are no more pencils in the inventory; this, at least, the chief told me. . . . By September 1914, already, I was using a pencil of the length of merely between 3 to 4 cm. I talked about it to a friend, and he advised me to attach a tendril made of bamboo to the pencil (it was done in the friend's workshop). Thus, at present, I could still use the pencil. On September 1, 1914, I was transferred to the station ticket office, but my pencil is now only 1 cm long . . .

Many greetings and respect from me, *Watjono*.[85]

It was a thin red line. We do not know what happened when Watjono finished his Johanfaber's pencil. Late in 1918, however, as World War I ended and as reports of an attempted socialist coup in the Netherlands reached the Indies, a radical dissident Dutch journal in the Indies, *Beweging*, "Movement," published a list of popular demands that, the journal felt, might be raised in case the revolution spread from the Netherlands to its colony. Besides doing away with the Dutch monarchy and other upside-downs, there would be, almost certainly, the journal wrote, a demand raised for "only one class to be installed on the railways, cheaper, and offering more comfort than it is now the case in the 3rd class."[86]

Revolution, however, like Godot, never seemed to come to the Indies. In 1921, in spite of a general decline in living standards, ticket prices on the Indies railways were raised. There were, still, the same three railway classes, with the third class as uncomfortable, and as native, as before.

The city of Semarang was a major railway junction in Central Java. (It was a place, also, close to Japara, where Kartini used to travel on the train most often.) By 1921, Semarang was called "The Red City" of the colony. From Semarang, in 1921, according to a historian, "Communist influences spread in its vicinity and to smaller towns along railroad lines—to Pati, Demak, and Purwodadi to the east; to Salatiga, Boyolali and further to Madiun and Nganjuk to the south; and to Pekalongan, Brebes, Tegal, and Cirebon to the West."[87]

These were the (railway) nerves exposed. On May 10, 1923, a railway strike broke out in Semarang, the biggest strike ever in the history of the Dutch colony. The strike spread through Java, as far west as Cirebon, a city and railway station only five hours by train from Batavia, and as far east as Surabaya, the biggest port and naval base and the most important strategic point of the colony next to Batavia. The rails and sleepers were torn off the tracks, and the battle cry of the strike was *spoor tabrakan*, "tracks crushing"![88] When a Dutch liberal journal, at the end of the era, in 1941, looked back and tried to identify one crucial moment when colonialism failed in its project of modernity, this event, the railway strike of 1923, was picked to be the point.[89]

Godot, anyway, did not come again. The smoothness with which the crucial moments were being passed over, in fact, seems to be the main theme of late-colonial history, and, perhaps, of Indonesian history beyond. The colonial project, in a sense, worked. There was a machine-like easiness of the past moving into the present, and the future. The railway strike of 1923 was over in a mere twelve days. The membership of the principal railway workers union behind the strike fell, in a single month after the strike, from 13,000 to less than 1,000.[90] The trains were running again, and, indeed, sharp on time.

In 1922, a year before the railway strike, Louis Couperus, the author of *The Hidden Force*, visited the Indies once more, and for the last time in his life. This time, he did not stay only in Java as usual. He traveled widely, through most of Sumatra, Java, and even to Dutch Borneo. The new late-colonial network of railways and roads made it possible by now. Couperus went by ship, train, and car. Some of Couperus' critics say that the refined author did not like the new Indies at all; the only true affection, they say he felt was for the Indies buffaloes. This is not entirely fair. Couperus clearly—and he was setting a new standard for the refined and anxious Dutch rulers, residents, visitors, and engineers as well—felt very much affection, also, toward the native drivers: "Both Javanese chauffeurs—there are two who will relieve each other on the long journey—stare straight ahead; sitting behind them we see four attentive trembling ears. I believe that could we see into their hearts we should find they are more afraid of *oranboenian* (ghosts) than of tigers."[91]

In May 1927, a new factory was opened by General Motors in Tanjung Priok, the harbor of Batavia. Already by early 1928, 5,732 Chevrolets left the plant for the markets in the Netherlands Indies as well as British Singapore,

Malaya, and Siam. As one of the papers for native Malay readers excitedly commented: "Now, we all can see how great progress and human endeavor is, and, with each day, how all this comes closer to ourselves."[92]

By 1939, there were 51,615 cars in the Indies, 37,500 of them in Java, 7,557 in Batavia, 4,945 in Bandung, and 657, for instance, in the Japara–Rembang region, where Kartini had lived and where she had died.[93] Most of the automobiles in the colony, of course, were owned by the Europeans. Unlike on the Indies railways, however, from the early times and increasingly, there were natives behind the wheel.[94] In the first three months of 1928, for example, in Surabaya, there were 524 serious traffic accidents, 42 percent of them caused by cars, 23 percent by trains or tramway, 17 percent by motorcycles. Among the drivers involved in the accidents, 11 percent were Europeans, 6 percent Chinese and Arabs. As many as 83 percent were natives.[95]

More than trains and tramways, also, the cars and motorcycles on the modern roads of the colony appeared vulnerable and bent on "going native." Initially, as should be the case, it was a matter of inventive engineering. At the historic first and big Amsterdam world and colonial exhibition, in 1883, for instance, a gold medal was awarded to a "Batavian one-horse hand-wagon on two wheels . . . Price 350 guilders . . . inventor—C. Deeleman."[96] By the 1920s, this very *deeleman*, which became *delman* or *dilman* in the process, was already fully taken over, and it belonged to the little people of the land. *Deeleman* became as familiar a name on the Indies modern roads as the names of the traditional horse-driven, ox-driven, or buffalo-driven carts such as *tjikar*, *grobak*, or *sado*. Gradually, and *deeleman* led the way, many of the *tjikar*, *grobak*, or *sado* began to use rubber tires. Toward the end of the Dutch era, some *deeleman* became motor-propelled.

It was engineering at a distance from a revolution. A battle much less visible, let's say, than the great railway strike.

In 1923, in the year of the strike, a long essay was published in a prestigious and scholarly Dutch Indies journal, *Koloniale Studiën*, "Colonial Studies," about what its author thought to be one of the most serious consequences of World War I in the colony: a historically new, and very fierce, competition between the automobile and the train. Before the war, the essay explained, the railways in the Indies had virtually a monopoly. The native animal-driven carts could not compete with the trains, in speed, in capacity, or in comfort. World War I, however, in spite of the fact that the fighting took place far from the Indies, caused a general disorder in the colony's economy and communications. In the suddenly emerging looseness, through fissures of the upset order, *vrachtauto*s, "lorries," appeared in the colony. The author of the essay made it clear that what he meant were native lorries, and that he saw them as a menace: "The state of tires on which the lorries ride leaves much to be desired. . . . Overloading of the lorries is a rule rather than an exception. . . . Public—very fatalistically inclined native public, naturally—makes quite some use, indeed, of this cheap opportunity for transportation."[97]

Two years later, in 1925, in *Koloniale Studiën* again, yet another many-page treatise was published on the subject. "Initially," this study opened, the role of the automobiles was thought to be in filling in only, where there was no railway available at the moment. "Passenger automobiles," besides, were meant to serve, largely, as a "*de luxe* means of communication." In Europe, according to the author of this second essay, the development of road traffic after World War I did not appear to cause a fundamental problem. In the Dutch Indies, however,

> as yet, it has been beyond us to build a railway network outstretching enough. . . . Therefore, *sado, deeleman, ebro, kretek, kossong, andong*, and whatever name all these carts might have—as the traffic intensifies, everywhere, and even in the largest cities—take on themselves, more and more, a role of the reigning means of public transportation.[98]

The European transporters in the colony, the study admitted, "have little understanding of the psyche of the native traveler." In the field of medium distances—and this was the crucial area—the "native transporters" (actually large numbers of them were the Indies Chinese), the "wild buses and trucks," were able to offer "faster transport, wider selection, with a greater flexibility, and along the routes the native travelers really desire to take." The adjusted animal-driven carts, and, increasingly, "native-owned" lorries and buses filled, crowded, a huge and growing "gap" in the colony's modern landscape. It was futile, and a little desperate, to argue, as the essay in *Koloniale Studiën* did, that this was not a modern technology, too: "These vehicles, as a rule, are second-hand, and, even, third-hand, and, very often, they are in a state of being barely able to hold together."[99] And, besides, they destroyed the asphalt.[100]

Language-game

> Certainty is *as it were* a tone of voice in which one declares how
> things are. . . . Doubt gradually loses its sense. This language-
> game just *is* like that . . . the *sureness* of the game.
> —Ludwig Wittgenstein, *On Certainty*[101]

People in America still ask me, when they read my texts and hear my accent, how difficult it is, for me, to write and speak English. I rarely tell them, but it is a fact, that, in Czech, my speaking and writing goes more slowly than in English. It is, perhaps, because Czech for me, more than English, is made of what Wittgenstein called "words with an atmosphere." For me, speaking and writing in Czech, more than in English, is like treading on expressions, each of which has a powerful yet uncertain memory, as walking on pebbles only loosely attached to the whole, in a state of equilibrium, so I still try to believe. My Czech, also, feels to me like a language of wet earth. The walking is soft and slow, each step is as if barefoot and expectant of an invisible root of a word that may halt the progress again.

My English, when compared to this, moves easily and fast. Its words are like blocks of stone in a pavement: cut and polished words, in my English, become usable tools. In English, in contrast to Czech, I am less expectant and less afraid that a word, like a stone, may come loose and, like a patch of plaster, peel off the language. That language, like the hard and clean surface of road, may crack and even, in a disaster scenario, fall apart.

Articulated in English, for instance, even my memories of Prague are like chiseled stones of pavement; or like the well-mashed corpses of prehistorical plants and animals that make much of the modern asphalt roads. It is how memory can most easily be stored. Uncouth pebbles are sunk into the once-organic matter. The language as asphalt can be spread out fast, and it can cement even the most incongruous things together. It can be, and it is, used to cover a large territory. My English is a language that is most handy against tardiness, and against the fear of language.

This is where the theme of this chapter, and my writing about it (done in America and in English), came most intensely together. I felt, and it was my imperative in writing, that there was a common potential and danger in both my English and the late-colonial language(s) of the Indies. Both, in order to simplify matters (being languages of strangeness, either of an exile or of an intervention), are eager to mash memories. Both entice to cruising. Both promise to be good against all stammering, but also against touching. Both are ready to help in building a shiny—and the most perverted—idea, that ruling a colony, as writing its history, may be done like hitting the road, a road that needs only to be smooth, and, in case of trouble, yet a bit smoother.

Magneet, "Magnet," first appeared at the end of 1913. It was the official organ of *Motor-Wielrijders Bond*, "Union of the motorcycle riders," and probably the first journal in the Indies devoted exclusively to life on modern roads. Reading through the first volumes of *Magneet*, one is impressed with how firmly its language-game was played. First of all, one is aware of smoothness.

Much of each issue of *Magneet* was taken up by various announcements and reports on *clubtochten*, "club trips": "Our aim . . . is first of all making collective club trips by members, with focus, first of all, on slow and careful driving."[102] *Magneet* described, for instance, the Wilhelmina Park in the center of Batavia, the heart of the colonial capital, as a scene of the greatest jollity:

Our first Club-trip

Wonderful Sunday morning, December 28 [1913] . . . The whole Weltevreden
is still half asleep . . . motorcyclists are standing in groups on the fresh green
grass of the Wilhelmina Park. . . . Just before half past six, our motors zoom,
and off we go from the meeting place. . . . What a delightful sight! . . . a medley
of little club flags . . . *ramêh* [Malay for busy, bustling, lively, festive, noisy], in
line . . . Photo-cameras click and clack. . . . Forwards! Past Rijswijk the cheerful
troops buzz, through the French quarter, via Tanah-Abang past Koningsplein
towards Matraman to Meester. . . . Past Goenoeng Sari the happy company is
drawing, attracting many eyes, then into the old town, and, out of there, still in
moderate pace to Versteeg, where cold beer is served. . . . The men are riding
through the busiest of neighborhoods, yet so considerately—it is proved by this
trip that it is not dangerous to take a couple of dozens of motorcycle riders and
let them loose on the street.[103]

The first trip of the union ended in the hotel *De Stam*, "The Trunk," in New
Godangdia, the newly built modern quarter of the city, with yet another cold
beer.[104]

Through the zooming of machines and the language describing it, the
colony's rural landscape was also pronounced safe. The motorcycle-club
members, as *Magneet* reported, penetrated deeper, from the center into the
suburbs, and inland. Joyful, excursion-wise, on bicycles, tricycles, with or
without sidecars, in groups or alone, they were—or were described as—
cruising. The union rented a "Club Café," *Buitenzorg*, southwest of Batavia
at the foot of the hills, and, in the hills, *Restaurant Rikkers* with a *Clubhotel*
near Cipanas, south of the capital. Other restaurants and hotels soon staked
out the club's most popular routes.[105] The language was powerful, and it is
still quite comforting to read about it in *Magneet*:

Along the Tangerang road

The calm and the darkness that rule all around are broken only by the hot-
tempered puffing of my motorcycle, and by sharp flashes of my motorcycle lamp.
There is nobody on the road except for a few drowsy watchmen—but they are
everywhere—and, now and then, there is a coolie trudging under his *pikoel*
[Malay for burden carried on shoulders]. . . . [Then] the somber tops of the
mountains . . . the air becomes rosy. . . .

The road takes me through a Chinese cemetery. There are graves on the left,
and graves on the right; a city of the dead—or so it looks to me—and it
stretches out in the vastness. Where the cemetery ends, there are *sawahs* [Malay
for wet rice fields], a bridge, then sawahs, and again a bridge.[106]

There were villages with nobody, as the club chairman described it in one of
his trip reports, "merely a couple of wretched huts, a market, and a chicken I
ran over . . . in a Chinese *toko* [Malay for shop] I got gas."[107]

Accidents happened, and the language of *Magneet* handled them well, too. A motorcycle crashed into a *grobak*, a native horse-driven cart. A native girl was killed as she crossed a road:

Fatal accident . . . A motorcyclist, in a cloud of dust, found himself, suddenly, against a native on a bicycle, coming from the opposite side. Various Europeans, men living in the neighborhood, declared that the native was *bingoeng* [Malay for confused], and that he zigzagged on the road in a most awkward way. . . .[108]

On the road between Klender and Bekassi, traffic was blocked by a *grobak* pushed from behind by three evidently drunk natives. We base the suspicion of alcohol on the fact that the cart on the narrow road was pushed in an extremely odd manner. On honking by a *Sacoche* motor-tricycle, the cart moved to the left, barely enough that it could be passed. But, when a *Sarolea* motorcycle began to honk, the fellows with the cart, who had to be absolute idiots, swung the said *grobak* in their good-natured stupidity to the middle of the road *in a diagonal direction*! Then, the scoundrels just stopped and waited. . . . All three natives have already been punished for their "performance."[109]

Collision. Mr. W. A. van den Capellen of Bekasi Road No. 3 was charged for killing with his motorcycle a native girl, called Moenah, of *kampoeng* [Malay for native quarter] Doereng III. The gentleman was driving without a driver's license. If he had had the license, the accident certainly would not have happened—[110]

Accident Chronicle . . . on an interior road near Modjokerto, a dogcart in which *wedono* [high-ranking native official] of Goenoeng Gendeng was traveling was hit by a car coming from behind. The *wedono* was thrown out of his little carriage by the impact, and he broke his leg. The car took him home.[111]

There were some hitches, inevitably. A certain Mr. Arriëns, *Magneet* reported, a member of the club, was beaten at a scene of one of the early Indies accidents. Evidently, he was considered by native onlookers to be the cause of the accident. He was attacked by stones first, as might be expected in the case of natives being enraged. Then, however, as *Magneet* reported, he was "worked over" by other means, too, including *Engelschen sleutel*, a monkey wrench. Fortunately, a Dutch official, an assistant-resident, came to the scene in time, and he took Mr. Arriëns away in his car.[112]

There were modern-road accidents, and, naturally, they multiplied. Yet, somehow, most of the time, they appeared to *Magneet* to be easily dealt with. In a sort of excursion-wise way, as all the rest of the news, fleetingly, the crashes on the road were made to read as a part of the club calendar. Placed always on the same page, among other regular features, they all seemed to happen on Sundays mostly.

Auto-terreur, "terror by automobile," breaking the rules by the natives themselves driving motor vehicles, was mentioned, in *Magneet*, but carefully and very rarely.[113] Natives driving on modern roads were overwhelmingly those in an animal-driven cart or—instant victims and the most ephemeral figures—on bicycles. Whatever might happen to them, articulated by *Magneet*, they were jocular. In the language of *Magneet*, with a preference, they were called *bruine broeders*, "brown brothers," or—in a mash of English (the *à jour* motor-club language) and Malay (the language of natives)—*would-be toeans*, "would-be-masters."[114]

The monkey wrench, with which Mr. Arriëns might have been beaten, in *Magneet*, disappears quickly amid, virtually, heaps of language describing other tools and other technologies. There were types of motorcycles described and depicted, a most impressive instrument in itself, as one advertisement in *Magneet* put it, "*Made Like a Gun*."[115] There were whole pages, naturally, on different segments of the motorcycle. As smoothness was the motivation, surface was in focus, and outer parts of the motorcycle were especially favored. Motorcycle reflectors were the most frequent and prominent presence; only the technology of honking, maybe, could compete with them. In these cases, the richness and extravagance of *Magneet* vocabulary peaked. Motorcycle klaxons, for instance, were *claxons, autofoxen, sirenes, gabriélles, olifanten-kasten*.[116]

In 1913, the magistrate of the city of Batavia decreed that noise and light by these very technologies had to be restrained or kept outside the city limits. Only "one-tone signal horns," for instance, were to be allowed.[117] To Mr. Lankhout, however, affiliated with the *Magneet* motorcycle club, this was not acceptable. He went all the way to municipal court to fight for his three-tone trumpet. His campaign was followed by *Magneet* in detail. The frivolous technology made for a frivolous journeying, and it was, justifiably, accepted as heroic, as a struggle for the smoothness of it all. As the matter was heard in the court of Batavia and *Magneet* reported it, Mr. Lankhout

> had all the native deputies on his side, as they understood nothing of the entire proceedings—somebody has forgotten to translate the debate into Malay, and they were simply saying "yeah" and "yeah." Only one *hoofddjaksa* [native prosecutor], who could understand Dutch, voted against. In this way, Mr. Lankhout could keep his trumpet.[118]

Nothing was said about the Dutch members of the court. Nothing needed to be said. Bemused, perhaps, by the native deputies' "performance," they voted for the klaxon.

Amusement (and inner calm), *Magneet* has shown, might be attained by quoting natives talking. A native policeman, in yet another *Magneet* story, decided on one occasion to stop a *Magneet* motorcycle club member who sported too big a lamp ("floodlight") on his machine. The burden of the story was the native talking: *Toean paké sekijnwerver tida bolé*, "Mister uses lomp [sic], that's not allowed." The hilarious spelling of the technical term—it should be *schijnwerper*, "light thrower"—was the trump of the language-game.[119]

As one might expect, racing was another important signature of *Magneet*. Even more than when on a club trip—beyond any discussion, and with absolute justification by the rules—when racing, or reading and writing about racing, one had an obligation to be impatient, and to have a zero tolerance for everything on the road that might impede the smoothness of cruising. The *bingoeng*, "perplexed," natives—simply, by purely technological means, by the very nature of the race—were excluded. The biggest race *Magneet* reported, the "Java Races," took place in 1915. The first prize was to belong to the motorcyclist, who, any time during September and October, could cover the distance between Batavia and Surabaya in the shortest time.[120] A *Magneet* club member, P. Heidsieck, made the trip on September 29 and 30 in thirty hours, and his report of the race was published in the magazine:

> Allah! My fears were well founded. . . . First, the damned baggage carrier! . . . Then: hey, all of you, the Dutchmen with some authority and with that "ethical approach" to it! . . . you are so eager to teach our brown brother to read and write . . . you are busily turning him into a Western dandy . . . if you can only teach him . . . to decently keep on the left side of the road as I am passing by on my motorcycle. If he just can learn this . . . he may, then, also become *my* brown brother. . . . I am truly fed up with all the *grobak* carts that compel me to reduce my speed. These fellows always wait until it is too late to get past them. You may think that they are afraid of klaxon. You should know better! Only my motorcycle *Fallot* lamp stirs some respect in them.[121]

Power in the late-colonial Indies, as *Magneet* language had it, could only be as potent as it was race-like, and racy. In August 1915, the Dutch governor-general of the Indies made an unprecedented inspection tour throughout an extensive area of Central Java. For the first time in the history of the colony, a large part of the tour was made in an automobile. The tour description spread over several consecutive issues of *Magneet*. The language of *Magneet* in the unusually long narrative was wholly uninhibited, and nowhere did it come to such fruition as in this case:

> *Central Java* . . . August 1915 . . . Hundreds of kilometers are being covered by the governor general in a couple of days, sometimes at a racing speed. . . .
> *Le circuit through 5 Provinces.* . . . an average speed 40 km/h! . . . we are in the old town of Cheribon. Why was the road not cleared before we came? Bang! Clack! Just in an instant, and I might have a foreleg of a *deeleman* horse in my car . . . anything like that should not happen, naturally. . . . The road along the sea [in contrast], is good, and we can travel at a steady speed. There are no carts moving, no, naturally there are none; . . . the order is well maintained. The road is entirely for us. Yippee, driver![122]

There was not much, in the smooth language of *Magneet*, about pavement or asphalt. It was too early for this, perhaps. Only on one or two occasions, the club members complained about the roads being "bad, and covered merely by that hateful rubble."[123]

In another Indies text four to six years younger than *Magneet*, and more on the serious, almost scholarly, side, the hardness of the road had already become the main issue. H.F. Tillema, the author of the text, came to the Indies at the end of the nineteenth century, and he became an owner of a big pharmacy in the city of Semarang. Early in the twentieth century, Tillema was involved in Semarang politics, and was elected a city councilor. He wrote extensively about the Indies. Among many other things, he was an organizer of some of the early motor races in the colony.[124] Tillema left the Indies shortly before World War I broke out, and, in Europe in 1920, he published his treatise on colonial roads.

Like IJzerman, the prospector of the Sumatra colonial railway, Tillema wrote as a pioneer, but unlike IJzerman, he had no patience with his roads' untamed parts. Tillema sensed the hardness of the modern roads and was on the hardness side. Like Kartini, Tillema believed in progress as cleanness; his, also, was a clinical cleanness—a pharmacy cleanness, indeed. Unlike Kartini, however, he was determined never to let himself be carried away by some boyish dreams.

Hardness was needed. "In Semarang," Tillema opened his article, the famous "Road of Daendels," the road built as far back as 1811, the pride of the Dutch in the colony, the most significant connection between the east and the west of Java, had become "one big latrine."[125] There was too much dirt on the road, and the dirt, Tillema carefully made it clear, came from the outside, from the beyond of the modern road. "The market people arriving from near and afar" brought the dirt on their feet. "Often," Tillema added, they defecated on the road, "sometimes in the very middle of it."[126] Day after day, Tillema wrote, "Siddin, so let us call the native garden boy," dumped the garbage from his master's house "from the kitchen and stalls," "instead of in the trash container," "over the fence," "with his gentle hand," upon the road.[127]

This was *stofjeremiade*, "dust jeremiad," Tillema wrote. Modern motor vehicles moving with some speed, and then using brakes, trams, trains, and bicycles, but most of all, "namely, the animal-driven *grobak* carts, buses and trucks, with their wheels and tires worn out and on totally off-centered axles"—vehicles so loved by the native Siddins, by the way, Tillema added[128]—they rode over "the feces of men, horses, and buffaloes, and made them into dust." The dust, then, got off the ground, all over the place, and it was *inwaait*, "blown into," open windows, especially of the houses near the modern road:

> It has been noticed by those who pass through the area between 5 and 6 P.M. that on the Boeloe section of the road, for instance, the layer of dust is several centimeters thick. As it is raised by bicycles, cars, and carts, people are infected by plague, cholera, typhus, and other things like that.[129]

It is widely known that the high child mortality, to a large extent, is caused by the dust, and that the inhaling of the dust, either organic or inorganic, causes serious throat, nose, and lung disorders . . . the dust of the road is a most serious threat to the health; it may be a matter of life and death . . .[130] Typhus spreads through the air. Pneumonia, also . . . and other pathogenic organisms as well.[131]

Sometimes, a little piece of stone, even, might be catapulted off the road by a passing vehicle, and the missile might be blown into a house, our house: "From what was said, it is understandable that the Europeans in the Indies consider the dust on the roads not just a great nuisance, but that they fear the dust for reasons of hygiene, and for other things."[132]

It was natural for Tillema, as a pharmacist, to appeal for modern roads to be taken care of, "healed," as modern man, in the same unkind tropical land of the Indies, was cared for and healed: "Just as we are ready to serve *obat* [Malay for medication] to the people here who are sick, so it should be the case with the sick roads. What castor oil and quinine make for the man in the Indies, water and oil residue can do for the roads."[133] Modern roads, wrote Tillema, should be soaked in water and oil residue so that the dust might not rise. Two photographs, of a "spray car" and "sprayer," a native with a "spray mechanism" on his back, were published with the article.[134]

Yet spraying the roads was not the real solution. Soaking might cause softness, and softness in the final definition, Tillema stated emphatically, was a threat. There was too much softness on the Dutch Indies roads, anyway. The tropical torrential rains soaked and eroded the roads constantly. Softness was everywhere. Even the stone most often used in the Indies to pave the new roads, *Indische grint*, "Indies gravel," as Tillema lamented, was dangerously soft: "The gravel is very porous, and it can absorb much water when it rains, which may be good. However, it keeps wet a long time, which is very bad."[135] What was the worst was mud!

Tillema mused about wooden-board roads, as were recently tried in Paris, France.[136] But something more Indies-like, evidently, was to be done in the Indies. What truly was needed, Tillema concluded, was a *bindmiddel*, "binding agent," and a *bindkracht*, "binding power."[137] Various oils might be used, he suggested, perhaps petroleum. Promising also, he offered, might be *melasse*, "molasses," the syrup produced by refining of sugar, the substance then being "mixed with slaked lime." This should work especially well, he thought, for the areas around the big Indies sugar plantations in East Java.[138] Yet another solution was mentioned by Tillema, termed "curious" by him, but, it seemed, not wholly discarded: "old shoes originating from the armies fighting in the past war [First World War] . . . 5 to 10% of powder . . . mixed with stone and asphalt; 57,000 pairs of old shoes [should be enough for] 1 km of road."

Street sprayer in Batavia, ca. 1916. (Tillema, *Een en ander*)

"It is a pity," Tillema added, "that most of the Javanese still walk barefoot."[139]

To feel safe on a modern road, went the message by Tillema, the road should feel hard and antiseptic. *Wegen-hygiëne*, "road hygiene," since the 1920s especially, became almost an indispensable part of any debate about the Indies communications.[140] In yet another article, Tillema again warned: "One has to be constantly aware that all improved traffic brings, also, more infection."[141] Tillema, sooner and more clearly, perhaps, than many others, saw that dust and mud is also a communication—a modern communication, indeed. Equally soon, he knew that, because of that, an overextensive contact had to be avoided.

"Whenever we go out, to see a movie, whenever we travel by a tram or something like that," one of several already existing Indonesian nationalist papers *Indonesia Raja*, "Great Indonesia," wrote in 1929,

> there are against us signs and billboards with "Native" written on them. This may seem to be no big matter, and not a reason serious enough to feel bad. However, when we think about it a little, it becomes clear that there is some deeper meaning beneath. The deeper meaning of those planks is that they divide "us" from "them."[142]

According to *Indonesia Raja* again, and also in 1929,

> Now we want to talk about a Dutchman as we can see him on a bus, or as we are moving through our lives on the same street. . . . There, the Dutchman is thrown into the maelstrom of the Indonesian people. . . . There, the Dutchman has to show his good intentions, there, on the street, the Dutchman, as now we can see him, the Dutchman on the road, and in the train. . . . There, it is there, that we are suddenly capable of thinking of "an intimate bond."[143]

Tillema's warning about infection, it appeared, was timely, and it was taken seriously. The point is, here, that the antiseptic quality went together with—and was as important as—grammatical correctness or correct driving. *Verminking*, meaning, in Dutch both "erratum" and "mutilation," was a standard term describing the natives in colonial offices misspelling Dutch telegrams they were to transcribe, or mispronouncing Dutch messages they were to convey on the telephone.[144] Natives, as the word *verminking* had it, were speaking and writing flesh and blood, or simply mud.

Wherever the natives went, and especially as they dared to approach a modern road, they were read and pronounced as carrying that soft stuff on themselves, on their tongues, on their feet, and on their wheels. Even by way of language, thus, they were made constantly to appear on the verge of slipping off the correct lines. There is a Javanese woman in Louis Couperus' novel, *The Hidden Force*. The woman is presented as, possibly, a dangerous native. She lives in "a half-marriage" with the resident, a retired Dutch official, and thus she is perceived as almost moving toward modern ways, toward modern roads, and into Dutch language. As the novel develops, the Javanese woman is encountered by a Dutch woman, Eva, the epitome of Dutchness, pure, correct, and, eventually, safe. Both women talk, and this is how, by the genius of the Dutch author of the turn of the century, the threat of the native is made to go away: "Beginning with a few words of Dutch and then taking refuge in Malay, smiling politely, she . . . did not know what to talk about, did not know what to answer when Eva asked her about the lake, about the road."[145]

Celeste Langan, in her study of *Romantic Vagrancy: Wordsworth and the Simulation of Freedom*, writes too about a passing encounter on a public road, and about common language. A bourgeois poet and a haggard vagrant seem to speak the same language on the public road. However, "The poet and the vagrant speak the same language only insofar as the common practice—the logic of infinite circulation represented as 'coming and going'—has detached their speech from the regional affiliations that might identify it as a dialect."[146] Langan also connects the encounter and the common language with "simulating": "simulating . . . the ethical ideal of liberal democracy . . . imagination of freedom: the right of individuals 'to come and go without permission and without having to account for their motives or undertaking.'"[147] The aim of this, Langan suggests, is to "attach" language "not to the soil but to technology," namely, the technology of "coming and going," the technology of the road.[148] "Old" language, whatever it might mean until now, is "expropriated" into a new common language of the road.[149] Through the new language of the road, then, an attempt is made to "abstract the world from its social conditions,"[150] to make it appear as if it is, technologically, suspended in the air.

Vaktaal (Dutch for technical language), the late-colonial language of the road, was hurled upon the native as the romantic language, in another time and another culture, was hurled upon the vagrant. Old language of the native, whatever it might have meant until then, was expropriated into a new language of the road and made as if suspended in the air.

There was much hazard in this language-game. In the *common* language of *public* road, a sense of ownership might easily slacken. The Indies was a very carefully ruled colony, yet by 1935, 150 Indonesian engineers had graduated from a colonial technical college in Bandung,[151] and in 1938, 13,500 Indonesians or so studied in the colony's low- and middle-level technical schools.[152] As Wittgenstein points out, in contrast to the "words of love" and other "words with atmosphere," technical terms are facilely exchangeable.[153] Unruly things might happen when native technicians and engineers encountered Dutch masters on the public roads of the Indies. In order to make the late-colonial technical language stick, to cling to ownership, atmosphere and love were to be faked.

There may be something, in every technical language, that makes this almost easy. As I clicked on "jargon," now, my *PowerBook Thesaurus* gave me back: "terminology, babble, gibberish, nonsense, balderdash, moonshine, palaver, twaddle."[154] Faking love and faking atmosphere, and thus trying to save the technical language of the public road for the masters, this was what the motorcycle club journal *Magneet* did so consistently. The amassing of the trivial terms of trivial technology, the personification of road machines, or the affectionate diminutives of spare parts (*lampje*, "little lamp," *claxontje*, "little klaxon")—all this was done to keep words by those who invented the game. The words, and the language with the words, were made smaller, into broken pieces, were gritted and mashed into semi-organic little corpses of memory.

Technical language could be befuddled. The main Indies magazine for police agents published a series of long articles in 1940 on the problem of modern versus unmodern traffic on the new colonial roads. A modern user of the road was not described in the article at all, was termed "general traffic," and was left to be merely felt, unaffected, and (thus, safely) at a distance. The unmodern was carefully articulated—and twaddled. He was a native horse-cart driver. In the otherwise wholly Dutch text he was called consistently in native Malay *Pak Koesir*, "Papa Coachman." Papa Coachman's position was set, and photographs were attached to show it, by a tennis-court-like design of straight lines running parallel or at right angles to each other. The lines were drawn on the road's clearly smooth surface, most probably asphalt, and they were marked *a, b, c*. . . . The horse, harnessed to the cart, stood calmly at Papa Coachman's side. The poor beast was labeled *p.k.*, for *paardekracht*, "horse power."[155]

Wittgenstein was also quoted as saying: "When attention to detail is abandoned . . . the real function of statements is missed . . . language *goes on holiday*."[156] The late-colonial language in the Indies, indeed, the technical language of modern road, as it was proudly trivial, was also imperiously vacational. It was an important rule of the game, especially as the colonial crises deepened, that attention to detail should, boisterously, be abandoned, and that the real function of statements should, buoyantly, be missed.

In the *Volksraad*, an advisory council of the governor-general of the late-colonial Indies, in July 1938, modern roads and railways came again to the debate. The next step, in the policy of the colonial government to improve the modern roads, a government spokesman told the *Volksraad*, was to re-model some of the features of the railway's first class. This rather curious priority met with a general approval of the Dutch deputies in the *Volksraad*, even while all the speakers admitted—they boasted actually—that there was not much in the first class to improve.[157]

The authorities, as some deputies suggested, it is true, might "take into account," in the future, the possibility of adding a few new carriages to the third class.[158] But the tenor of the debate was markedly elsewhere. Madame Deputy Razoux-Schultz, very outspoken on the issue all the way, talked about the bad ventilation and "odor" in the third, native, class. Now and then, she said, so she heard, it was not even possible for the service (a waiter?) to reach the third-class passengers: "In the 3rd class, the service, truly, is not always equally good. Especially during the rush hours, the travelers of the 3rd class might not even get any service at all, as even the ordering may be difficult."[159]

The burden of Razoux-Schultz's speech, however, was in her ardent plea for tourists. They, she said, traveled on the railway's weekend lines, to the mountains and the spa resorts of the colony. Their bikes in particular should be transported by the Indies railways for a radically lower price than was the case at the moment: "One feels so strongly for the bike parties, which have to use trains on their longer trips, and it is not easy."[160]

Like the first-class train carriage, the structure of the late-colonial Indies modernity was to be constructed: layer upon layer, in a multitiered and upholstered realm that would rest upon what, underneath, a long way down, at the very bottom perhaps, might be the salt of the earth. In the same railway debate in the *Volksraad*, yet another outspoken Dutch deputy, Van Helsdingen, agreed with all that had been said and aggregated the argument by adding, playfully, one little complaint:

> At present, the accommodations in the 1st and 2nd class leave very little to be still desired. The cooling in the new carriages is outstanding, and sitting space is so excellent that it is an undiluted pleasure to be in it. Sometimes I even wonder whether we did not go too far here. The springing of the seats is so good that one actually cannot peacefully read any more. As you settle down with your book, you are swayed and rolled up, down, all over, and all the time, so that one of the greatest benefits of the train travel might be taken away.[161]

A blanket of jargon, vacation language, and twaddling modernity was being laid over the Indies. Mas Marco, again, soon detected the binding power and the muffling power of the triviality—sooner, in any case, than most among his brothers, the soilsons, the natives, who were thus being veiled and silenced. If we leave the mushy substance on us, untouched, Mas Marco warned, it will harden into an unbreakable crust. In 1914, Mas Marco's early Indonesian journal *Doenia Bergerak*, wrote:

> Automobile, in our time, is a vehicle best loved by all the dignitaries and the capitalists. . . . Nowadays, indeed, what is considered most powerful, and toward which all the strengths and all the time are exerted, is improvement of roads. Big roads become better, more beautiful, straighter, and smoother every day. Measure by measure, as the roads are plastered, covered with concrete and gravel in the most advanced manner, the mountains of gravel also grow. . . . It is certain that this is the future; even the village roads, and all the hamlet roads—no road is too

small for it—will be made bigger, wider, more attractive, and smoother. The mountains of gravel will grow higher. In our forests, we will keep on digging for the gravel, and the mountains will grow still higher, higher than the real mountains. . . . Already, peasants dig for the gravel, instead of digging in their fields. . . . It is certain that there will always be enough gravel . . . and, at the end, no car will ever break down on the good roads.[162]

The best of the Dutch engineers in the colony knew as much as Mas Marco knew. In 1919, for example, a a successor of Dr. IJzerman of sorts, a well-known and much respected railway engineer, Rietsema van Eck, defined the Dutch civilization in the Indies as "an upper or surface authority laid over, and covering a native authority."[163] About the same time, another outspoken Dutch technician in the Indies, engineer-architect Plate, in a lecture before the learned Indies Society in Batavia, described the Dutch rule over the colony as *cultuurloos maar gewapend. Cultuurloos* is "culture-less" and *gewapend*, either "armed" or "reinforced."[164] Plate was an engineer. Something even more advanced, a crust even harder than of gravel or asphalt, might be invoked here by the engineer's jargon. Engineer Plate certainly was well aware that *gewapend beton*, "reinforced concrete," even as he spoke, was becoming, in the East as well as in the West, in building up as well as binding in, the most trendy thing there was.[165]

This was the power of the Dutch rule and late-colonial language—not a capacity of their inner form to emanate power beyond themselves, but, power resting in (hopefully) being impregnable. Several decades after the Dutch colonial rule ended, in the post-colonial Indonesia of the early 1960s, talking to a graduate student from Cornell University in the United States, a very old and most revered Javanese scholar, Dr. Poerbatjaraka, still seemed to recall exactly this. The beauty of Dutch, so the student interpreted what the old man said, was in the language's "invulnerability to Javanese: In it one could say anything about Java and Javanese culture no matter how sacred the subject."[166]

Bahasa Indonesia, "Indonesian Language"

At the end of the monsoon season, in 1928, "Mas Arga, a journalist, and a native son of this land, started from Tjilegon on an expedition around the islands of Java and Madura. He travels on a motorcycle *Indian Scout*, with 1 guilder (one roepiah) in his pocket."[166] The image of the Dutch rule, covering the Indies evenly, like asphalt, must time and again be qualified. There was always, and increasingly, a number of natives who made it onto the modern late-colonial roads and who did not slip off their smooth surface. There were many, even, who moved faster and faster.

One has to be skeptical, of course. This was a very unequal situation. It was easily possible that the sort of encounter described in Celeste Langan's story of nineteenth-century England was being repeated in the Indies:

> I put my hat upon my head
> And walked into the Strand,
> And there I met another man
> Whose hat was in his hand.[168]

The meeting of a bourgeois and a beggar through the common language of the public road might be as relevant a model for capitalist England as for the colonial Indies. A picture, to paraphrase Langan, of the beggar as a "subject without property" being turned—through the culture of the public road—into a "subject without properties."[169]

Mas Marco, again, appears to be the best author to read when looking for answers. He has been called the first modern Indonesian journalist and writer. He almost never wrote in his "mother," "native," "old," tongue, which was Javanese. Virtually all the time he wrote in Malay, the lingua franca of the archipelago throughout the centuries of the Dutch colonial period. Malay, as the first language in the late-colonial Indies, was spoken merely in parts of Sumatra, and a dialect of Malay was spoken by some in the big cosmopolitan cities, primarily in Batavia. For most of its users, Malay was a practical, and not very warm, medium of communication with strangers.[170] In other words, it was a language of the road.

For Mas Marco, beyond this, Malay often also appeared to be a language of exile, a "linguistic exile,"[171] a voluntary exile from an easy sense of home, and a simple shifting between places. Words of his sentences, it often seemed, were not made to fit together readily. Frequently, unexpectedly, he broke his language, fissured it, with an utterance as from another world. Dutch words and technical jargon he used often, overused even. Sometimes a Javanese word appeared as an incongruous flash and remained as it was, unaffected by its verbal and syntactic surroundings, occasionally even sticking out from the newspaper page in Javanese script.

While the Dutch Indies police magazine, as we have seen, called the native coachman's horse *p.k.*, for *paardekracht*, "horse power," Mas Marco, with the pleasure of a linguist and a rebel—or so I would very much like to believe—made the Dutch distinguished *Welvaartscommissie*, "Welfare Commission," into *W.C.*,[172] or sometimes made the word by which he called his brothers, the people of the land, the sons of the soil, *boemipoetra* (*boemi* in Malay, is earth or soil, *poetra* in Malay is son) into technical, farcical, and biting *b.p.*[173]

Instead of Tillema's *Kromoblanda*, a hybrid that should denote a calm and happy land of "native(and)Dutch," Mas Marco used *kromolangit*, "our skies," *kromoremboelan*, "our moon," and *kromobintang*, "our stars."[174] He used expressive Malay-Dutch hybrids of his own, neuters of modernity—so again we like to believe—like *rasaloos*, "feeling-less," or *maloeloos*, "shame-less."[175]

He was said to write *koyok Cino*, "like Chinese."[176] The language of Mas Marco made uncertainty. It might confuse the order that was a must of *Kromoblanda*—across racial lines, for instance. Mas Marco wrote in a language that could easily be mistaken for, or thought identical to, what was described as Sino-Malay, the language of Indonesian Chinese. Both were often called, and misnamed, "Low Malay," "Batavian Malay," or "*bazaar* [market] Malay."[177]

Mas Marco seemed to master scratching and cracking whatever might be polished and smooth in the colony. The Dutch authorities, at least, thought so. Repeatedly, Mas Marco was arrested for what he had written and sent to colonial prison. Mostly, his crimes were defined—appropriately, given the guarded impregnability of the late-colonial roads and language—as *haatzaai*, "hate-sowing." Properly, so it also seems, given the culture, Mas Marco died in 1932 in the notorious Dutch colonial internment camp on Boven Digoel, New Guinea, thus in exile and on the road.

Mas Marco's language was justly described as "witty" and "sharp," but, also, with an equal justification, as "speedy."[178] This brings to mind again a very disquieting thought. Was not Mas Marco's language—in spite of his scratching and kicking, in spite of his rebelliousness and love for his country—the new kind of language Celeste Langan is writing about, a "language disfigured by a logic of infinite circulation," a "peddler's" language? Is not Mas Marco the first great Indonesian writer, actually, by installing this language? Did not he tie his writing so well to the road that—in the weird company of the writings by IJzerman, and Tillema, and Kartini, too—it became suspended in the air?

In 1928, all the principal Indonesian nationalist youth organizations convened in Batavia, and, in the Youth Oath, the young people pledged their lives to "One Archipelago, One People, and One Language."[179] The language they meant was *Bahasa Indonesia*, "Indonesian." It was Malay, in fact, but correct Malay, "High Malay" actually, as standardized by the Dutch colonial government and government-owned *Volkslectuur*, "People's Reading," or *Balai Poestaka*, "House of Book," publishing house. Some even called it "Ophuijsen Malay," the name of the Dutch government language engineer chiefly responsible for it.[180] This language, now, was newly conceived, or constructed, by the emerging generation's sheer will to become independent of the Dutch.

In 1938, *Partai Indonesia Raja*, "the Great Indonesia Party," the biggest indigenous political party of the time, instructed its members to use only Indonesian in all their public statements. In the same year, even the native deputies in the *Volksraad*, the most conservative wing of the nationalist movement, decided that, forthwith, they also would use only Indonesian, when speaking in the council.[181]

A young, brilliant, and now forgotten Indonesian intellectual, Soesilo, in

May 1939, explained in an article in a liberal and Dutch journal *Kritiek en Opbouw*, "Criticism and Construction," how daring, indeed hazardous, it was suddenly—by political, cultural, linguistic will—to decide for, and to cling to, one's own chosen mode of expression: "We must struggle against everything; constantly, we have to perform and to keep on performing an acrobatics of brain! Yet, we have to have the language if we want to build a unity in our Indonesian struggle."[182] And, in spite of all the acrobatics and daring, Soesilo wrote, one whole region of language, culture, and hope had to be left out: "the field of technical sciences, all the matters that one can grasp only through the purely Western scholarly concepts. This is necessary, otherwise our own cultural values might suffer."[183]

This appeared to be just another kind of making a language in layers. For the sake of struggle, the "language of unity" was being engineered with little time and strength for nuances and accents. It was a potential asphalt language of a weak and emerging nation. Whoever might look for another way was in a danger of appearing to have mud on his feet.

Sarmidi Mangoensarkoro was a teacher in *Taman Siswa*, "Garden of Pupils," a system of Indonesian nationalist schools built up as a modern, yet Indonesian, alternative to Dutch and Western education. In quite an emotional series published in 1939 in a school journal, *Keboedajaän dan Masjarakat*, "Culture and Society," Sarmidi described—similarly to Soesilo—how difficult it was to speak out and to write out. Unlike Soesilo, against a polished and carefully uniform language, he pleaded for *mengeltaal*, *Mischsprache*, or *gado-gado*, "mixed-language," "blended-language," potpourri language," a language of not just a single taste:

> Some people, so it seems, close their mouths and keep silent, just in order not to speak the *gado-gado* Indonesian. Some people, it seems, are afraid that if they spoke the language, they might lose the respect of others. . . . Yet, whoever studied the history of the English people, for instance, knows that theirs, too, is a mixed language, *mengeltaal*, *Mischsprache*. And, who would argue that English is not a language to be respected?[184]

According to Sarmidi, Indonesian should mature, be spoken, and be written naturally:

> *First*: The progress of such language should happen not merely by intentional care, but, also, and very much so, by the language's own potency. . . . *Second*: The progress of such language should happen continually. . . . *Third*: The path for the progress of such language should not be staked out by rational calculus, but the progress should be driven by fate.[185]

This was a courageous, but, perhaps, a rather desperate position. In this moment of history, someone—Dutch as well as Indonesian—may acclaim "traditional culture" and get away with it. To write, however, about an "unstaked path" and "progress driven by fate," one placed himself on the modern road, in the middle of it, and, inside the language of the road, one stood there, not too much different from the Papa Coachman of the late-colonial stereotype, barefoot, *bingoeng*, confused, bewildered, panicky, perplexed, laughable amid the general traffic.

Sjafroeddin Prawiranegara, a Sarmidi partner in dissent, as he wrote about the Indonesian road to modernity, and about the language, even spoke of grass! Sjafroeddin, in the late 1930s, had been a student at the exclusive Batavia Law School and, also, an Indonesian-Islamic student activist. In an article for a student journal published, like Sarmidi's series, in 1939, Sjafroeddin started by warning that "A language must not become a machine." Something bad was going on, he wrote. "In fact, now, language is often called an *instrument*, and even a commodity." This was how the market was getting its sway: "The West speaks too much."[186]

Sjafroeddin was a Muslim and a patriot, and he genuinely resisted the Dutch rule as it developed in the Indies. He tried to ascertain a position for the Indonesian people in the modern world, and he rejected the Dutch idea that

> through the centuries, the [native] people completely lost a memory of its origin, [and that] it was purely Western and foreign scholarship, linguistics and ethnology first of all, which retrieved the [native people's] connection with the past and, thus, the only concept on which Indonesia might be based in the future.[187]

Like Sarmidi, Sjafroeddin accepted the rule of the game, stepped upon the modern road, and spoke of the language of the road. He was a modern man. History, and the history of his nation as well, Sjafroeddin was convinced, moved by the force of, as he put it, contradictions. Contradictions, as Sjafroeddin defined them, "dry out like a banana." "Contradictions," he wrote further, "are solved thus, that the old vanishes into the ground, and it leaves just a trace on the ground." Whatever humans and their scholarship ever did, he wrote, had been merely to put a light touch on the pliant face of the earth: "Already before Copernicus the Earth was round. Only people thought that it was flat. Since Copernicus we have a round Earth and a totally *changed and broadened* world view. . . . However, the Earth remains the Earth as it had always been, and it will remain the earth."[188] Sjafroeddin offered an unacceptable sense of culture, and Indonesian culture as well, as dirt, as flow of water and blood, as roots of grass breaking through armed or reinforced and culture-less modernity. The Indonesian blood, Sjafroeddin wrote, had always been, and always will be, like the Earth and the Earth will always be. "Always, it will flow, and, where it will not be able to stream, it will creep, like the roots of *alang-alang*, coarse tall grass."[189]

Dissidents in every movement are like a broken accent. In the Indonesian nationalist movement, too, there could be only very little tolerance for them.

Steamrollers of the Colonial Public Works. (KITLV)

Unbroken velocity, rather, speed and/or belief in speed, made Indonesians into a mainstream. From here, also, came the idols. This is what *Soeloeh Indonesia*, "Torch of Indonesia," the most influential among the Indonesian radical nationalist journals, wrote in 1927, a year before the Youth Oath. The author signed himself or herself as *Setijogroho*, "Passive Resistance":

Self-knowledge, self-confidence, and self-respect
. . . Comrades . . . Know yourself. Respect yourself. Believe in yourself! . . . This is from where our power stems! Learn by heart what Henry Ford is telling you.[190]

Against the full-blown Dutch rule in the Indies, a full-blown Indonesian nationalist movement was rising. With impressive speed, the movement set for itself a clear-cut choice—in fact, so clear-cut a travel plan that it had nowhere else to go. Soedjojono, a young promising artist and an Indonesian nationalist, who after 1945 was to become the celebrated painter of the revolution, wrote a series of articles in 1939 about how the Indies of his time was to be seen and painted. Not anymore the lush tropics, with the half-naked women. Soedjojono demanded, instead: "sugar factories and starved peasants, automobiles of the rich, and pantaloons of the youth: shoes, trousers, and gabardine jackets of tourists on asphalt roads."[191]

Modern roads and railroads, or so one tends to expect, were the veins and the arteries of the movement. The movement's pain and hope, as well as its gathering sense of revolution, were supposed to concentrate on the modern roads.

Sopir, "Chauffeur," was a little Indonesian journal that belonged both to the movement and to the modern road. It was published in Yogyakarta, Central Java, in the early 1930s, through the time of economic depression by a cooperative, and then a trade union of Indonesian drivers. Taxi drivers and hired chauffeurs, so it was stated by the journal, made up the bulk of the union membership and of the readers.

In virtually every issue of *Sopir*, there was one article of a series running through the journal's existence: *Nasib kaoem Sopir*, "The Fate of Drivers." The same illustration at the head of each article showed a driver behind a car wheel; the car was modern, the driver was an Indonesian, and he wore a *pici*, black velvet skullcap, since the 1930s the most visible marker of radical Indonesian nationalism. The fate of the Indonesian drivers, the series wrote, should be cause for grave concern:

In the early days, and, if we are not mistaken, until 1908, the profession of drivers in general was considered a good one, and it was highly valued. This could be seen from the fact that salaries were high, and the Dutch, the Chinese, and the Indonesians all liked to be drivers. Gradually, however, the esteem for drivers declined, and, at present, there are rarely Dutch who may like to become drivers, not many Chinese want to do it, and, indeed, only we, the nation of the Indonesians, remained with it. We, so it appears now, hold a monopoly.[192]

Solidarity among the Indonesian drivers was most energetically encouraged by the writers of *Sopir*. What should keep the Indonesian drivers together, according to *Sopir*, was the drivers' professional pride. The most impressive leading articles were "Why do some motors not run even when ignited with starter or handle?" and "What we need to know about the speed on the road": oxcart made 4 km/h, horse cart 5 km/h, regular bicycle 18 km/h, tram 35 km/h, ocean liner 40 km/h, horse at gallop 45 km/h, stop train 45 km/h, fast train 60 km/h, express train 75 km/h, car on a medium-good road 60 km/h, and airplane 150 km/h.[193] *Sopir* also gave advice to its readers on how to buy a second-hand or third-hand car, and how to start their own taxi business, for example.[194]

Kesopiran, "driver-ness," was a word that appeared in *Sopir* as often as "nationalism," "fate," or "future." Indeed, they seemed to be merely various forms of the same idea. Driver-ness, according to *Sopir*, was the capacity of drivers to find their place in the world. The essence of driver-ness was to care for the car.

As the editorial of the first issue of *Sopir* declared:

> this is the first magazine about the knowledge of motor, published by our nation itself. This is the aim of this new journal, to spread an awareness about the matters of motor car. . . . In connection with this aim of ours, we hope that this magazine will be useful not merely to our drivers, but also to all those among us who like motor cars and motorcycles.[195]

There were, according to *Sopir*, two fundamental obstacles in the Indonesian drivers' way toward a better future: "I. the number of drivers is larger than the number of motor cars; II. the drivers as yet are not fully aware of their driver-ness."[196]

Time and again, *Sopir*, and its series on "The Fate of Driver," explained that driver-ness was driver's awareness of oneself as driver, and that it meant:

> a. to care about the car thoroughly, diligently, industriously, and cleanly; to correct everything in the car that is not right, to maintain painstakingly and to protect the car as a whole, and all its parts . . .

> b. to take full responsibility for the car's safety [*keselamatan*, Indonesian for happiness, welfare, safety], and to drive the car in accordance with law.[197]

This, again, was why *Sopir* was here, "to unfold and to explain to the nation of Indonesians the significance of motor."[198]

In April 1932, *Sopir* published the statutes of *PCM, Persatoean Chauffeur Mataram*, "Union of Mataram (Yogyakarta) Chauffeurs," the drivers' union in the region that had been founded the previous November.[199] In section 5 of the PCM statutes, driver-ness again was defined from both the point of view of nationalist politics and traffic rules: "Everyone can become a member of the PCM, who is of Indonesian nationality, originally from the region of Djokjakarta, and who owns an automobile driving license."[200]

In the same issue, the emblem of the new trade union was shown and explained. The shape of the emblem might vaguely resemble *Garuda*, the mythical bird-vehicle of the Hindu god Vishnu, not an unusual symbol in the Indies and in the Indonesian nationalist movement as well. But it was no Garuda. It was, as it was explained by *Sopir* at length, an image of a car:

Purpose And Aim Of PCM Emblem

The emblem of PCM represents a car as seen from the front, looking toward the mask of the radiator. On the left and right sides, there can be seen fenders and automobile lamps. Above the radiator *PCM* is written in distinct letters. Below, a *bumper* is visible, with the full name of our association painted on it. . . . Now, what is the meaning of all this?

First: The car radiator, in fact, is a device to cool hot air, so that the motor can work properly. For the PCM, this means that all lowliness, animosity, conceit, in short, all that is not proper and good for man, can be caught and cooled.

Second: The two lamps mean that the PCM will shed light of conscience on everything that otherwise might remain dark.

Third: The *bumper*, with the full name of the PCM painted on it, means that the PCM is capable at any time of robustly shoving off everything that might impede or disturb the happiness, welfare, and safety of the PCM.

Fourth: The wings on the left and on the right [there is a pair of wings painted as growing out of the car] wait impatiently to carry the PCM high and still higher.

Fifth: The name *PCM* painted above the radiator, in the center, suggests the will of the PCM to sit high in the driver's seat.[201]

There were reports, in *Sopir*, in this bad time of economic depression, of other drivers' unions "busily" emerging also in other parts of the Indies.[202] A letter to *Sopir*, from the Minangkabau region in West Sumatra in November 1932, spelled out, if this was still needed, why trade unions like this were being founded:

Before HCM [Himpoenan Chauffeur Minangkabau, "Minangkabau Chauffeurs Union"] Existed

You can easily imagine what might happen to a driver in Minangkabau before there was the union, and unity. When such a driver found himself on the road, in the middle of the road, in danger, when his car broke down, what could have saved him was only himself. Even worse, when he did not know what was wrong with the motor, or when he ran out of gas, or when he happened not to have the

A railway and a pedestrian bridge, East Java. (KITLV)

right tools with him, wrenches, and things like that, then he might be forced to
sleep in the open, on the roadside.[203]

Now, however, the letter explained, there was the union, whose members
were ready and willing to help. At some points on the roads in West Sumatra, small union repair shops, and even gas stations, were set up or
planned.[204]

This was an ideology, engineering, struggling, and dreaming in an asphalt
way, and in layers. *Grobak*, *tjikar*, or *sado*, the native animal-driven carts of
the Indies, were mentioned often in *Sopir*. The intensity of the driver-ness
demanded that the native vehicles and their Papa Coachmen be referred to
with an intense sense of apprehension—often in a state of high irritation.
Coachmen were not drivers! They were unruly, the cause of accidents, and a
threat to drivers. A brother magazine of *Sopir*, *Motorblad*, "Motor Magazine," lamented under a headline, "BUGGIES MUST HAVE LICENSE": There are
countless examples to prove this. Virtually every motor-car driver suffers, and
not to a small extent, by the brutality of the coachmen. They frequently and
with great indulgence linger in the middle of the road, and, in all possible
manners, impede the general traffic."[205] Everybody driving on a modern road,
the same journal suggested, should be obliged to carry a written report of
one's recent medical examination.[206]

At certain moments, to recall Wittgenstein again, words can be used to replace feelings. "Hurts," for instance, can be used instead of weeping. The
word "revolution," similarly, can be used to replace a sense of struggle and
change. The word "revolution" no longer describes revolution but replaces
and displaces it. Thereby it creates a new way of behaving. The word comes
between us and the world "like a sheath."[207]

There was something irresistibly raw in the battle cry of the big Indonesian railway strike of 1923, *spoor tabrakan*, "tracks crushing." It truly sounded
like trains, and words, crushing against each other. *Spoor*, "track," was Dutch,
and *tabrakan*, "crushing," was Malay. Terrorism is never nice, but this crash
was like touching. *Spoor* was Dutch, but in fact it was Malay, too, as the
appropriation of the Dutch word by the local population was very far advanced at the time. *Tabrakan* was Malay, and in fact it was becoming Indonesian, too, as the language was just being born out of Malay. This was one of
the brief moments in history and language when many possibilities seemed
open. Or, perhaps, the striking railway workers spoke so beautifully because
they did not have enough time to express themselves.

In 1939, sixteen years after the strike was defeated, the railway workers, of
course, and even some of their trade unions, were still there. The best known
among them, *Persatoean Boeroeh Kereta Api*, "Union of Railway Workers,"
published its own journal, *Sinar Boeroeh Kereta Api*, "The Ray of Railway

Workers." In May 1939, the journal published a report on the activity of the branch in Cirebon, West Java, one of the union's largest subdivisions. The language of the report, as of all the other texts in the journal, is very correct Indonesian. The strike, indeed, had failed long ago. The activity as reported is smooth. The meetings in the report are fleeting encounters at the railway stations.

It was, so it seems, the only kind of optimism that the culture and the language could give. The union ran as well on time and oily, as the colonial trains did:

At 2 A.M., train no. 2 from BTB-SB has arrived. The comrades were already waiting on the platform. But the train in the direction of KTS was 15 minutes delayed. . . . We missed the connection to BL and TA, the regular departure at 6.30 P.M. We arrived at TA by the next train, yet it was too late [for the meeting], and, therefore, we had to continue directly to Trenggalek in order to catch the connecting train. We could not stay long in Trenggalek, and we left for the *Westerlijnen* [the Western Rail Lines], stopping on the way in TA for about 1 1/2 hours to hold a meeting with the comrades at the TA branch. . . . At 11.29 P.M., we boarded the train to KTS, and there we had to wait for 2 hours for the connection. . . . [The next day] at 3 P.M., we boarded train no. 17, toward DK, and we arrived at DK at 6.30 P.M. As it was already late, it was the rainy season, and we had just 1 1/4 hours left, we could not make it to the meeting with the comrade chairman of the branch in DK. . . . Next day in the morning, February 23, 1939, we boarded train no. 23 to PWT. . . . At 4 P.M., we boarded train no. 19, back to Cheribon . . .

Warm greetings. Kartodiwirjo, Chairman of the Cheribon branch[208]

TWO

TOWERS

Towers hold bells, display clocks, provide targets for the eye,
furnish lookouts over prospects, proclaim power, mark authority.
They may be free-standing, but more commonly they will be
incorporated into a building that otherwise follows the
normative standards of a type.
—Van Pelt and Westfall, *Architectural Principles*[1]

Homes on Wheels and Floating Homes

THE ALLUSIVE virtue of style is . . . a matter of density," Roland
Barthes wrote.[2] Culture is a vessel, a receiver always full with a most
diverse mass. The densest and heaviest sediments of the mass concentrate always close to the bottom. To mark the density of culture, at the
crucial, highest-density spots, people erect stones and build up monuments,
cathedrals and temples, steeples, spires, and towers. Towers are anchors.

If towers mean anchor, then, in the modern Indies, remarkably, there was
a pervasive sense of anchorlessness, of floating—or wheeling at least. In
1842, one of the earliest Indies magazine, *Kopiïst*, in the same series in which
it wrote about railways,[3] reported at a great length on a "movable iron house
in Belgium":

> Thus far (with few exceptions), man built his abode firmly in the ground, and he
> fastened it tightly to the spot he had chosen for it. Affection for a house worked,
> also, as an affection for the place on which the house stood. Since now (if the
> change is generally followed) this would no more be the case. You may choose
> your house, now, in a certain town, without being compelled by this act to stay
> or, when deciding to leave the place, to leave the house behind you to someone
> else. You may like it in A. no more, so, you just take your house with you to B.,
> and then to C., and so on. . . . For the slightest reason, now, you may abandon
> the place where you seemed to be so firmly established; you may go without a
> worry that you would have to build a new house to your taste!—
>
> The invention is by *Mr. Rigand* of Brussels. He has constructed a movable
> iron house that resolves it all. The walls of the house are made of a double (iron)
> coating, with a space between; into that space, the warm air, as much as is
> needed, is brought from the kitchen of the house, it can be preserved, there, and

its flow can be regulated by the means of valves. The house has two floors above the ground, and it weights 810,000 Dutch pounds. Moving the house from Brussels to, say, Liège, Ghent, or Antwerp would cost as little as 300 guilders.[4]

Certain qualities, and this seemed to be the essence of the house, were needed for anybody to live in the dwelling without a foundation—first of all, perhaps, as one stayed, a yearning to be somewhere else.

The house without foundation was specially outfitted, as for a trip. The idea was not as exceptional, at the time, as it might seem. The urge for a trip was a part of the culture. The house on wheels, and its equipment, as *Kopiïst* described it in great detail, fit what, in the Netherlands since the mid-nineteenth century, was a widespread sense of life, a style called often *het lelijke tijd*, "the time of ugliness":

> The artists drew their inspiration from all sorts of historical styles: the style of Romans, Mediaeval gothic, seventeenth-century Dutch renaissance, rococo, baroque, empire. They picked now from this and then from another style. The houses were filled with new objects in old styles, best if varied to the extreme. . . . Moreover, no model from the past should be copied exactly, but it should be adjusted according to modern needs and methods of construction. For instance, garden furniture was made of wrought iron and in gothic style, or a Mediaeval castle was filled, among other things, with thirty bathrooms with running hot and cold water.[5]

To unburden time and space by quotation marks. In a typical time-of-ugliness painting, a bourgeois family in mid-nineteenth-century Utrecht, Holland, is seen relaxing in the small backyard of their house: "Father is shown sitting on a wrought-iron classic-style garden bench. The chair of madam is made of lacquered rattan in gothic style."[6]

Two centuries after the Golden Age, the seventeenth-century era of Dutch greatness and world supremacy, the little Netherlands of the nineteenth century was still a colonial power. To keep a balance, the Netherlands tried to be a little distant from its actual European petite self, not to be fastened tightly; to be ready, or to appear ready, for a trip—a trip to the Indies most logically. The "gothic" table of the middle-class Dutch family in the ugly-time painting just mentioned was of lacquered rattan. It was a common trend. At the industrial fair in Haarlem, in 1861, "lacquer products of the Amsterdam firm of Widow F. Zeegers & Son" were much applauded, "indistinguishable from the real Indies [sic] thing."[7] The "best Eastern carpets" were said to be made in a carpet factory in Deventer, northwest of Amsterdam.[8]

Those Dutch men and women who actually moved to the Indies took their ugly-times with them. The wheeling, floating, and faking remained. Just the perspective shifted. What was an exoticism in the Netherlands became the way of life in the Indies, how to keep in touch with things back home, there, equally far away, over the ocean:

The nineteenth-century upper-class and middle-class townhouses [in the Indies] revealed a mixture of Javanese and neo-classicist elements. . . . Sometimes, the houses were built on an elevated place like an ancient Greek temple. . . . Another variant of neo-classicism [in the Indies] was a *Waterstaatsstijl*, "Department-For-Water-Regulation Style." . . . the builders from the [colonial] Department for Water Regulation . . . combined . . . classicist elements with embattlements, oriel windows, and gothic arches. . . . The "Department-For-Water-Regulation Style," was, then, being transformed into other, and yet other, mixed-style forms.[9]

There was, also, a difference in intensity. While in Europe, the bourgeois house was equipped to permit its owner to sense the possibility of an occasional excursion to another place and time, a temporary reprieve, in the Indies the house was being built and equipped as for a permanent trip.

While in Europe, also, by the late nineteenth century, the ugly times were passing away, in the Indies, with the intensifying sense of modernity, they grew pervasive. While in the Netherlands, the functional and efficient twentieth century was setting it and making the ugly times into expressions of nostalgia and cultural margin, in the Indies, the modern twentieth-century homes—the more modern the more so—were still ugly-built and filled with *kunstnijverheid*, "arts and crafts," Eastern and Western, ancient and avant-garde, all of them if possible, and in maximal quantity. One author, in 1916, described what was offered to be used in the Indies house:

> huge number of samples of Eastern articles of all tastes: imitation tortoiseshell, imitation gold and silver . . . Eastern stuff, much of which are cottons printed in Twente . . . brutally glazed yellow, monotonous, and gross copper . . . ludicrous copies of ancient Greek vases embellished with figures from *wajang* [Javanese puppet-shadow theater]—![10]

The same author also explained why genuine stuff was not necessarily good, and might, indeed, be dangerous to use:

> I am well aware that a full-blood Dutch man can appreciate, and with some pride even, when his wife wears Dutch velvet blouses batikked in a Javanese way . . . he also much admires batikked covers on his European table, books, and pillows—Yet, with a manifest air of disdain, the same Dutch man will look down on an authentic Javanese batik cloth. Something like that worn by his wife, for instance, might be confused with the *sarong* his Javanese maid likes to wear.[11]

In order not to appear like the "people of the land," the Dutch in the Indies were overequipping themselves.

In a guide for the young Dutch housewives newly arriving in the Indies, *Ons Huis in Indië*, "Our House in the Indies" a well-known Dutch woman writer, Catenius-van der Meijden, described in 1908 what she believed was a typical Dutch house in the colony at the time. It is striking, in her description, how barren the house appeared to her, and how urgently she felt the need to redress this "fault," and, intensely, to "equip" the house:

> In Europe, in the very old as well as in the most contemporary houses, one finds a great variation of halls, large and smaller rooms, squares, octagonal, and other shapes, deep and shallow window embrasures, many kinds of galleries and penthouses, all making it easy for the owner to rest in and to enjoy a cozy corner or some other stylishly furnished space. In the Indies, on the other side, one has to do, time and again, with the same set of a front verandah, an inner verandah, a hall in the center, an always equal number of the rooms on the left and on the right, and then the back verandah—all rectangular.[12]

The cavity of the house, Mrs. Catenius urged, should be quickly filled with objects and shapes—good-for-filling objects and shapes in fact: "in one corner, place a piano . . . in another, a boudoir lamp . . . music cabinet . . . Behind hang some draperies made of light stuff, and, at the side, install a knick-knack, a novelty piece of furniture [*fantasie-meubeltje*] of some kind."[13]

The objects in the Indies house were to be carefully maintained "throughout the stay." They were to be arranged as if to keep them quickly on hand when one "might go on a trip again." The accumulation of objects in the Indies house was geared toward a display. This is one reason, perhaps, why there appears to be such an affinity, during the late-colonial period, between the Indies house and the fairs and exhibitions.

Walter Benjamin called the exhibitions and industrial fairs in Europe of the nineteenth century "the sites of pilgrimage." Bourgeois Europe got on the move, Benjamin wrote, attracted by "the commodity fetish."[14] Indeed, there has always been a connection between the exhibitions and journeying. The first big Dutch world and colonial exhibition, in Amsterdam in 1883, was also the first really significant occasion when the Dutch railway companies, on their lines to and from the exhibition, offered a discount.[15] As the contemporary press remarked, the exhibition made the Dutch, for those few weeks, into a "nation on board": "daily, thousands of every rank and status were carried from all the provinces of the country to Amsterdam. The bond between the capital and the rest of the country strengthened."[16] There was an order based on transience. Like in the Indies house, the display exuded temporality. Visitors came to "dwell" just for a little while. They were attracted to the exhibition, as to the house, indeed, by the fact that all the accumulation of objects had just been brought from elsewhere and soon would leave again.

Colonial exhibition was the epitome of the colonial house. The permanence of the exhibition was neither the visitors nor the objects, actually, but a future exhibition. The early arts and crafts exhibition in Amsterdam, in 1877, for instance, contained many objects from the Indies, and it became the core of the arts and crafts museum in Haarlem, which, in its turn, supplied the best objects to the Amsterdam exhibition in 1883.[17] This exhibition ended, or lived further, in several newly established museums of Amsterdam, and so on.[18]

Passage, also, was the essence of the exhibitions, and the central piece of their architecture. The main entrance to the Amsterdam exhibition of 1883 was by a monumental throughway under the fresh new *Rijksmuseum*, "State Museum" (all the Rembrandts and Vermeers were just being moved in at the moment). The throughway was a tunnel that everybody coming from the city to the exhibition had to use:

> The throughway is a wide road under the ceremonial hall of the Museum, it is the largest entrance to be found anywhere in the Netherlands, Belgium, or Germany, and it is two hundred Netherlands ells long.
>
> . . . As we enter the throughway and move toward the exhibitions grounds— astonishing sight!
>
> As if by a powerful charm, we are moved out of the Netherlands into the heart of Asia.[19]

At the end of the tunnel, the exhibition grounds themselves were also floating. As on a trip between the Netherlands and the Indies, displayed to the visitors, a model of a habitat, as a well-conserved ugly-times dwelling, as a contemporary Indies house. For a few hours, even several days, a visitor could "stay" at the exhibitions and browse through a space completely filled with oil paintings of Javanese landscapes by Antoine Payen, stuffed animals, models of "native Indies" temples, and "historical Dutch" ships, puppets, clothes, tents, and jungle boots,[20] "a model of the monument built in Aceh [Sumatra] in memory of the fallen in the struggle for the Fatherland and the King, a glass house for tropical plants, a hut out of which authentic Indies music could be heard."[21]

After the Amsterdam event, the exhibitions became enormously popular in the Indies. In the big cities like Semarang, Surabaya, Medan, and Bandung, and in the capital of the colony, Batavia (often with the help of rich Chinese),[22] the fairs became regular and, often, the focal events of the calendar. They were a passage easily accomplished over the vast expanse between the metropolis and the colony; a person being Dutch in the colony could dwell in the same (exhibition) space in the West as in the East. The Dutch in the Indies might feel, while at an exhibition, a functional and convincing temporality. They might feel more at home than in their own house.

There was yet another space in the Indies that expressed the late-colonial culture of habitation very deep in its essence. *Soos* is an acronym for *societeit,* "society," or "club," in Dutch. *Soos,* by the late nineteenth century, was being established in every large city, and in many small towns of the colony as well. The Batavia *soos, Harmonie,* "Harmony," the oldest one, existing since 1814,

Social club 'Harmonie' in Batavia, ca. 1890. (KITLV)

was the most prestigious and the model throughout the Indies.[23] According to the last *Harmonie* statutes from 1935, besides the rules such as "In the billiard and card room, as well as at the bowling court, as these are used, gentlemen can take off their jackets. . . . It is forbidden, during the dances and other festive evenings, to switch off the fan, except when all the parties present agree,"[24] it was clearly stated: "The Society is called 'Harmony.' . . . It was established for a cohabitation among its members. . . . For all times, it will maintain its Dutch character."[25]

This was the *vast clubje*, "steady club," with its *eeuwige brandy-soda*, "eternal brandy soda."[26] The late-colonial *Harmonie*, like the late-colonial exhibitions, became crucially important by means of its flair, that of a place one visits, browses, and belongs to, significantly but lightly. Hotels were another model home, and a model for home, in the Indies. As time went on, in the words of an observer in 1920, "A large proportion of the Europeans [in the Indies] lives . . . in 'villas' . . . in little pavilions glued to the 'villas,' or simply in hotels and pensions."[27] The same observer noted how, among the Europeans in the Indies, there had been a rapid increase of what he called *uithuisigheid*, "gadding about." It did not mean some new-quality enjoyment of the outdoors. Rather, there was an increasing drifting between house and house: "The old-fashioned dinner parties at one's home are ever more replaced by dinners at the Club or a restaurant. The costs of what earlier was a welcoming table at one's house are diverted to the rising budget for wives' toilettes."[28]

Like the "movable iron house" of Mr. Rigand, the modern Indies houses were to be close to the roads. As the means of transportation developed, and as there were faster ships sailing, and steaming, between the Netherlands and the Indies, the exhibition-like, *Harmonie*-like, and hotel-like houses in the Indies, for the Dutch in the colony, became a norm: "With the periods between the times on leave shortening, indulgence has been encouraged, but conviviality diminished. Now, one has to save in order to have a fine time off, and for the Indies, increasing numbers are content with just staying in a little house or a hotel pension."[29]

And there were real homes on wheels, and homes really floating. From their inception through the end of the colonial era, like the objects at a colonial exhibition, the railway coaches of the Indies were brought piece by piece from elsewhere: "motor from factory Hohenzollern in Dusseldorf, teak wood carriages from factory of Beynes in Haarlem, switches and semaphores from factory of Grutson in Magdeburg."[30] Nowhere, perhaps, was there so much of the ugly times kept alive. Only the carriages of the third class, the *kambing*, "goat," class, for the natives, were allowed to remain barren, just a transport.

The carriages' exteriors were sleek. But the interior was what mainly mattered. Extreme care was taken to equip the carriages inside. When the Dutch pavilion, at the last and the greatest colonial exhibition, in Paris in 1931, burned down, among the "national jewels" lost in the flames, as it was lamented and listed by the official report, besides ancient Javanese sculptures and seventeenth-century Dutch Golden-Age navigator's and wall maps and paintings, was new furniture by Lion Cachet and a few other Dutch designers, made especially for the Indies railway carriages and ocean liners.[31]

Lion Cachet, and also the architect Penaat, who specialized in the Indies railways,[32] belonged to the most respected figures of the late-colonial culture. They helped significantly to make the idea of the late-colonial Indies house tangible. The philosophy of the rattan-like gothic ugly times lived in their work naturally, as rarely anywhere else. For Cachet and Penaat, the use of the most expensive and most strange-to-place materials was typical. They used deliberately complicated techniques. It was said of Lion Cachet that he "sought . . . diversity in subjective feeling [and] worked from a position of idealized intellectualism."[33] The interior of a train carriage or ocean liner designed by Cachet or Penaat might be

> paneled with oak wood and gilded in golden colors in the *genre* of Renaissance. . . .[34] each panel contains a different Eastern ornament . . . the walls are inlayed with various metals. . . .[35] . . . there is a glass mosaic . . .[36] . . . balustrade of the railing has bronze crowns . . . there are red and brown tinted golden reliefs of lions . . .[37] part of the interior are [copies of?] the reliefs from [the ninth-century Buddhist temple] Boroboedoer, and various other artifacts from Bali and other Ancient-Indies locations.[38]

It was more than an accidentally chosen metaphor when Mrs. Catenius, in her *Our House in the Indies*, wrote: "A good journey overseas feels like a pleasure trip in a movable hotel!"[39] The Cachet's and Penaat's Indies trains and ocean liners, in fact, might be even better than *Harmonie*-social clubs, hotels, and exhibitions at providing a model. They were floating homes and homes on wheels, truly. In them, more than anywhere, as long as the wheels turned and the ship sailed, there was almost perfect late-colonial homeliness.

In 1905, the *Rotterdamsche Lloyd* commissioned Henri Borel, one of the best-known Dutch writers of the time, to travel to the East on one of the newly built company liners. It does not matter which way you go, Borel wrote in the travel guide that resulted from the commission and the trip: "I have described the trip from Batavia to Rotterdam, but those who make it from Rotterdam can use the guide all the same. They just would read it from the end to the beginning; the places just pass by in a different sequence."[40] Thus the ship sailed smoothly one way or another, and the ship's inside was the main thing that mattered. As it might be expected, Borel praised the *Lloyd* for an excellent job: "in the music salon, the smoking salon, or in the reading salon of 'Sindoro,' one completely forgets that one is on the sea, and imagines himself to be on the land."[41]

Whichever way the passengers went, as the ship sailed, the airs of home went with them: "LUNCH. Peasant soup. *Entrecôtes* beans. Baked potatoes. Prawn salad in curry sauce with rice. Desserts and fruit. Coffee. DINNER. *Julienne* soup. *Cromesquis.* Roast beef and peas. *Chicorée—flamande.*"[42] This was as modern a space as one might imagine at the time, and naturally technology was in focus:

> Bathrooms and water closets can be found at the middle of the ship on both lower and upper decks, and they are outfitted with all the newest inventions. The ship kitchen, the ovens, and the electrically powered kneading machines are of the latest models, and they can take care of 600 people. Two ship machines, one on the stem, the other on the stern, work entirely independent of each other.[43]

The ship, Borel wrote, sailed "calmly like a Friesian clock."[44] Nowhere in the late-colonial empire, indeed, and Borel made it very clear, did a home—a Dutch home—truly appear more real. It resonated deeper with the sense of home than that of the ugly times. It approached the highest standard of home, as every Dutch schoolchild knew it from the stories and paintings of the Golden Age. As on the rich seventeenth-century still lifes, the ship, the home was floating: "In the ice room, there is a constant temperature of -3 C. . . . fish, fowls, ± 25 cows, and several pigs hang on the hooks, and in the cool room there is milk, fruits, and canned food."[45]

It was a home, rising and falling softly on the waves, so that one would hardly even notice, when, at a point, the journey ended. As the ship arrived, if one happened to go the eastward way, the passengers were awaited in "*Hôtel des Indes* in Weltevreden . . . the best hotel of all Batavia." It was the best hotel in the whole colony, possibly, and it was situated next to the best of the social clubs, the *Harmonie*. The hotel, also an ideal of sorts, and yet another late-colonial high way of dwelling, was floating itself, fully defined by the road, by the journey, and by the ship: "each Tuesday evening there is music, as the departure by the 'Rotterdamsche Lloyd' back to the fatherland is scheduled for each first and third Wednesday of the month, and by the 'Nederland' for each second and fourth Thursday."[46]

The idea of floating homes and homes on wheels was open—it was made to be open—to all kinds of technological improvements. If there ever was an even more home-suggestive image than that of an exhibition, club, hotel, train, or ocean liner, it was—in the last few years of the colony, as the great economic depression and final crises of the colonial system set in—the image of an airplane.[47]

It is "lunch time," a typical advertisement folder for the new *KNILM*, "Royal Netherlands Indies Airlines Co.," wrote in 1932. "We are in the plane interior, and the steward brings a cup of tea." On the folder's photographs, there is, undoubtedly, calm on the outside. White skies, like the unruffled surface of the sea, are seen from that airplane window. The sign above the window is suggesting: "Without a reading lamp, nobody can travel well."[48]

Nothing, in the late-colonial Indies, was more *comfortabel,* "comfortable," than traveling by plane. This was a soaring house:

> Repeatedly, the travelers who think of a flight to Holland ask us whether the change from the tropical heat to the Dutch cold is not something unpleasant "especially because high in the air it is so cold." . . . In fact, there is no difference with what your friend feels, when he steps into the cold out of his house, which is usually well heated. . . . It is the same. You just put your coat on, there is no difference from your friend in Holland . . . the airplane cabin is at least as well heated as a room in a Dutch house.[49]
>
> Already more than 2,200 times, comfortable planes have risen high in the air, and as many times the passengers have been brought back, down to the land of the Indies.
>
> The only irritation sometimes still mentioned, namely by those of the weaker sex, is that there is no special KNILM flight toilet in the plane!
>
> . . . Each of the planes' three motors has 225 horsepower. In all combinations of just two of the motors, these great transport planes can fly on, undisturbed, and they can even ascend. . . . *Travel with us.* COOL, FAST, COMFORTABLE!! *Become the Air-Traveler.*[50]

The Cities

In 1919, a long essay, written by one of the first Javanese with Dutch training as architect, Noto Diningrat, Engineer, described the layout of a traditional Javanese house, *oertype bij alle standen,* an "archetype for all estates," of the population.[51] The "unity of the way of life" of the inhabitants of the house, Noto Diningrat wrote, was "anchored" in one particular room, "deepest" in the layout. In this innermost chamber, "messages could be received from the world of souls and spirits."[52]

The article was remarkably modern, technical, and gloomy. In the more recently built Javanese houses, the author wrote, during the past few decades, as the "longer-lasting materials" began to be used, a "hygienic point of view" began to be respected, and "combining of parts" was becoming a matter of skill, the Javanese began to look more toward the outside for models. "What examples can they find, now, except the European building in the Indies?"

> In the old type, the core of family life took a place in the center of the house, and the center was the symbol of depth and soul; in the new type, family life drifts away from the core.. . .
>
> Earlier, there was an equilibrium between the outer parts of the house and its area for the retreat (it was parallel to an equilibrium between the outer and inner states of mind). In the new types of home, there is much less of that; in some houses, even, the shrines have disappeared.[53]

Javanese houses, Noto Diningrat's article concluded, were becoming functional in a new way. They began to reflect new pulls and pushes in society. The houses were being pulled and pushed; they were made into parts of the new Indies architectural space. Noto Diningrat wrote:

> Departmental-services buildings, assembly buildings . . . etc., are built for the government; churches, cathedrals, monasteries, and so on are built for religious service; schools . . . libraries, museums, laboratories, theaters, concert halls, etc. are built for the arts and education; markets, abattoirs, health and hygiene centers, factories, workshops, banks, hotels, and so on are built for everyday cohabitation of the city; railway stations, post and telegraph offices, warehouses . . . and so on are built for commerce and transportation.[54]

The new functionality, expressed most distinctly in the new Indies cities, sucked the life and the meaning from the inner space of the native homes. Javanese, especially the "modern natives" among them, those few well-off who could afford it, were pushed and pulled by the power of the new functionality out of the depth of their houses. They became inattentive to the inner space of their houses, and their houses became shallow. A fake affinity emerged between the rulers and the ruled. Ki Hadjar Dewantara, a Javanese aristocrat, national educator, and respected Indonesian nationalist leader, described in 1929 how his modern countrymen lost their freedom. They, he wrote, "lost (their) free national spirit (*roch kebangsaan merdeka*) and (their) own ideas died. [Now, they lived] as if lodging in someone else's hotel, [content] just to eat good food and sleep."[55]

Nellie van Kol, a well-known Dutch woman writer mostly of children's books, lived in the Indies. She was the wife of the already-mentioned prominent Dutch Indies engineer and Socialist International colonial expert, Henri van Kol. She also very early sensed the new and disturbing phenomenon of the late-colonial city emerging in the Indies. Already in 1884, she wrote how sad she felt that there were no real villages in the Indies. After giving it some more thought, however, she qualified the statement: "yet, there is Buitenzorg, . . . I call it village, indeed, and I spell it as a name of honor . . . the park that is the jewel, the crown, and the heart of Buitenzorg . . . that freedom . . . coziness . . . the holy *waringin* tree . . . that harmony . . . oh, Buitenzorg, where are my childhood dreams."[56]

Buitenzorg, today Bogor, was the summer residence of the Indies governor-general, just before the cool mountains, away from Batavia; its Dutch name meant "no worry." The park in Buitenzorg was the world-famous botanical garden, with ponds built in, landscape fixed, and trees, including "the holy *waringin*," probably transplanted from elsewhere. One might plunge into the yearning and never hit the real. It was like a soft waving of the sea.

Pasar Baroe, "New Market," in Batavia, ca. 1920. (KITLV)

A modern city nostalgia was constructed for the late-colonial Indies as carefully as the floating homes and the homes on wheels.

There was a discomfort among the Dutch in the Indies as early as the time of Nellie van Kol. In 1892, the most prestigious colonial journal, *De Indische Gids*, "The Indies Guide," expressed almost alarm about how often the Dutch in the Indies, instead of just living, felt an urge to "fortify themselves." One method of fortifying, *Gids* wrote, was "by overeating, and by exciting in oneself desire for further eating that overburdens the organs of the body so that it could take no more."[57] The journal compared the discomfort to that of an insecure traveler. The Dutch in the Indies, the journal wrote, ate as much as if they still were on the ship or in a transit hotel on their trip to or from the Indies.[58]

Urbanization made the insecurity more distinct, as the modern natives, pushed and pulled out of the innermost parts of their homes, were increasingly living uncomfortably close to the Dutch. This was to be a complex culture of feeling to be out of place. One of the most widely used primers for the Dutch children in the Indies schools, in the early decades of the twentieth century, characteristically bore the title *Ver van Huis*, "Far from Home."[59]

In *Het Leven in Nederlandsch-Indië*, "The Life in the Netherlands Indies"—a candid, often brutally self-searching book by a Dutchman, Bath Veth, about the Indies, published in 1900—(it was a best-seller, which meant that it touched sore nerves of not just a few) a sense of exile was prominent:

> all Indies places have something of a hotel. They accommodate the guests, who have just unpacked their bags, and who are already thinking about leaving. We, in the Indies, are like traveling salesmen. Those of us who are not rich, and who struggle through their careers, are forced to spend years in the hotel places, and they absorb much from the character traits of head waiters and porters. They get a flair of strangers, and they come to be part of the ways of the hotel. . . . [For those of us who already became rich, there are] clubs, receptions, after-dinner parties, assemblies, dilettante choirs, and theater associations—this is into which the social life of the "hotel guests" dissolves.[60]

This is a book on floating and wheeling, moving and shifting, yet there is no transport, and even very little detachment or tranquillity: "There are places, in the Indies, where the newspapers, besides the column of train and passenger-ship time tables, contain nothing but a section on *reception days* . . . family A receives on the 2nd and 4th Monday of each month; family B entertains on the 1st and 3rd Thursdays."[61] Life, according to Bath Veth, fell into the hotels in the Indies, as into a black hole: "An Indies hotel is the abomination of all Indies abominations. . . . Much too large a portion of the European population shut itself into these places. What is more, the life of the Netherlands Indies is a reflection of the Indies hotel life. Poor life as well as rich life."[62] European mankind in the Indies, according to Bath Veth, degraded itself into "a stream of hotel guests."[63] By the same token, Bath Veth declared, the Indies were truly *hotel-thümlich*, hotelish:

> It is night, the Indies night. It is night in the Indies hotels also. . . . All are asleep. Out of a sudden—boom—the cannon shoots five o'clock; *poekoel boem* ["time is boom"] as natives say. . . . Whispering in Malay can be heard. Then first soft claps and knocks. *Tŏk, tŏk, tŏk*— pianissimo; a two-minute pause. *Tŏk, tŏk, tŏk*—forte; a two-minute pause. *Tŏk, tŏk, tŏk*— fortissimo. *Tŏk, tŏk, tŏk*— furioso. The doors begin to open from the inside. This is the uprising! Uprising of black he'taeras, *baboes* [maids], prostitutes, *njai* [concubines], who, in the evening, had slipped into the mean rooms of the male guests of the Indies hotel. . . . The doors open just a bit wider. And an ebb tide, the black venereal dregs, flow and stream out of the hotel.[64]

The threat, especially as modern cities were emerging in the Indies, was in fluidity. In semen and blood untamed, as Bath Veth described it, in water— polluted, dripping, leaking, or flowing unregulated. We have seen what water could do to the modern roads in the Indies. To rule the colony, to become modern there, to stay, meant to confine the flow.

The problem of water was always felt in the Indies as an agrarian problem, mainly of irrigation.[65] By the late nineteenth century the problem of water was felt acutely and essentially as a problem of urbanization. In the most widely read Indies magazine, *Nederlandsch-Indië Oud en Nieuw*, "Netherlands Indies Old and New," in 1918, an author listed the pipelines for drinking water in the Indies cities among the cornerstones of colonial power, side by side with "bridges, roads, and railways." Above them, in fact. Water regulation, the author wrote, was more "directly felt," and it affected Indies life more intimately.[66] In 1918, also, an early, ambitious—and never fully realized—project of a colonial city was announced. The regulated water was a crucial part of it. Two new modern quarters were to be built in the city of Batavia, *Menteng* and *Nieuw Gondangdia*: "They are designed as city quarters specifically for the better-off"; "the pipes for collecting refuse in the settlements are planned to be at least six meters from the limits of each of the inhabited compounds."[67]

H.F. Tillema, the author of *Kromoblanda* and a Dutchman in the Indies who did much thinking about asphalt, as we have seen, devoted much of his time to the problem of water. There are lengthy passages and many illustrations in Tillema's books and articles on the Indies colonial water regulation in design and action. The images are strictly polarized in clusters of "there" and "here," "before" and "after." "Before" and "there" is chaos, or, at least, messy and smelly space, of *kampongs*, the "native quarters." "After" and "here," there is water tamed, dammed, canned; this is the realm of the Dutch, of colonial pipes and dikes.[68] Tillema was one of the greatest dreamers of the empire. He liked to call himself "engineer of health and hygienist."[69] Others described him as a "propagandist for a clean water."[70]

There was some progress—and drifting—on the water distribution and the *rioleering*, "sewerage," in the Indies as the years passed. In 1927, for instance, the inspector for people's health in the city of Batavia stated that "The need to care for good drinking water is being recognized more and more. People begin to see that carefully draining off the feces may well pay back in reducing the occurrence of infectious illnesses."[71] In 1938, still another decade later, and three years before the Dutch rule came to its end, another Dutch leading expert again "foresaw" that "the ever denser building over the urban space, and an increasing concentration of new houses, will inspire an improved care being taken of canalization."[72] First of all, this was a commerce in dreams, or plans. The particular expert just quoted made it clear also that he still spoke, of course, merely about the "better off" inhabitants of the Indies, thus almost exclusively the Europeans and, predominantly, the Dutch.[73]

The water ordering, in dreams, plans, and actuality, in the Indies—in a special sense—worked. The water channeling and its culture was evolving into a usable system of making the colony both dirty and clean. In many reports by contemporaries, the state of affairs was described. On the one side, there were the Dutch and European, or Dutch-like and European-like, city quarters, with sewage regulated and clean water running on tap. On the other side, there were *kampongs*, the native quarters as one Dutch author in 1917 depicted it in modern negatives, "places filled with huts made of bamboo, wooden planks, and woven mats, with shutters but no windows, with no floors but earth, with no bathrooms, no washing place, and no water closets."[74]

Yet the system never became perfect. As is natural for water—semen, blood, and slime, too—fluidity persisted in the colony, and so did the threat. This was a place of tropical rains, and also of natives, modern natives in particular, who seeped into the late-colonial modern urban space. One could feel it, even in the imperfect language of the water ordering. In a thoroughly matter-of-fact study on the clean water problem, for instance, written in careful Dutch and filled with international technical terms, of course, the one single word that was not translated (was felt to be untranslatable?), the singular native word left glaring and ominous in the text, was Malay, and Indonesian, *bandjir*, meaning "flood."[75]

The late-colonial water regulation might turn into a dangerous business. Even a process of learning—by the natives about clean water, for instance— might turn into a *bandjir*, a flood of sorts. In 1914, an early Indonesian journal edited by Mas Marco, the already mentioned *Doenia Bergerak*, "World on the Move," carried a big headline: "*Clean drinking water needed. People get sick and die.*" There had always been epidemics in our land during hot seasons, the journal wrote; there had been typhus, cholera, and plague, and, of course, all intelligent people knew why—because of bacilli carried by rats and fleas. Intelligent people knew very well, also, how this all might be dealt with. In Madiun, East Java, for instance, at the time of a recent plague epidemic, "*five wells* with clean water were opened; there were just five of them, however, for a vast area, and *no pipes were installed to bring the water to the distant places*."[76]

We Indonesians, *Doenia Bergerak* wrote, know well what is right and wrong. So why we are still dying in the modern cities of the Indies? Why, moreover, now, in modern times, unlike in *tempo doeloe*, the "times past," are we Indonesians dying by the scores not just during the hot seasons, but during the cold seasons as well?![77] The answer was simple, according to *Doenia Bergerak*. In their effort to kill the bacilli, the Dutch authorities spread "artificial fertilizers containing sulfuric acid and highly concentrated phosphate" around the sources of clean water. When the rains came, as they regularly do in the Indies during the cold season, the chemicals got into the wells and streams, and they killed more than even cholera and the plague could.[78] In the Netherlands itself, the Dutch did not appear to cause problems like that. "Possibly," was the journal's conclusion, in their own land, the Dutch handled water scientifically.[79]

It was a divided society, yet, with a stinking lavatory often in the middle. Here there might repose a point of resistance. In an article in 1927, the radically nationalist *Soeloeh Indonesia*, "Torch of Indonesia," a decade after Mas Marco's *Doenia Bergerak*, used a motto from Tillema: "pray that the great Indonesia does profit from the benefits that only hygiene can bring." The article dealt with the fact that, among Indonesians, there were still many who "prefer a shanty to a house, and there are some Indonesian servants, even, who can foul a water closet when the mistress of the house is not looking."[80] This was not nice at all, *Soeloeh Indonesia* wrote. It was as disgusting, almost, as the notorious Dutch public latrines in Amsterdam, where, until recently, "special walls had to be built, slanting low above the holes, so that the people were forced to sit down when using the place."[81] Civilized men and women, the article ended, should do their best not to behave like these people in Amsterdam.[82]

The shit universalism can become a powerful argument in any battle for equality. In the late-colonial Indies, namely, privies, outhouses, and wells, the space supposedly and manifestly antiseptic, became the point where a political debate developed with an extraordinary clarity. In another *Soeloeh Indonesia* article in 1927, the colonial times were depicted as water running—running out: "Older Soerabayans still remember well the style in which water policy was made. There were festive openings, speeches, light shows, and big headlines: *Tap water is clear like crystal, and it costs nothing*."[83] Even some public fountains actually appeared, and they worked for a while. Then,

the population watched with sad eyes how the number of both the festivities and the public fountains was being reduced. . . . Eventually, the remaining public fountains had opening times cut to merely few hours a day. Finally, as a crown to the whole edifice, there came—a shutting down of the public fountains altogether. . . . Instead of water on tap that cost nothing, the guilds of water bearers and water sellers appeared on the streets.[84]

The very place where so much Dutch ideology and feeling had been invested became most open to assault. There were separate toilets in the Indies still, in the late 1920s—some for "Europeans," some for "ordinary Asians," some, occasionally, for "higher Asians," "modern natives who made it closer to the European sphere. "Those of us who wear a native scarf, or a Moslem skullcap," *Soeloeh Indonesia* wrote in 1927, "have to relieve themselves in ordinary latrines."[85] As late as March 1941, a few months before the end of the Dutch empire, another Indonesian paper still found the matter unsolved and characteristic of the whole system: "in a small place in the interior, at the local bank, on the doors leading to the 'inevitable installation,' there are four tags hanging with the message: 1. Bosses; 2. (White) Staff; 3. Asian Personnel; 4. Clerks and Others. *Good heaven*, friends! Even in such a situation, discrimination?"[86]

In the Indies, big cities grew fast: between 1900 and 1925, Batavia by 130 percent, Surabaya 80 percent, Semarang 100 percent, Bandung 325 percent.[87] Even in the Indies, some authors, writing about architecture and urban planning since the early twentieth century, wrote about *revolutiebouw*, a "revolution in construction."[88] They meant a different revolution, in a sense, an acceleration of untamed flow. As all the statistics showed, the native population of the colony, in villages and small towns, was rapidly increasing. It trickled and it streamed into the modern cities. As the colonial system grew complex, the number of Europeans in the cities of the colony also rapidly increased.

In 1913, at the International Housing Congress that took place in the Netherlands, H.F. Tillema, the propagandist for clean water and asphalt roads, was the main speaker about the Indies. Tillema warned that, because of the influx of people into the cities of the Indies and the soaring prices of land, most of the houses recently built for the Dutch in the colony were so close one to each other that not enough light and fresh air could get in. With a sense of urgency, Tillema also spoke about a necessity of *kampongverbetering*, an "improvement of native quarters."[89]

It was like the story of the public fountains quoted above. A great deal of publicity was given to the problem, especially in the beginning. In Batavia— not too urgently, it is true, six years after Tillema had called for it, in 1923— a plan for *Taman Sari*, "Fragrant Garden," *modelkampong*, "model native quarters," was announced. The natives who had made it to the city, it was planned, were to be moved into seven new large residential buildings, each with a public bathing and laundry place, and, altogether, with two "universal shops." Like the public fountains in Surabaya, the experiment never got off the ground. It faded away while still in the planning phase "because the building costs, it was found out, would be too high."[90]

As water could not be purified for everybody, so the line had to be drawn with modern housing. There were yet two other housing congresses, this time convened in the Indies, in 1922 and in 1925. Papers were presented, and the necessity of doing something radical in urban planning in the colony, namely, in the native urban quarters, was emphasized. There was a third congress on the problem planned for 1942, but, in this case, not even papers were read and plans made, because in 1942 Dutch rule was already over and Japanese armies were in the Indies.[91]

Assaineering, "sanitation," or "slum clearing," as well as *gezondmaking*, "sanitation," of the cities, was talked about continually and, as a rule, animatedly, since the early twentieth century and throughout the late-colonial era. An "urgent need" to launch a special field of study at the technical college in Bandung was debated, of "engineer-hygienist" or "health engineer" to encourage "rationalization" of the late-colonial city growth.[92] All this—the intensity and futility of it—was highlighted, typically, at colonial exhibition fairs. The booths and panels on "city hygiene" (and quarantine)[93] were among the most elaborate and best visited.[94]

In Europe, through the interwar period, wealth could still be thrown around widely, and new technologies of habitation could be applied almost at will. In the Indies, in the revolutionary twentieth century, as one observer put it in 1917, "the Europeans, forced by necessity, were learning how to accommodate themselves to *smaller* houses, yards, and gardens, to a lesser space and less air; this was what was meant by the word 'modernity.'"[95] The same Dutch observer described the modern city of the colony, also, from the point of view of the Europeans, as "long, eternally long streets. . . . Between the streets, street-less *kampongs*, of the natives."[96]

The Camps

In 1938, Sukarno—a native architect[97] living in exile at the time, a leader of the Indonesian radical nationalist movement, and the future president of post-colonial Indonesia—was asked by a leading Indonesian literary journal *Poedjangga Baroe*, "The New Writer," for his views on the tradition and the potential of the Indonesian architecture. Sukarno, Engineer, answered:

> In general I am of the opinion that Indonesian architecture does not know three dimensionality. For me personally this amounts to a "shortcoming." We are very skilled in the area of two-dimensional decorative art, but very "meager" when three dimensions are involved. Indian and even Islamic influence were not strong enough to provide three dimensionality on a grand scale to Indonesian architec-

ture. People will be able to understand this opinion of mine if they compare our architecture to that of foreign countries.

The Mendoet temple, the Pawon temple, and the small temple in the Sewoe temple complex satisfy me in their overall conception (which is of modest proportions), but the Boroboedoer [all the edifices are in Java] is only satisfying in its details. Moreover there is a problem: is not the overall conception of the Boroboedoer Temple (although it recalls a Stupa) a failure?

I am not a pessimist. But I think the view I hold is correct. I am an optimist, also in the field of architecture. On the basis of comparisons with other societies it is possible to conclude that Indonesian architecture can become three-dimensional. There are signs which confirm that optimism of mine!

Respectfully SOEKARNO.[98]

Intensely modern, Sukarno reflected the prevalent architectonic idea of empire. Norms of construction and space were inevitably to come into the Indies from the world. Intensely nationalist too, Sukarno of course did not draw, not explicitly at least, the logical conclusion to the same idea: whoever among the natives might wish to be part of the new Indies architecture was very welcome to participate, meaning (at this stage of "learning," "becoming modern," and "catching up") to provide ornaments.

Here is, for instance, a description by a Dutch architect of the building of the headquarters of a Dutch Indies railway company in Semarang, East Java, in 1917:

> Except for the rough stone used in the main walls and in the foundations of the building, and except for the largest pieces of the carpentry timber, all the building material has been brought over from Europe. . . . All had been carefully designed, modeled, and pre-made over there, including about 350 cubic meters of granite stones, readied so well that not a single piece of the stone had to be re-cut when it arrived in the Indies.[99]

The article is accompanied by a photograph—the construction site and the design well visible, and hundreds of ant-like natives on the scaffolds.[100] "Building in the tropics!" another Dutch observer and architect in the Indies wrote, at the same time as he described another construction, the tramway company headquarters in Tegal, Central Java: "what an extraordinary charm that, here, awaits everyone of us, the builders! To build in an undiscovered land! The climatic demands . . . the completely new contingencies of the architectural composition. The different sunlight . . . The foreign people . . ."[101]

There were shades of difference among the Dutch architects in the Indies on how to handle the strangeness. One group had argued most vocally since the 1920s for what it called a "Eurasian style." The prominent man of the group, architect Maclaine Pont (just quoted writing about the "charm" of building in an "undiscovered land") to prove his point

Villa Isola. Architect Wolff Schoemaker. Ruins, photographed ca.
1947. (KITLV)

went to the archeological excavations in Trawulan (East Java) to study the origins of the [ancient Javanese] Kingdom of Majapahit. . . . In 1929, on the basis of his research, he built of bent bamboo an experimental cupola for the archeological museum in Trawulan, and, later, he followed this by building a hexagonal cupola of bent and laminated wood.[102]

There are some buildings (but neither of the two cupolas) designed by Maclaine Pont in the Eurasian style still standing in Indonesia today. The most striking and best known among them is the technical college in Bandung, historically a cradle of the Indies late-colonial modernity of sorts, and also the alma mater of Sukarno, Engineer. Maclaine Pont finished the building in 1920. It has heavy Tudor-like interiors of wrought iron arches. Above the construction, in which nothing is non-European, there are, as if affixed in an afterthought, native-like, constructionally non-functional, and richly ornamented roofs.[103]

An opposite school of Dutch Indies architects worked to prove that purity mattered, that only imported notions of construction and space might work in the colony. According to the leading figure of the group, C.P. Wolff Schoemaker (a teacher and mentor of student Sukarno, too),

> there is not anymore on Java a native builders' tradition that might serve as a model for any modern Indies building style. There are some good examples of building traditions still to be found on some outlying islands, like Nias, Sulawesi, and in parts of Sumatra, but these specimens cannot be expected to find a fertile soil [in the most developed part of the Indies] on Java.[104]

Wolff Schoemaker and his followers eagerly absorbed all the modern and avant-garde impulses of the time, and "from the world"—Frank Lloyd Wright from the United States, *Bauhaus* from Germany, *Nieuwe Bouwen* from the Netherlands.[105] Intensely and narrowly, they focused on building unmistakably modern sites in the Indies. Schoemaker, for instance, built in 1920 the exhibition grounds for the great annual fairs in Bandung;[106] in the same city, in the years that followed, he designed the Grand Hotel Preanger,[107] and, most strikingly, *Villa Isola*, a private residence in the hills above the city made of reinforced concrete and glass.[108] Schoemaker and his colleagues built colonial modernity straight. When, in one of Schoemaker's creations, *Concordia*, the European social club in Bandung, an image of the Hindu-Javanese god *Kala* (most frightening, human-eating monstergod), somehow got placed above one of the entrances, Schoemaker called it a "slip of pencil."[109] Schoemaker and his colleagues were convinced that it was imperative not to waver. Any "experimentation with a Hindu-Javanese" or other "Eastern" art forms, they warned, even the ornaments, might "afflict the building with a broken accent."[110]

Some were more hesitant than others. But there had been general agreement: one could play with the line, but any crossing of it implied danger. H.P. Berlage, the main arbiter of what was good and bad in Dutch architecture in the Netherlands and also in the colony, warned that the Eurasian style in the Indies might decay into a "slack renaissance . . . a failure by attempt to give a building a local-related character through Hindu-Javanese ornaments."[111] *Eenige Hindoe-Javaansche versiering,* "some Hindu-Javanese decoration,"[112] perhaps, Berlage accepted. But the substance of the building was to be kept clean.

Berlage found some examples of great native beauty in the Indies—*pondok,* a traditional Javanese construction, most simple, made just of a roof on pillars and without walls. It was graceful, but it could hardly be used as a norm for a modern city. The real, undiluted good of the Indies architecture to Berlage was in constructions he saw in the city of Semarang. "The loveliest town of Java," he wrote, "creation of Plate and Karsten . . . of Klinkhamer and Ouëndag . . . the new city quarters. . . . I will call Semarang the city of Karsten."[113] This was in praise of the Dutch architects who came to the Indies and managed to remain themselves. It was a formula that might be used indefinitely. Bandung was the city of (Dutchman) Schoemaker, Batavia was the city of (Dutchman) Cuypers, Surabaya was the city of (Dutchman) Citroen.[114]

This architecture and urban planning, again, was a technology for how to get across the ocean, on the foreign shore, and not to notice. In 1923, on his first and only trip to the Indies, as he was driven from the harbor in Batavia to the city, Berlage passed, he wrote, "along a canal that almost could be mistaken for a canal in Holland."[115] "An hour after the arrival," Berlage "sat down to drink tea . . . on a terrace of a cozy house in the new city quarter Menteng, which felt like relaxing on a terrace of a cozy house in Baarn or Hilversum, just an hour or so from Amsterdam."[116]

This was an often recorded impression of the Dutch visitors and settlers. Wim Wertheim, who came with his wife in 1931, remembered: "We stayed one week in a villa quarter that made us think of Wassenaar."[117] Another Dutchman, M.W.F. Treub, who arrived in the Indies in 1922, as he saw the town of Siantar, in Sumatra, "thought that he had arrived in a replica . . . of Bussum."[118]

Whatever the changes, however advanced the gadgets were, as in the ugly times, the architecture made one think of being, or about to leave for, someplace else. The spot the house or the city quarter stood on seemed to carry little weight. The space built over seemed empty. There is a chilling description of the center of the city of Batavia, in fact, the center of the colony, by Berlage. The main square of Batavia and the colony, *Koningsplein,* "King's Square," as Berlage found it in 1923, was "sliced [by railway tracks] . . . it suggests that the railway is not in the city, but the city is on the railway . . .

the square's character is mended by several amusement parks and sports fields, whose reconstruction, moreover, is suspended now, as it is a time of malaise."[119] The Dutch in the Indies might evade the sense of being surrounded, prickled, and imperiled by strangeness, if not just their houses, but whole cities were built as if floating.

For the same reason, a neighborhood—any neighborhood, as it implies stickiness—was undesirable. The system should work calmly and smoothly, without a hitch, "like a Friesian clock." Willem Walraven, a Dutch journalist, who said that the Dutch were living in the Indies "like flies upon milk," wrote about the Indies in 1942 as

> a land, merely, of . . . how-are-you's and *how-do-you-do's*, [humans] who, in reality, are worthless either for companionship or for citizenship. A man whose children play with my children, who borrows my newspaper, whose wife comes to borrow ten potatoes *in case of emergency*, this man, at any moment, can spread gossip about me, or accuse me of a most serious crime.[120]

Another Dutch author, Walraven's friend, H. Samkalden, wrote this in 1939:

> The European community here [in the Indies] constitutes an extremely heterogeneous group. The elements of the most diverse strata of the society in the motherland, here [in the Indies]—*par droit de peau*, "by the right of skin"—are brought together, having in common merely being the European *Oberschicht*, "upper layer."[121]

The Dutch community in the Indies, Samkalden wrote, was *sterk atomistisch*, "strongly atomized": "the people live, here [in the Indies], side by side, one along the other, one near the other, but without an organic bond. Hence the reliance upon formalism that is so typical of this society: where one's place is more or less accidental, it is left to bourgeois flattery and disparage to shape the social order."[122] The Dutch within the Dutch community in the Indies took care not to touch each other. If they ever did, Samkalden wrote, it was a touching by "slander," "rumor," or, most of all, "curiosity": "curiosity that never becomes a sustained interest. . . . Just talking about others, gluttonously reporting facts, and little facts, and the littlest facts about others . . . feeling that one is a citizen as long as he or she continues with this kind of conversation . . ."[123]

An Indonesian architect Soesilo, one of the rare exceptions, a native with a Dutch engineering college degree, agreed with Samkalden. He added, however, an even more sobering note. Not merely the Dutch, he wrote in 1938, but all the modern city dwellers in the Indies, Indonesians including, fundamentally missed something: "One lacks, in the Indies towns and cities, the social homogeneity that exists in the West since the urban communities emerged."[124] This, in Soesilo's view, blocked the natural growth of the Indies society and of each of its constituent parts. The problem of fluidity was in the foreground again: "For a freely flowing social evolution, intense processes inside the society are inevitable. For this, in turn, it is necessary that the society is fully differentiated in every one of its layers, and that the layers, all together, continually grow up one out of the other, and fit, each of them into all the others."[125]

This was the problem of flow, and touching, again. Soesilo, logically, writing about the Indies late-colonial architecture and society, wrote about "blocking of natural growth." There had been as little of a neighborhood in the modern cities of the colony, and for exactly the same reason, as there was little of trees. Trees, naturally growing trees, that is, in the late-colonial Indies cities, were becoming something of a sign of imperfection.

Trees, in fact, increasingly and with much fervor, were written about in the Indies journals, technical as well as home-fashion or literary ones. Trees were given a part to play, as "schemes of trees" recommended to the Indies city builders as well as city dwellers. These new trees of sorts appeared to exist as much as they were printed—contours and gadgets—on the journals' pages. The one important "natural" quality left to them was that they would grow fast, but not beyond the guidelines of the plan. They were display trees, and they made one think, again, of colonial exhibitions. Manifestly temporal, and brought from elsewhere, they were supposed to give a new meaning to the colonial streets and squares.[126] Other trees were cut in the process:

> nowhere [in the Indies cities] are there the lavishly laid-out parks and gardens as one can find them in so many places in Europe; there are [in the Indies cities] just plain, large, open spots, grass fields mostly, used as a grazing ground for cattle during the rainy seasons, and left, burnt by the sun and barren, during the rest of the year.[127]

Less and less one found, in the Indies cities, "the shadows of the galleries . . . a sheltering *waringin* tree." Instead, there were ". . . rows . . . almost treeless, shadowless. . . . If there are some trees planted in the new town and city quarters . . . then it is done in a hurry, and the trees are the fast-growing acacias or other of the so-called rain trees."[128] Willem Walraven, as usual a Dutchman extreme in his views, or perhaps ahead of his time, saw it as an ultimate failure. "It is no more thinkable," he wrote in 1941, about one of the most modern urban communities of the Indies, the port city of Surabaya, "that a tree might still flourish on Pasar Besar, Gemblongan, Toendjoengan or Simpang. . . . Not merely that the trees can no more live and breathe in such places, but such places have no right to have trees, equally as a roofless house has no right to be a home."[129]

The first decades of the twentieth century in Europe, as mentioned already and repeatedly, were a time of the most marvelous dreams. A new city dreamed of or planned was to be the fullest expression of ideas that were both pure and simple. In the view of van Doesburg, of the Dutch avant-garde, in 1923, for instance,

> Town is a tension in the length, and a tension in the height. Nothing else. Two correctly made contacts of an iron wire make a town. Each urban individual lives through attempting—by feet, train, tram, or by tearing things apart (replacing the future)—to find the crucial middle point where the tensions mate.[130]

In the Indies, at the same time, a culture was emerging that increasingly looked like a caricature of the same dreams or plans. Or perhaps there was a stronger premonition in the Indies of the global modernity crisis. The colony, perhaps because more visibly sick than the metropolis, sensed more acutely that the modernity of the age should, and did, look hard for a purist and escapist solution.

Parallel to the pure-energy, pure-technology, pure-esthetics modern and avant-garde ideas of habitation in the West—the "geometric cities" or "linear cities," as they were sometimes defined[131]—there was, in the Indies, a distinct and increasing emphasis on putting matters, sort of, in line too. Since 1909, and with great strictness, all the designs of government buildings and sites, schools, pawn shops, post offices, and prisons in the colony were subjected to an official standard: "For each type of building, a standard was decreed, the so-called 'Normal Plan,' from which only variation could be permitted. . . . For a coed school letter D, for instance, the *Normal Plan* prescribed a complex made of five buildings: a gallery was to run around the compound at each place 2.50 meters high."[132] Gradually this became the new *tropenglorie*, "glory of the tropics," as one Dutch author put. Everywhere in the colony, "except the remotest corners of the archipelago . . . in all places, the greatness now was in channeling, staking out, *normalization*."[133]

Life in the Indies was increasingly made in line and by line. We saw it already: the "long, eternally long," Dutch-like, European-like streets with the streetless native quarters in between. There were also, in a sense, "geometric cities" in the Indies. "Sweeping" geometry gained its place in the Indies. The problem of cities, increasingly, was debated and solved as a problem of *slordig straatmarkering*, "sloppy street marking." There was a copycat order on one side; on the other side, the disorderly urban rest was to be swept away or under:

> The level of architecture in the Indies is low because of the extensive use of master-builders [instead of architects]. . . . Even those who are sometimes addressed as architects are, in fact, only supervisors with good connections. . . . With little individuality . . . the parts of houses, paragons of ledges, pillars . . . balustrades, everything is copied from model books, so that the buildings are the same to the smallest details.[134]

In November 1937, a long article was published in the journal for the Indies technicians, *Tecton*, under the title "Should Native Quarters in Towns be Retained and Protected, or Should They be Pushed Off and Destroyed?"[135] *Kampongs*, the low-class, native quarters in the Indies towns and cities, the author began, were first of all "sources of danger." Fires and epidemics start there ("flies, mosquitoes, rats, fleas, stench, etc.") and spread beyond the *kampong*'s limits; thieves come from there. It would not do, at the same time, just to push a *kampong* away, to sweep it beyond the town's limits. Servants lived there. Of course, the servants earned so little that their employers would have to pay for their transportation by autobus, jitney, dogcart. If the servants were made to walk, they would arrive too tired to work. Or, much worse, the servants might to decide to sleep somewhere on the way, along the road. Nobody wanted more tramps in the streets.[136]

At this point, the echo of the European avant-garde was heard again in the Indies: the idea of "housing blocks." It was popular, at the time, among the most progressive social architects of Amsterdam, Rotterdam, and other European cities, to house the working-class people in a most modern, efficient, and safe way. The authorities in the Indies, *Tecton* argued, should also be radical. They should also build the poor quarters in the cities from the ground, completely anew. They should start

> entirely simple, sober large-size buildings . . . shelters . . . on a permanent basis (fire proof), and they should place *kampong*-dwellers in the buildings. In this, we can learn from the West, and we can look closely at how, in the European towns and cities, the question is solved. The Westerners have, for that purpose, a kind of house-barracks [*woonkazernes*]. They would satisfy us, in the Indies also, only they should be designed modestly, and also otherwise adjusted to the Eastern circumstances.[137]

Building up modern Indies cities was done by way of cleaning—and, increasingly, clearing out. Like the trees, some people—and this became an extensive movement with the passing years—were pushed off, or pulled out, of the cities themselves. As in so many other matters in the late-colonial period, and as in dreams, it was difficult to distinguish between a design and an obsession, between another attempt at cleanliness and an infection. The very name of one very striking, and increasingly massive, clearing out, in the Indies, suggested this: *bungalow-itis*, meaning a "bungalow epidemic."[138]

Bungalows, whole cities of bungalows, tiny *poppenhuisjes*, "dollhouses,"[139] and bigger "summer" places were built in the Indies countryside, in the hills and mountains especially, around and above the big cities like Batavia, Semarang, Surabaya, or Medan. Since the 1920s, the bungalow cities also grew around and above smaller Indies towns.[140] The bungalows were built by the people, naturally, who could afford them, almost exclusively Europeans. Especially with the outbreak of World War II in Europe, and with the evident approach of the war to the Indies, as an Indonesian nationalist paper noted, late in 1941, "man discovered that air is good for health."[141]

With the crises, the prices of the bungalows should have collapsed. They soared enormously. To build a bungalow, it was clear, one had to believe strongly that this was some sort of fundamental solution. As Wim Wertheim, a young Dutch professor at the Batavia Law School (who was also "infected"), remembered, in order to buy a bungalow, one had to change a great deal of one's life: "We had to buy a second-hand car for Fl 500. It was in March [1940]. The car was cheap, but it needed so much gas. Besides, because neither my wife nor I could drive, we had to enlarge our domestic staff by an expert, a chauffeur called Kawi."[142]

Some suggested that it became difficult to breathe in the Indies, if one did not succumb to *bungalow-itis*. Eddy Du Perron was a Dutch writer born in the Indies. He came back to the Indies in the mid-1930s, escaping from Europe in the shadow of fascism in search of a freer space in the Indies, his "land of origin." He found himself, as his correspondence shows, moving through the colony with diminishing hope, and, most of the time, between bungalows:

> Tjitjoeroeg is located really high. And it has better climate even than Bandoeng. It is cool, pleasant, and less humid.[143]
>
> . . . sometimes I walk a little, toward the *Villa Panorama*, a pension in the neighborhood from where there is a divine view.[144]
>
> . . . It can be breezy here, and as a whole, for the Indies, there is an excellent climate, and the natural beauty here is amazing.[145]
>
> Our little house in the mountains (4,200 feet, about 1,350 meters high) is primitive yet wonderful: walls of bamboo, mats as the floor, etc., but we are completely at bliss that we have a W.C. with "chaise."[146]
>
> At present, we live in a place that is about sixteen kilometers outside and above [the town of] Soekaboemi, and it is so cool, here (1,100 meters) that we have to sleep under blankets. This is a bungalow system . . . *clean and healthy*, and the surrounding is marvelous. . . . By the way, the bungalows' names are *Utrecht, Gelderland, Zeeland*, and so on like that. *Groningen* is presently under construction. We live in the just-opened *Drente*.[147]

Even when kept at a distance and tiptoed over through the bungalows, the Indies was distressing. Disgusted with the colony, Du Perron returned to the Netherlands in 1939. He died of heart failure in Bergen, on the Dutch-Brabant coast (in a sort of a bungalow, in fact) on the day the German armies invaded the Netherlands.

The problem of Indonesian history, however, is that it rarely comes to an all-absorbing drama and to a flash of recognition. The strength of the colonial culture, and all the more so as it was approaching its climax and end, was in trivialities. Bungalows were toy-like cities in the (real) mountains. The architecture was a notion of the ultimate survival being merely toyed with.

Instead of the model cities of Europe, models of cities, dolls of houses, of streets, and of squares were increasingly typical of the Indies. Miniature became a significant way in which the Dutch Indies architects increasingly worked. Images of models of cities, by the mid-1930s, as there had been increasingly less of newly built real building, filled the pages of the Indies technical journals and popular family magazines. Between 1936 and 1939, most strikingly, plans and models of the main square of Batavia and colony, *Koningsplein*, of the future square, better to say, became the leading feature in the most prestigious *Ingenieur in N.I.*, "Engineer in Netherlands Indies." In endless variations, the models, plans, drawings, and photographs of the models were presented and most hotly debated.

One could see in the models an almost perfect, cute in its model form, and largely empty square, dotted now and then with icons of a future little sport stadium, little railway station, little chief colony's offices, little radio headquarters, little governor-general's palace.[148] It was also a little future jumping-off place: "it makes one feel good to listen to Wouter Coolop, Engineer, as he imagines the Koningsplein in the years not far ahead; then, he says, the city with the colony's central offices would need the square, also, as a landing space for little airplanes."[149]

Something in the Indies came yet closer to the idea of the ultimate late-colonial city, closer even than the clean-air bungalows and toy-like models. Professor M.W.F. Treub, an influential spokesman for the empire, did not find very much to his liking as he visited the Indies in 1922. He was particularly impatient with what he saw as a lack of colonial daring. When shown some of the Indies *dienst parks*, "service parks," however, that had just began to appear in the Indies, he became truly excited.

The Indies service parks were built first on plantations, and then by other colonial corporations for their officials and workers. Service parks, as Treub found and described them, were "compact," "calm," and "roomy." About one of them, the service park at Djatiroto, a sugar plantation in East Java, Treub wrote: "the housing for the head administrator, administrators . . . and some other personnel . . . impressed me as a well and orderly established villa park."[150] In Cepu, the place of large oil fields, also in East Java, as Treub saw it, "The whole town is in the shadow of oil. . . . The housing of the oil-company personnel, however, looks comfortable and neat. The compound of the administrator . . . in one word, is princely."[151] Treub moved in the Indies between service parks and was increasingly happy. One of them, in Batavia, a service park of the *KPM*, the Dutch principal shipping company, was built for predominantly native personnel of sailors and dock workers. Treub called it a *kampong* in his report:

The *kampong* makes a good and robust impression. The houses look tidy, and the *warongs* [Malay for stall] for the inhabitants is clean. The terrain is spacious. . . .

The whole territory . . . is a town by itself. One has all one needs: a bathing place, goods, food and drink. In the end, there is a ship that one can go on with, toward the next step of one's career. *Mein Liebschen, was wilst du noch mehr?*[152]

Treub's was not a long book. But yet another service park was included, this time a park for the native workers of the Deli Railway Company, in East Sumatra. This *dienst park* is also called *kampong* by Treub, and it is equally highly praised: "This *kampong* may justifiably be called a model . . . its houses . . . are more spacious . . . than . . . the ones in [the workers' original] homes . . . here, the *kampong* is full of light, spacious . . . hygienic."[153]

Treub's was not a unique voice. H.F. Tillema, the man who so often set the tone for the late-colonial Indies culture, was no less positive about the service parks. As Tillema portrayed them—and this is, again, why he is the late-colonial Indies classic—the service parks were happy prefab places, made for seasonal dwellers to stay at and to move on from.

Throughout his *Kromoblanda*, Tillema lamented the demise of what he called the old Indies way of life. He found the way in which space was being lost as cities got overcrowded, and as the price of land soared, especially deplorable.[154] In the service parks, as he presented them—as he led his readers to them—the space between dwellings was ample. It was more, it was flagrant. It stood up, as if at attention during roll call. Messiness was gone, and order was made into geometry. Tillema's photographs of the service parks, as one turns the pages of his books and articles, look increasingly like—and many, indeed, are—photographs of models and drawings. Best of all, technical drawings. The images are as if they were diagrams of pure ideas, in black and white, sharp lines, linear, and, in the Indies sense, avant-garde.

The Indies service parks, and the Tillema images of them, were a part—caricature, omen, or both—of a contemporary, modern European, and global culture: "Town is a tension in the length, and a tension in the height. Nothing else. Two correctly made contacts of an iron wire make a town."[155] The Indies service parks were undoubtedly modern, well-organized display cases. Even a newcomer to the Indies could not but be impressed.

In the spring of 1942, the armies of the Japanese Empire conquered the Dutch colony. Their architects and their engineers were also designing and dreaming up a new space in the Indies for themselves and for the natives. At least some of the Japanese planners and dreamers found it perfectly logical to place their newly established concentration camps for the European men, women, and children, into "just a little adjusted" (more people to a unit, straighter lines, sharper edges, more flagrant space) Dutch Indies late-colonial mountain hotels, bungalow towns, exhibition grounds, and, certainly, best of all, the service parks. "If the Deli Railway Company had foreseen," a

Public housing project near Bandung, 1920. From an album offered to the mayor of Bandung, B. Coops, on the occasion of his retirement to Europe in August 1920. (KITLV)

Dutch engineer, employee of the company, and then an inmate of one of the Japanese concentration camps, wrote, "that Pulu Brayan [the company's service park] would be made a women's internment camp in 1942, the company would certainly have pushed some of the already planned improvements of the service park through, and it would have made sure that they were accomplished on time."[156]

The Towers

According to Heidegger, "Mortals dwell in the way they preserve the four-fold in its essential being": house, ideally, is a unity of mortals and gods, and skies, and earth.[157] Also, according to Heidegger, "A boundary is not that at which something stops but, as the 'Greeks' recognized, the boundary is that from which something *begins its presencing*. That is why the concept is that of *horismos*, that is, the horizon, the boundary."[158]

There had always been an awareness, in the Dutch architecture and urban planning, of a tension between the outside and the inside of a house. The Dutch model, most celebrated and reproduced in countless images since the seventeenth-century Golden Age, had been that of an energetic harmony between the two realms. The model Dutch house, in these images, is strong enough, or "full" enough, in Heidegger's sense of the word, that it is able to shelter both humans and their gods. The gods, as Kierkegaard put it, are still there in the house; they have not yet "taken flight," and they have not "taken the fullness with them."[159] The model house in the Dutch tradition, in other words, is bold enough to face the outside, the street, the square, and the town: "In the traditional Dutch townhouse the front room or *mooie kamer* (literally beautiful room) represented the symbolic presentation of the home to the street. . . . The curtains of the living rooms of Dutch houses, which traditionally face onto the street, are often left undrawn and the rooms brightly lit."[160]

The Dutch architects and urban planners in the early decades of the twentieth century were at the head of Europe in the efforts, in the changing times, to retain the order and harmony of a house facing the town. Much of their best effort, still, was to dynamize the interior of a house, to keep the house "full," so that it remained strong in order to stay open to the world.

The walls of houses were crucial. H.P. Berlage, the most influential Dutch architect of the time, wanted house walls that "came [as he believed they had] from the hanging carpets of nomadic tribes . . . [to remain] a colored plane rather than a solid structure."[161] Berlage preached "plurality and tolerance" in modern society. House walls, according to him, were to be a means of connecting rather than separating different rooms inside as well as outside a house. Brick was the material favored by Berlage—best of all, exposed, unplastered bricks, the same kind of brick walls on the inside as on the outside: "For Berlage, brick symbolized the relationship of the individual to society."[162]

In the Indies, too, a struggle between the interior and the exterior of a house had been going on for centuries. From a Dutch point of view, it was summed up in a lecture by B. de Vistarini, Engineer, an Indies architect, giving a talk at the Bandung technical college in 1938. At the start, in the seventeenth century, de Vistarini said, the houses in the Indies were built by the first Dutch settlers as if out of fear of the newly conquered, foreign, and inhospitable land. The first Dutch houses in the Indies had "little windows . . . exposed gables . . . and virtually no regard for the climate."[163]

In the centuries that followed, as the Dutch began to feel more at home, "colonial styles" developed that dared to open a colonial house a little more to its environment. There was, eventually, "a large compound, galleries running around the house, an inner gallery as a shelter from the direct daylight, and several outbuildings in the compound."[164] Toward the end of the nineteenth century, however, as the modern era fully touched upon the colony, according to de Vistarini, "the circle was closed." The parts of the colonial house that tended to open the home to the outside began to shrink: "First, people began to let the galleries decay; then, as if by design, the front and back galleries began to be limited in size, and a new type of house emerged, around 1900, in which the front gallery, typically, was merely yet another room, only missing its front wall."[165] Another twenty years later, in 1920, according to de Vistarini, "We can find the *mode* of the open front galleries no more. Most of the rooms, now, have a size no larger than 16 square meters, the height is limited to merely 4 meters, and the whole compound [formerly an open space] is built over as much as one-third of its whole area."[166] As in the beginning, it was becoming typical for the modern Indies houses to have "unsatisfactory regard for the climate . . . windows behind jalousies . . . exposed gables."[167] The very newest Dutch houses, according to de Vistarini, Engineer, had "no fundamental advantage over the first Dutch houses built in the Indies."[168]

It was like disproving Berlage's beliefs. But this was not modern Europe. It was what Europe might become in the decades ahead. The backyards and the gardens at the sides and behind the Dutch houses in the Indies, as Eduard Cuypers, a prominent Dutch architect in the colony, wrote in 1919, typically became "closed by walls built around the whole building, with just a single entrance left in the front, and another one in the back."[169] Thus, the front of the houses, just by using *draaibare jaloezieluiken*, "venetian blinds," "could be shut off."[170]

Some among Cuypers' and de Vistarini's contemporaries realized that by looking at the "closing" and the "covering" of the modern Dutch houses in the Indies, they, in fact, were witnessing increasing emptiness. If the function of the front of a house is to face the world, the faces of the late-colonial Indies houses were going blank. Willem Walraven wrote about this in 1933:

> People now build houses that could be closed at the front side. Such houses have a bell by which a visitor is expected to report.
>
> . . . Many houses now would be more habitable if they were turned around a hundred and eighty degrees; thus, the outbuildings from behind would be placed to face "the main street," and what is now the front-gallery entrance would be turned to what is now the backyard.
>
> . . . Listen to the architects as they are giving us their advice: fortify the front of your house, close it, hang on yet another bell at the entrance, and a mirror, perhaps.[171]

The traditional space where the Indies houses and their inhabitants used to meet their neighborhood, the front galleries, Walraven wrote in yet another article in 1941, "became too small even for making a turnaround properly."[172]

In contrast to everything Berlage might preach in the Netherlands, the excellence of a modern Indies house was increasingly in its capacity for closing. Walraven, writing about the Indies neighborhood, inevitably turned to the Indies doors:

> It is the *mode* to have three double doors on the front gallery, one leading to the room on the right, the other leading to the room on the left, and the third leading to the inner hall and toward the back gallery. Even more so, double doors are becoming increasingly fashionable, opening to yet another double door, a *porte-brisée*, with its wings of course made so that they open in the opposite way.[173]

Even the smallest house, if its owner wanted to keep up with the Indies times, had to have an imposing set of doors:

> All the doors are double-wing doors, even garden gates are *portes-brisées*, . . . often they become so small that, in order to make it possible for a person of European posture to get through, *both* wings have to be opened each time.[174]
>
> . . . There are houses where one can find three pairs of doors in one wall opening: one pair of jalousie doors on the outer side, one pair of glass doors on the inner side, and the third pair, called *tochtdeuren*, "storm doors." None of the three pairs appears fully satisfactory by itself, yet their combination neither appears perfect. There has to be a special scheme in place to maneuver these door schemes, and most of our servants never learn that.[175]

This is how we get, slowly, to the idea of towers. The modern Indies houses, with their increasingly blank faces and increasingly heavy doors, were also pulling off the ground. It was a very striking change, again, and it was often remarked upon by contemporaries. Instead of houses that were open and sprawled amid their tropical surroundings, shaded by trees and galleries, now a second floor, and even a third floor, appeared in the Indies. Already in the guide for the Dutch housewives in the Indies by Mrs. Catenius, in 1908, almost all the houses photographed as modern were houses with at least two floors above the ground.[176] To live in a house with an upper floor, Catenius wrote also, was something in the Indies "quite new." It provided the colony, she believed, with an "utterly new architectural order."[177]

This of course was a continuation of the tendency toward floating—logical, if there were not so much heaviness in it at the same time. Some among contemporaries were proud, others were puzzled. People had to be very well off to get into trouble, as one Dutch architect in the Indies pondered in 1916: "To decide on such a costly house, one truly has to have a passion for climbing stairs and for dealing with the heat in those tropical bedrooms, high under the European roof."[178]

As the twentieth century opened, H.P. Berlage, in the Netherlands, expressed a confidence in the art of architecture as "the mother of all arts." He defined the architecture he believed in as "a principle of decency (beauty [that] is efficiency)."[179] For Berlage, a modern building was a "total, organic complex . . . ,"[180] and, in his view, architecture should "shape a space into an organic whole. Every single segment should invoke the others, and every single segment should mirror the whole; nothing, in the construction, can be missing."[181]

The Dutch modern and avant-garde architecture, through the first three decades of the twentieth century, followed Berlage on this principle of organic totality. No detail would make sense if it did not serve the whole. The Dutch, avant-garde De Stijl, "The Style," movement sought an architecture of a "classical purity which would abolish the distinction between art and utility." De Stijl architects wrote enthusiastically about "the similarity between a Greek temple and the modern functional design of an ocean liner."[182] With the other avant-garde architects in Europe, they dreamed about and planned houses "as characteristic industrial products of the twentieth century like the motor car."[183] Or, as Le Corbusier put it, "The house is a machine for living in."[184]

There was an echo of this, again, in the Indies. Technology and a (kind of) principle of organic totality had always been important in building and defining a modern house in the Indies. To build and to define a house in the Indies, sooner perhaps, and more intensely than in Europe, technology of wheeling, floating, and soaring had been invoked. Ocean liners and trains, as we have seen, long served as models for the Indies late-colonial house. With the late-colonial era evolving, the notion of technology that might make a home into a machine in the Indies maybe, again, sooner than in Europe, became acute.

Air conditioning is one example. Very much beyond—and, often, not very much related to—its practical use, the idea of air conditioning got very early, and forcefully, into the late-colonial Indies culture. Eventually, close to the end of the empire, the idea became one of the dominant themes of the debate about the future of the Indies.

It had long been an accepted view in the colony, as the most prestigious *Indische Gids* wrote in 1891, that heat had an adverse effect on *geestkracht*, "mental stamina," of the white people in the Indies, and that *koude*, "the cool air," was desirable, at least to "the parts of body that were directly exposed— such as mouth, nose, eyebrows, and, also the lungs." "The health of a European," according to *Indische Gids*, was endangered, and the capacity for work diminished when the heat had its way for months at a time.[185]

Some kind or idea of air conditioning by mechanical means, too, seemed to have been around for a long time. In 1892, a year after the just-quoted *Indische Gids* article, *Tijdschrift van het Kon. Instituut van Ingenieurs afd N.I.*, "Journal of the Royal Institute of Engineers, section Netherlands Indies," wrote about air cooling abroad. Characteristically, this early report was on cooling of "homes on wheels": "On the *Rohilkhand Railway* [in British India] . . . the first-class carriages were equipped with *punkah* fans that are set in motion, through a cog transmission, by a person who is located in a separate part of the carriage."[186] In Switzerland, and also on the Pacific Railway, in the United States, iced water was provided to passengers, and much more was planned:

> According to *The Scientific American* of May, 1881, a plan is announced in Philadelphia for installing a ventilation system consisting of an iron bar, placed on the outside of a train, under the roof, at the whole length of each carriage. Short iron wings will be fixed on the bar at regular intervals, and in alternate positions. The bar will be made to spin through a cog wheel fixed upon the roof of the carriage, which, in turn, will be propelled by a stream of wind as the train travels forward. Inside the carriage, moved by the iron wings on the spinning bar, air will begin to circulate as well.[187]

Something like that, the journal suggested, was badly needed in the Indies as well, namely, on the notoriously warm lines of the Indies railways in Central Java connecting Semarang, Surakarta, and Yogyakarta.[188]

Not everybody accepted it. For some it seemed to be too artificial an idea not to be funny. To cool real houses, not carriages, was open to ridicule. In the *Indische Gids*, next to the reports just quoted, an author joked about a project that proposed "through technical means" to "fight high temperatures" and even "make them harmless to our white race." There are some people, the author wrote, who fantasized about how "cool stations would be constructed, out of which one might draw fresh air" for "cooled houses": "These people, it seems, can see no financial problem . . . they may admit that a man cannot stay inside the cool walls all the time, but they argue that he can do so at least at night, lunchtime, and dinnertime."[189] The author just laughed and laughed at how these people believed that the air, before it is let in a house, can be cooled, and at the same time (this had always been an important part of it) disinfected: "What sense of reality! . . . the air will be pressed through filters so that nothing but purified air is let into inhabited parts of the house. To keep our segment of the house free of undesired air, one will merely have first to pass through a disinfection room, then take a bath and change clothes. How practical!"[190] Imagine: "A gentleman relaxes and enjoys some reading and fresh cool air under the auspicious working of a cool station. The man calls his servant, the servant steps in a disinfection room, takes a bath, and changes his clothes—Not bad as an All-Fools joke."[191]

Essentially, the chuckling and laughing author was right. Extremely few houses in the Indies before 1942, before the end of the era, were "air conditioned through technical means." Yet, exactly because of this, the theme gained its extraordinary strength. This was yet another powerful because pure idea of dwelling—and staying—in the Indies. No more a joke.

In March 1935, a new and costly—lavish, given the economic situation of the colony—"laboratory for technical hygiene and sanitation" was opened in Bandung. Most of the research in the laboratory shifted almost instantly to air conditioning. In August 1938—the laboratory was already an "institute" by the time—a big international conference on air conditioning was convened in Bandung under the institute's sponsorship. There were papers read, at the conference, on "air conditioning and comfort," "air conditioning and malaria," or, for instance, on "temperature of whites in tropics and in tropical air-conditioned places."[192]

The matter was considered to be of grave importance. One speaker put it in almost a Heideggerian way: "man is a cosmic resonator,"[193] he said. Upon the Dutch nation, also according to that speaker, a special burden and a privilege rested in this particular task, if only because it was the Netherlands that had always been in the forefront; was not it in Holland, namely, in the seventeenth-century Golden Age, that the thermometer was first used as a tool in medicine?[194]

Air conditioning in the tropics, as was the conference view, should first be installed where the Europeans were *meest kwetsbaar*, "most vulnerable," when, namely, they were "in a sleeping state," which meant in the European houses' bedrooms.[195] Air conditioning, also, according to the conference, was especially important in the coastal areas of the colony, for the modern people who could not, at the moment, take refuge and relax in the naturally cooled hotels and bungalows of the mountains.[196]

Some papers dealt with air conditioning for hospitals. But, "of course," in those difficult economic times, it was impossible to think about the matter except for a very few of the most expensive and exclusive clinics in the Indies.[197]

The Indies movie houses appeared to be the only places frequented by natives en masse, and considered for the new technology of habitation:

> We surely can get closer to the European level. Theaters and movie houses, for instance— when they are made air conditioned—what a wonderful impression it will make, at the entrances, in the halls, as one sits down.[198]
>
> The Maxim Theater in Soerabaia is already adapted for air conditioning, so that all the public of that city, as moviegoers, can familiarize itself with climatization.[199]

On second thought, the air conditioning in the late-colonial Indies, rather than exclusive, was theatrical.

Air conditioning, like the floating houses, became emblematic of the culture. By a certain moment in history, no one could talk about a modern house in the Indies without mentioning air conditioning. Air conditioning had become a dynamic, and often defining, part of the culture. One of the speakers at the Bandung conference described European colonial office buildings in the Indies of the future:

> Cooled office buildings will be built lower; or better to say, in the buildings of the same height there will be more floors. . . . When inside the buildings, one will not notice any difference whatsoever from Europe. There merely will be more doors against the draft and the inner partition of the building will be more concentrated and condensed than in Europe. The climatized office buildings will have one great advantage above all: even in the buildings that will be situated on the most busy streets, one will work in calm because through the windows—they will naturally be closed all the time—the noise from the streets will penetrate only very feebly, if at all.[200]

Because of air conditioning, the limits and the edges of the modern house, logically, would have to be most sharply designed and respected: "Movement, especially in large buildings, is still the most important problem that has to be solved . . . inhabitants . . . have to be protected against recurrent and possibly too sudden temperature shock as they will walk around, enter, and leave the building."[201]

Walls of the air-conditioned buildings would have to be carefully erected and strong indeed, otherwise catastrophe might occur: "How painful it would be, for instance, if the air from toilets were to penetrate into bureaus and suites."[202]

Air-conditioned houses in the colony, by their very nature, had to be strong, "more solid than it is usual in the Indies today."[203] "It speaks for itself that the outer walls of the air-conditioned buildings will have to be further sheltered against an unrestrained sunlight, possibly through overhanging roofs, built-on penthouses, additional walls, and so forth."[204] Air-conditioned houses in the colony, by the merit of technology, proudly, on their own principle of organic totality, also had to have heavy doors and windows: "There will have to be double doors and double windows in the air-conditioned buildings . . . the doors and the windows will have to have a greater capacity to isolate, and, let us hope, more solid materials could be found to make them than it is the fact today."[205]

The idea of air-conditioned houses in the colony—and, again, this was why it was so intensely talked about—appeared to offer a new, most modern, avant-garde chance for the Dutch in the Indies to float and to stay:

> By the air conditioning of houses, life and work in the tropics will be made much more effective than it has ever been possible to make it. It is probable that this will become a means by which the European community here in the Indies, as it has always been so much desired, will be enabled to keep its character. Already we can see how some persons in leading positions who normally would have to repatriate, now, thanks to climate regulation of their office or home, are beginning to think about extending their sojourns in the Indies.[206]

Purely on technological consideration, by the laws of modernity—or so it now could be made to appear—the air-conditioned homes in the colony might be designed and built huddling together more than ever, against the uncooled rest of the colony:

> A house built in a row of houses, touching upon the house next to it, with one or more floors above the ground, probably in a Dutch style, will have a multiple advantage for air conditioning. Because of this, the increasing introduction of such houses in the Indies should be encouraged. . . . These buildings, as we will undoubtedly soon encounter them here in the Indies on a much larger scale, will appear to us much flatter and taller from the front view, they will have small rooms, even smaller than it is common in the contemporary modern little houses in the Indies today.[207]

"Still much denser a construction" was foreseen for the colonial Indies of the future, "acclimatized building blocks," dwellings "small but healthy."[208]

The climax to this, again, as to almost everything in the late-colonial Indies, came with the Second World War. Wim Wertheim, the young Dutch professor at the Batavia Law School, remembered how in 1940 the Netherlands was invaded by the German armies. All connections with the Indies were cut, and the Dutch in the Indies were getting ready for a truly long sojourn. Wertheim and his wife could no longer send money home to Holland, as they had done all the years before, to support their parents and to invest for their own European retirement. Completely in the Indies at last, Wertheim recalled, "We installed a modern air conditioning system in our bedroom. Without being able to send the monthly sum of money to Holland anymore, this little luxury became possible."[209]

As it was increasingly troubling in the colony on the outside, the air inside the modern Indies house was, increasingly, conceived of and handled as a technological substance. Very seriously, it was written, for instance, about a possible health effect of *luchtelectriciteit*, "atmospheric electricity," in the colony's households.[210] As the final crisis of the empire approached, more than ever—and more than in any dreams or plans of the European avant-garde—the modern house appeared in the Indies as a laboratory, and as a machine. As the Dutch author just quoted put it in 1938: "equipotential layers of atmosphere rest upon the upper layer of the houses. . . . One can think of a house as of a Faraday cage."[211]

When Professor Treub visited the Indies, he was correct and perceptive when he pointed out—aside from service parks—electric kitchens as modern hearts of the colony. Treub was truly excited about one particular house in the Indies, that of tea magnate K.A.R. Boscha—a house, as Treub saw it, a dream, and a future. The house stood in the middle of Boscha's own plantations, high in the mountains above Bandung. Correctly and perceptively, Treub defined the Boscha house through strawberries, caviar, and oysters. They were served in the tropics as a surprise, he wrote.[212] And, it was comforting, the idea being that modern man was able to build a source of electricity somewhere in the house to feed a refrigerator, and, thus, in the midst of the Indies, create a world for himself.

In Europe, in the Netherlands, namely, soon after the First World War, a movement started, *Vrouwen Electriciteit Vereeniging*, a "Women's Electricity Union." The movement was inspired by Dutch women's sudden enthusiasm for electricity—even better, electricity in the house. The Dutch Women's Electricity Union had, in the early 1920s, about one thousand permanent members. It organized lectures and excursions, and it published a monthly *Bulletin*.[213]

In the Indies, a magazine closely resembling the Dutch *Bulletin* existed. It emerged, however, not as in the Netherlands, on the high tide of general optimism after "the last of all wars" was over in 1918. The first issue of the Indies journal was published in June 1931, at the moment of the impact of the great economic depression in the colony. The name of the magazine, *Alles electrisch in huis en bedrijf*, "All Electric in House and Business," was changed after a year into the more homey *Alles electrisch in het Indische huis*, "All Electric in the Indies House." While in the Netherlands at most about one thousand copies of the *Bulletin* used to be sold, in the Indies—with a much smaller number of potential Dutch women readers than in Europe, of course—*Alles electrisch* sold, in the first year, 25,000 copies, and, beginning with the second year, 30,000 copies every month!

The glossy magazine was made largely of photographs, captions, and headlines of various sizes. The first issue of *Alles electrisch* announced, for instance: "Do you want to be ill?! Illnesses cost money! Health, therefore, means saving. . . . Try the General Electric cooling box. . . . The new brochure 'Nutrition Hygiene' can be sent to you through International General Electric Co. in Soerabaja. Telephone South 2157."[214] From the first issue, too, comfort in the house was a theme: "*Alles Electrisch* can help. Even long-standing consumers of electricity can be brought more up to date about what ELECTRICITY can offer them in the field of EFFICIENCY, HYGIENE, and COMFORT."[215]

This, in fact, in the Indies, was the climax of the ugly times. All could still be saved, in comfort, as long as the house was crowded with equipment. Keeping up with progress, this meant electric equipment: milk cookers, irons, pans.[216] The house was as self-contained as it was electric: "Bake, cook, and iron on electric AEG machines . . . 1001 Watt *Slender-Line Lady* by Mr. Duske . . . *Thermoroller* . . . *showroom* of Gebeo . . . *Siemens* goods."[217] The house promoted was simply called "Electric House": "from the front galleries . . . one enters the electrically equipped living rooms, eating rooms, bedrooms, kitchen, and bathroom."[218]

As long as the electricity worked, and as long as one kept on reading the magazine, it was cozy in the Indies house. Inside such a house, if only it were electric enough, even an electric native might be placed. There was a short article, rather exceptional but to this point, with several pictures and a poem published in the September 1931 issue of *Alles electrisch*. Under the headline "*Washing Girl and Electric Iron*," a native *baboe*, "maid," was shown inside an electric Dutch house, ironing. In the staged photographs and the poem, at first, there was an incredulity expressed by the girl, as she spoke to herself in a funny mix of crude Malay and clumsy Dutch:

't is terlaloe kras-, "'s too hard."

It is too difficult, at the beginning. But as she learns, she goes on, and indeed she finishes her job, okay,

in minimum van tijd, "in minimal time,"

Geen vieze boel en narigheid!, "no mess, easy."

Elated, the "maid" exclaims, still in half Malay and half Dutch, as crude and clumsy as before:

Djempol!—"Wow!"

zooals dat ijzer glijdt!—"how that iron glides!"

Dat doet de E-lec-tri-ci-teit!. "This is, what that E-lec-tri-ci-ty does!"[219]

She is (safely, let us all hope) plugged in.

A Viennese author, Robert Musil, in his *Man Without Qualities*, wrote early in the twentieth century about a "social *idée fixe*" of a modern European of his time:

a kind of super-American city where everyone rushes about or stands still with a stop-watch in his hand. Air and earth form an ant-hill, veined by channels of traffic, raising story upon story. Overhead-trains, underground-trains, pneumatic express-mails carrying consignments of human beings, chains of motor-vehicles all racing along horizontally, express lifts vertically pumping crowds from one traffic level to another—At the junctions one leaps from one means of transport to another, is instantly sucked in and snatched away by the rhythm of it, which makes a syncope, a pause, a little gap of these intervals in the general rhythm in which one hastily exchanges a few words with others. Questions and answers click into each other like cogs of a machine.[220]

Then, Musil added, as an important part of the idea: "And elsewhere again are the towers to which one returns and finds wife, family, gramophone, and soul."[221]

What are these towers? There certainly has never been a European city, even a village deserving its name, without some sort of skyline. Towers make the skyline. Skyline, for a European city, has been a necessity, a city identi-fication, a pointer to its character, an anchor. The Europeans, it would seem, had to bring some sense of skyline with them to the Indies. Then, in design-ing and building up modern cities in the Indies, what was the skyline? What, in the Indies, were the towers?

In 1936, A. van Doorn published quite a bitter article in *Tecton*, an Indies magazine for technicians by profession, on "how man sometimes builds" in the colony. He chose what he believed was a manifestly bad example of the modern building: "there is absolutely no connection between the surround-ings of the house and either its style or the material it is built of."[222] The house, van Doorn wrote, did not take the least notice of the location where it stood. The garage was dominant as one approached the house from the main street. The front of the house, the main entrance on the side of the garage, and the windows of the finest rooms, in their turn, faced nothing but a big, ugly water tank.[223]

Reading van Doorn's article, however, one has a feeling that in fact this was, in the late-colonial Indies, as it should be. This indeed, so it seems, was the culture, the style, and the search for meaning. This also was the skyline. The garage should be the dominant, and, even more so, the house should face the water tank.

The circle, indeed, was closed. The electric equipment crowded the houses as for a trip. The energy was in wheeling and floating. The house was, al-most, self-contained. Electricity, the newest helper of the colonial culture, was bringing water into kitchens, bathrooms, and lavatories, thus the ulti-mate, it seemed, coolness, purity, order, and calm, inside the house: "taking into account the violent tropical rains, to secure movement of water into and inside the house as feasibly as possible . . . water is led into reservoirs that are placed as high as possible, in towers. That solves the problem."[224]

Cultures always giggle, when one looks long enough. Here were, perhaps, the towers for the Indies. According to Eduard Cuypers, Berlage's contemporary and, also, an immensely influential voice in the Indies late-colonial architecture, urban planning, and culture in general, "Many among us, in the Indies, are quite conservative, as far as home building is concerned; very often, people here seem to be satisfied with an imperfect and imperfectly equipped house, just because they feel that their stay here is temporary."[225] This flabbiness, Cuypers wrote, had to be countered with all strength:

> Namely, by adequate water regulation, by means of an air pressure water reservoir that allows the owner of a house to devise a bathroom and water closet with an amazing supply and outlay of water. In the system of the air pressure water reservoirs, the clean water is pumped from a well (and there is a well in every compound of the Indies), is forced into a water reservoir or a kettle, and, from there, it could safely be distributed throughout the house.[226]

This, Eduard Cuypers knew well, was the crucial idea of the empire. A possibility still open, perhaps, to prevent leaking and sinking of the colony. Quite inevitably, Cuypers, who knew this, wrote about the water towers as they grew above late-colonial Dutch Indies houses during the last decades of the empire, as if he was writing about spires of cathedrals:

> The layout of the fourth type of house (D) is based on a completely different outlook than the types A, B, and C. The compound of the house, in this plan, is of an exactly equal-sided octagonal shape, and it is closed on all sides: by the mass of the main building in front, by three service buildings on the next three sides, and by four covered galleries on the sides that remain; the whole complex is interconnected through inner passages. . . . Looking at the front, toward the right side, above the bedrooms, there is a little water tower. There, the reservoir is located, which provides the house with its own water. This little tower is brought into the plan for practical reasons. Yet it gives the whole complex, as we look at it from the outside, approaching the front, a spirited silhouette.[227]

THREE

FROM DARKNESS TO LIGHT

The City of Light

> But it is fruitless to ask whether the Greek language or Greek
> economic conditions, or the idle fancy of some nameless pre-
> Socratic, is responsible for viewing this sort of knowledge as
> *looking* at something (rather than, say, rubbing up against it, or
> crushing it underfoot, or having sexual intercourse with it).
> —Rorty, *Philosophy and the Mirror*[1]

In 1883, the islet of Krakatoa in the strait between Sumatra and Java erupted in flames. Seven years later, in 1890, an account of the event was published by R. A. van Sandick, a former chief engineer in the Netherlands East Indies. The book, *In het Rijk van Vulcaan: de uitbarsting van Krakatau en hare gevolgen,* "In the Empire of Vulcan: The explosion of Krakatoa and its Consequences," became an instant best-seller, both in the Netherlands and in the colony.

The native peoples of Indonesia, van Sandick wrote, writing about the Krakatoa explosion, lived through this disaster (70,000 people died from the explosion) as they had through many other shakings of their land. The Sundanese of the west part of Java, for instance, always believed that their island was precariously borne upon the back of a buffalo:

> From time to time the animal moves: hence, the earthquakes. If the weary beast were to decide to shake off his load, the whole island of Java would disappear. Therefore, whenever the natives feel the earth is shaking, they throw themselves upon it, kiss the ground, and shout loudly: "aya, aya." With this they try to say: "we are still here." They hope, because buffalo is a domestic animal, that it will understand, it is its duty to keep on hauling Java on yet for a little while.[2]

There was, as we read van Sandick, strikingly little fundamental difference between the native peoples of the land and the Dutch masters of the colony. To the Dutch in the Indies, too, the land appeared to behave like an animal. It might shiver, but it could gently be touched and talked into behaving well again. The recent modernity, lighthouses, were the first thing swept away by the wave of destruction coming from the volcano. Telegraph poles and wires, where they were already installed, were torn out and apart. The tragic story of what happened next, still, so it seemed, could be told in a tone of an almost morning-after tenderness. There was, also, a great deal of mud.

On the eve of the eruption of Krakatoa, on August 26, 1883, in Anjer, a small town on the coast of Java facing Krakatoa, thunder was heard and lightning was seen. Telegraph connection between Anjer and the nearest provincial town of Serang was broken. However, by half past nine in the evening, both the thunder and the lightning weakened, and "the whole Anjer went to sleep." Things like that had been happening all the time:

> At six o'clock in the morning of the 27th August, the employees of the post and telegraph service were busy reinstalling the telegraph wire. Some Europeans had already gotten up and wandered in their pajamas or in sarong and kabaya [Malay for shirt or blouse] over the yards, or were taking a bath in the *mandi kamer* [Dutch-Malay for bathroom] behind the house. Others were still in bed.
>
> Toward half past six, the flood came. . . . Most of the residents did not even see the wave coming, others did not have time to escape. The water soared, and, in an instant, it drowned the town. . . . Some buildings of Anjer were carried away, the lighthouse broke in half. . . . The jail disappeared with all its inhabitants, prisoners as well as the guards.[3]

In an eerie way, this was a natural scene. Above Batavia, the capital city of the colony, almost a hundred miles away from Anjer and Krakatoa, flocks of birds were seen moving toward the east.[4] Technology did not work: "The magnetic instruments at the meteorological institute in Batavia did not register anything out of the ordinary."[5] When the people in Batavia, however, went down on their knees and pressed their ears to the ground, they could hear the rumbling.

There struck darkness and light. Offshore on ships, "Fireballs were seen . . . moving with a great speed and disappearing in the water. Upon the ships' masts, blue flames flashed."[6] Sun appeared by fits and starts, blue and green, clouds were red and yellow.[7] In the streets of Batavia, the most modern city of the colony,

> gas lamps suddenly went black, and doors and windows banged.[8] There was darkness . . . gas factory could not send enough pressure into the system. . . . The natives downtown, squatting in front of their houses, calmly waited for the events to take their course. They chewed their *sirih* or smoked their straw cigarettes. Merely *pelitah*s, little oil lamps, flickered on the ground next to them, as ever.[9]

In the van Sandick account, there is no mention of a disturbance in a political or criminal sense of the word. The police are mentioned, once, intercepting a letter from "a high priest in Mecca." The letter is said to call upon the people of the land to deepen their Islamic zeal, and to accept the eruption as a warning that the end of the world might be coming.[10] There is one more reference to the police. A police brigade was dispatched to some of the worst affected spots. The task force left Batavia onboard the steamship *Kedirie*, September 3, 1883, and it is described, in some detail, by van Sandick. It was made up of three military officers, ten police agents, and twenty coolies. On its journey, the mission is described as acting intensely, but mainly as trying not to get lost in the devastated landscape. As far as the mission's policing went, there did not appear anything, throughout, to be investigated.

Fatima was a native, and she was eleven years old when Krakatoa erupted. We know it because the book about her murder—published by F. Wiggers in 1908, and another Indies instant best-seller—had a title page arranged like a police form. The victim's name was printed at the top: *Fatima*; below, religion followed: *Islam*; cause of death: *murdered*; date of murder: *February 22 or 23, 1908*; place of murder: *Pasar Senen, Batavia*.[11]

Like van Sandick's book on Krakatoa, *Fatima* was rich in texture, full of sounds and trembling. The earth did not shake as in a volcano explosion, but birds were clearly heard.

This was the story. Before one sunrise, in 1871, as *tjoelik* birds began to sing in the trees, Dasima, a middle-aged and childless village woman, dreamed about magnolias. As she woke up, she decided to see a witch about it and was told that a child would be born to her, a girl, happy at the beginning, but in about fifty years' time yielding to an unspecified yet dreadful fate. On the night of the Javanese-calendar day of *Rabo wage*, a girl, Fatima, was indeed born.[12] The family moved to Batavia and stayed at a place called Djangal near Pasar Senen. There Fatima grew up, and she became "the rose of the neighborhood." "White men, yellow men, short men, all were trapped if only her eyebrow moved a little. Young men, the whole neighborhood, everybody, might be said to be crazy about her."[13]

Fatima married, divorced, and married again. She remained childless: "Her hair was still fine and black, the shape of her body was even better than when she was young, only her teeth grew bad."[14] She ventured into a jewelry business and became rich. She lived alone, in two rented rooms at *Kampoeng Baroe*, "New Quarter," and frequently she had visitors.

In 1908, Fatima was forty-nine.[15] On Sunday, February 23, 1908, the police were called to investigate at a murder scene. Fate had struck Fatima:

> The corpse of Fatima was sprawled near the wall of the room, the planks of the wall were smeared with blood, and on the floor there were bloody puddles. In the room there was one table, one clothes hanger, one bottle of port wine standing on the table, and two glasses also on the table from which, evidently, somebody had drunk. In the room also, there was found a piece of something, like

Dutch chalk that is used in school for writing on the blackboard. It was suspected to be poison.[16]

The police concluded that Fatima was killed "between six p.m. on Saturday and six a.m. on Sunday"—this meant, when there is dark in the Indies. "Her neck was cut by a sharp device." It appeared that "Fatima had had visitors in the house, they drank port wine, and, when Fatima got drunk, they forced her down on the floor and killed her like a goat is slaughtered. There were at least two persons besides her, one did the holding, the other did the cutting."[17]

There are police. Yet birds sing through the story; the earth shudders as the body of Fatima hits the ground. The police, at this early age, appear almost as physical as the victim herself. They also, so it seems, can still be closely watched.

The final part of the book, the section that every true crime novel enthusiast would expect to be its climax, the last fifth to be more exact, is fully taken up with very detailed information about the detective in charge. His name was *Toewan Schout Hinne*, "Mr. Sheriff Hinne."[18] He was born, we learn, in Sambas, West Borneo, on September 4, 1852, and we are also told that it was on Saturday:

> At the very beginning, by the decision of the resident of Rembang dated November 27, 1867, he worked as a clerk-accountant in the assistant-resident office in Bodjonegoro . . . his salary was Fl 25 a month. . . . By the decision of the resident of Rembang dated July 18, 1871, no.369, . . . he was promoted to become scribe in the office of the resident of Rembang . . . with salary Fl 25. . . . By the government decision of July 25, 1904, no.15, his salary was raised to Fl 500, and, by the decision of September 20, 1905, no.7, in addition to this, he was awarded a bonus allowance of Fl 100 a month.[19]

The police inspector, Mr. Sheriff Hinne, was well paid and clever, but also pure and wise:

> Great number of dark cases could not be explained until they fell into his hands. Then they became brilliantly clear.[20]
> He was just, wise, and good . . . Even when talking to a humble person, his language was courteous and noble. Thus, it came as no surprise that all the little people honored him and asked him for his help.[21]

The reader is not told about the detective's hair, and whether his teeth were still good, but he is told almost as much. Mr. Sheriff Hinne "did not know what fear is . . . people made all efforts to avoid being arrested by him personally." Once, when he was apprehending a criminal, "he was wounded, and the medical costs came to as much as Fl 322.25."[22]

Naturally—and "naturally" is the fitting word, here—the sheriff found out who had murdered Fatima. It was in the chirping in the air. He just let his agents "listen to whispers." At the new Batavia electric tram central station, at last, inevitably, they overheard the truth.

Sherlock Holmes, at the time, had already been known in the Indies. Yet, Sheriff Hinne did not appear to care about the wonderful man. He certainly did not work from the shadows, and—optics being yet another theme of this chapter—he did not even use a loupe.

There was a touching optimism in Wiggers' *Fatima*. The murderers were grabbed, and another case was made brilliantly clear. It was a modern book. Facing the title page, there was a full-page, crudely drawn, askew, yet evidently lighted, and enlightening, petroleum lamp.

Wiggers, the author of *Fatima*, believed in progress. Possibly a Eurasian, he became the chairman of *Maleische Journalisten Bond*, "Union of Malay Journalists," founded in Batavia in 1906.[23] Besides *Fatima*, he wrote other thrillers, like (together with a Chinese colleague, a journalist we know only by the initials Y.L.M.) "A Poem about the Java-Bank Robbery." But he also wrote or translated into Malay "Statutes of the Pasteur Institute in Weltevreden," "On Rabies in Dogs," "Book for Accountants in Orphanages and Auction Houses," "On Bankruptcy," and "On Criminal Court."[24]

His was an optimistic time, or so he believed, and also the time when all sorts of light were rising. The same year *Fatima* was published, a new political "light" was lit up in the Indies. In Sheriff Hinne's precinct, in fact, next to Pasar Senen, *Boedi Oetomo* was founded in 1908, arguably the first modern political organization of the native people of the land. It was the time when new "lights" and lights appeared on a daily basis. It might seem that murders like that of Fatima would never happen again. As Fatima lay in blood, her Chinese landlord, we read in Wiggers,

> was embarrassed that such a wicked thing could happen without him knowing anything about it, and in his own neighborhood, which was so lively. . . . Ach, what a misery, and all this, just at the time when, in order to improve the neighborhood further, he planned to install public gas lighting on the main road.[25]

The optimism of van Sandick's *Krakatoa* was less easy—as it was less easy to tame a volcano than to catch the men who killed Fatima. However, van Sandick, too, was a man of progress. Chapter XII of *Krakatoa*, a climax of this story of sorts, described a visit to the fire mountain, three years after the eruption, by a famous Dutch botanist of the time, the director of the Botanical Gardens in Buitenzorg, Dr. Melchior Treub (a brother of Professor M.W.F. Treub, whom we met in the previous chapter). Dr. Treub climbed the quieted volcano, took samples of the soil and vegetation back to his laboratory in Buitenzorg, subjected the samples to a microscopic analysis, and published the results in the *Annales du Jardin botanique de Buitenzorg*. This was described by van Sandick in the greatest detail.[26] The message still reads very clearly. Whatever the volcano might do, now it had been placed under the Dutch (magnifying) glass, and all that is needed will be found out.

Both Wiggers' *Fatima* and van Sandick's *Krakatoa* became Indies best-sellers, meteorically successful, but they were gone fast. For reasons I will try to explain, they did not become the classics defining the upcoming age. This, in contrast to Louis Couperus' *De Stille Kracht,* "The Hidden Force," and Raden Ajeng Kartini's *Door duisternis tot licht,* "From Darkness to Light."

Couperus, whom we have already met, was a descendant of an old Dutch family in the Indies. He spent his childhood in Batavia and, in 1900, now in his thirties and a famous Dutch writer, returned for a one-year visit. A novel resulting from the visit, *The Hidden Force,* was published first in 1900, and then reprinted many times.

Couperus was considered decadent, oversensitive, bizarre, and, often by the same people, prophetic. In the mood of the turn of the century, of Zola's scientific naturalism, among other things, Couperus was also praised as an observer with qualities of "camera obscura," or "photography."[27]

In a sense, *The Hidden Force* was an even more optimistic story than either *Krakatoa* or *Fatima.* Nobody was murdered in Couperus' novel, and a Dutch community in a provincial Javanese town, the subject of the novel, however much it resembled the ill-fated Anjer of *Krakatoa,* was not drowned in the mud. There merely was a jungle of unexplained sounds, an unknown, and a twilight around Couperus' Dutch men and women in the Indies: "in the long, broad galleries there was . . . barely light . . . A stream of moonlight floated over the garden, making the flowerpots gleam brightly until they shimmered in the pond."[28]

There was no corpse, but there was also no tenderness. In a curious way, the smell of blood was sadly missing. Instead, there was a high-strung nervousness, a stiffness between light and darkness: "a hundred paraffin lamps . . . Waltzes and Washington Posts and *grazianas* were danced . . . a handsome villa adorned with slender, fairly correct Ionian pillars of plaster and brilliantly lit with paraffin lamps set in chandeliers. . . . This was the Concordia Club."[29]

The Hidden Force was sort of detective story, too. But, to use words of Wiggers from *Fatima,* no case here was ever made "brilliantly clear." In *The Hidden Force,* all the time, something had been "discussed in whispers" or in a "muffled" voice.[30] If not a native, one was encircled by "rumours . . . anonymous letters . . . dark . . . palpable falsehood."[31]

When it came to the climax of this novel, only a "little nickel lamp" illuminated the scene. The wife of the top Dutch official in town, the resident, was attacked by "the hidden force" as she took her evening bath. Her naked white body was polluted by never-explained red stains, blood, or betel-red saliva, from the darkness.[32]

An enlightened space in *The Hidden Force* had painfully sharp contours. As the resident, in the opening of the novel, takes his evening walk, the torch light carried along by his servant picks up from the darkness around only what safely, without an anxiety, could be seen: "First the resident passed by the secretary's house, notary's house . . . a hotel, the post office . . . criminal court . . . Catholic church . . . railway station . . . European toko [in Malay: shop] . . . open space, in the middle of which was a small monument with a pointed spire, the town clock . . ."[33]

As Couperus understood it, where there was a light, there was "our" space in the Indies. As Kartini understood the same thing—wrongly, as history would prove very soon—where there was the light, there was modern Indies for all.

In 1900, Kartini, the daughter of a native regent under the Dutch supervision in Japara, was twenty-one years old. She wrote, in 1900, to a friend: "Couperus is still in India. When he is back in the fatherland, I believe that a brilliant book about my country will appear."[34] This was in anticipation of *The Hidden Force*.

Besides trains and clean and hard roads (as we have seen), Kartini believed in brilliance, too. Her own book, an edition of her letters to Dutch friends and mentors written at the same time as Couperus was in Java and published a decade later, carried the title *From Darkness to Light*. In a sense, the book was nothing but a brilliance. Kartini believed in a mission and duty for herself and those in her position in the Indies, "to put on the light in that poor dark world."[35]

Throughout Kartini's book, there was a "trust in the sparkling sunny path"; the "golden fields [that] bathed in glow"; a "nature [that] was a smile and light," "jubilation and splendor of the sun":[36] "the silver moon shines, and gold and silver flickers on the surface of the sea . . . the stars glitter gently in the azure sky."[37] There had been sunlight, moonlight, and starlight. But, as Kartini believed in a mission, and as it was a modern time, the book first of all was full of man-made light.[38]

Kartini could not be "grateful enough that we have good eyes."[39] Her eyes moved with an intent, and fast. They followed "the clouds as they journey through the vast heaven till they vanish behind the waving green leaves of the palm trees. My eyes then pick up the glistening leaves painted with the gold of the sun."[40] Kartini's watching, full of quick curiosity, was also well organized, and—her little drawings witness to it[41]—remarkably akin to the well-organized, informative, and eye-pleasing drawings and early photographs of "sights" that filled the Dutch magazines of the time.[42]

Kartini, a true daughter of a Javanese aristocrat and official, understood the power of watching:

above all what I have always liked doing is observing our own people. . . . We [Kartini, her sisters, and her father] try to come into contact with the people as much as possible. When we go out, we always stop and visit the village huts along the way. At first, the villagers would stare at us somewhat strangely, but now that is not so.[43]

Kartini was a strong and modern woman. Visiting villages with her father, she especially liked when either her father or one of his *toekang*s, "craftsmen," took photographs: "it is very difficult to make even one single photo in a village . . . [there is] the superstition . . . [they are afraid that] their lives might be shortened if they allow us to take a picture of them."[44] Most of all, Kartini wished to have a camera herself: "I wish so often that I had a photographic machine and could take pictures of our people—as only I could, and as no European can. There is so much that I would like to make into words and pictures, so that the Europeans might acquire a pure image of us Javanese."[45]

Kartini is known for liking to present her Dutch friends and mentors with samples of her own, or "her people's," skill. Often, among the things she gave away as gifts or sold (fire screen, file cabinet, postage stamp receptacle, table top), were "portrait stand . . . box for storing photographs . . . with various wayang [Javanese shadow puppet theater] figures carved . . . wayang photo box."[46]

With a passion not always easy to understand in our age, Kartini sent, received, collected, and exhibited photographs: "Dad and Mary . . . portraits, they stand in my album."[47] She reported on the exact placement of the photographs in her rooms: "how wonderful . . . surprise . . . portrait of Your highly esteemed Husband . . . the crown of it all . . . the main attraction of the bookshelf . . . the greatest jewel . . . precious souvenirs. On the top shelf . . . between the flowers."[48]

Like the anxious Dutch community of Couperus' novel, Kartini propelled herself toward the light. Often, much too often, Kartini sent photographs of herself:

> What do You think about this portrait . . . June of the last year, we went to the photographer in Semarang.[49]
> Keep this portrait . . . Poor clover leaf [Kartini and her three sisters] . . . pulled apart . . . the best portrait . . . at Christmas time . . . our last portrait taken together as [unmarried] girls.[50]
> We send You hereby the latest portrait of us both. What do You think about me as Her Highness?[51]

She sent photographs to show where, how, and why she, her sisters, and "her people" lived: "Recently, I acquired a couple of nice pictures of sawah rice fields . . . and this is what I dream about, and it is such a nice dream: to get these pictures to Holland, and to let them be copied and published."[52] When the first of her sisters was being married, a camera was not at hand. But Kartini dreamed again, and she wrote to a Dutch friend again, about the dream, that she might have taken the pictures and sent them: "So veiled and crowned, it was as though she had stepped from a page of the Thousand and One Nights . . . like one of the fairy princesses. The costume was very becoming to her. What a pity she could not have been photographed![53] You should have seen sister as she sat there before the *kuwadé*. She ought to have been photographed."[54] Kartini learned to call Japara *vergeten hoekje*, a "forgotten corner," almost as if it were her birthplace's alternative name.[55] We feel it was a forgotten corner also, and perhaps essentially, because there were not enough cameras in it.

Kartini was a revolutionary. She fought to reach toward the kernel of what she believed was the modern world and a source of light. She could not have been happier when one of the Dutch magazines of the time published a few photographs of her and her Japara work: "I have a wonderful letter from . . . *Eigen Haard* . . . the copies . . . of the pictures are nicely printed . . . on fine paper."[56] Her handicraft might even get to the great Paris world and colonial exhibition of 1900: "When [the colonial director of education and industry] saw our work . . . he asked if it would be possible in a year's time for us to have an exhibition. . . . 'You must send a great deal of what you have shown us.' O, Stella, I could not speak; I turned to him and to [the director's wife] with tears in my eyes."[57]

Kartini was a revolutionary and her fate was tragic. Not just because she died at the age of twenty-five, just after giving birth to her first son. Thinking of Paris, she reminds us of "certain Communards." "Certain Communards," Roland Barthes wrote in his *Camera Lucida*, "paid with their lives for their willingness or even their eagerness to pose on the barricades: defeated, they were recognized by Thiers' police and shot, almost every one."[58]

On November 19, 1859, the first concession by the Dutch minister of colonies was given to install a public gas lighting system in Batavia and environs.[59] By 1883, 1,270 gas meters worked in Batavia and 384 in the port city of Surabaya.[60] In 1897, electric power was made available in Batavia, as the electric company spokesman put it, "to answer a call for more light."[61]

In 1865, there were only three lighthouses in the whole archipelago.[62] In 1900, there were twenty-six of them, including the one destroyed in Anjer by Krakatoa and then restored, and another one in the harbor of the town where Couperus' *The Hidden Force* took place.[63] In 1912, an average distance between the street lamps in downtown Batavia was 60 meters—shorter than in downtown Amsterdam of the same time.[64]

Willem Walraven, a Dutch writer who chronicled the late-colonial history of the Indies better than anybody else, first arrived in the Indies in 1918. As a soldier, he served in several towns of Central Java. Two decades later, at the end of the era, on a journalistic assignment in 1938, he visited some of these places again, among others Rembang, the town where Kartini was married, and where, in 1904, she died:

Javanese pharmacists in the private hospital of Dr. Bervoets.
(*Nederlandsch Indië Oud en Nieuw*, 1917–1918)

How dark Rembang is in the evening! . . . There is no daytime electric power supply and only a very meager one in the evening. . . . Shops and Hotel Loeberg have just the Nulite-burners. One uses the electricity as little as one can, only during meals, and in some guest rooms. . . . One has no trust in Rembang, and Rembang has no trust in electricity![65]

In 1938, when Walraven wrote this, only 9 percent of the population of Java, the best illuminated island of the Dutch colony, lived in areas where electricity was theoretically available. Merely 5 percent actually used it. There were, in 1930, 76,000 natives in Java who used electricity—less than 0.2 percent of the Javanese native population. The number even decreased, to 54,000, in the next four years as the depression set in.[66] The minimal costs, in 1938, were calculated as "25 cents a week for two lamps of ten Watts." No one in the Indies doubted, as the same source noted, that this was "too much for the little man."[67]

An Indies civil-engineering student journal wrote in 1935, with only a little exaggeration, that the colony "held a record, the highest tariffs for electric power in the world."[68] The director of the state-run electric company pointed out in 1938 that the "cardinal fault" of the Dutch colonial energy policy was "that we, here, in the Indies, amidst purely an Eastern society, build up an electric technology on an exclusively Western basis."[69]

In many countries on the Pacific rim, according to this expert, in Japan, China, Australia, or the United States, electric energy distribution networks were laid out "simply," even "primitively": "high-voltage and low-voltage lines are carried over one and the same row of poles, while transformers are placed directly on the poles." Why we, the Dutch in the Indies, the expert asked, "do not follow the example of so many . . . why do we remain so stubbornly attached to the perfectionist system of network construction?" Why did the Indies have the stiffest electricity regulations imaginable?

Every electricity expert who has been in Japan [for instance] . . . would tell you that the electrification of that country extends into even the poorest and most remote regions, and that [in spite of benevolence of regulations] there still are many people in Japan who were not electrocuted, and that, on the contrary . . . the number of fatal accidents . . . is extremely low.[70]

Why did the Indies stick so tightly to its designs of luxury?

As a good example of this, there are two villa-park neighborhoods, Berg and Dal, in the hills near Bandung. For just twenty households, there are four transformers connected by 6,000 volt cables that, in large part, are laid over four separate and parallel rows of poles.[71]

There was a fascination, in the late-colonial Indies, increasingly noticed by contemporaries, with "colored-glass fillings," "leaded glass," and "cathedral

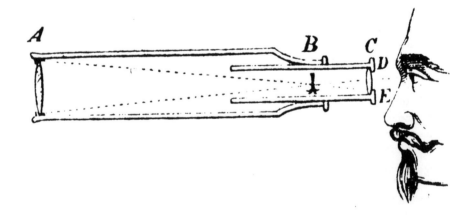

Dutch telescope. (Oudemans, *Ilmoe alam*)

glass." Visitors and architects wondered about "the sickness."[72] "Ventilation is unsatisfactory . . . the fashion of leaded glass is spreading like an epidemic. To disastrous excess, it floods and drowns the windows, the doors, and all the other houses' openings."[73] Walraven, again, in 1939, saw "neat and little villas with little terraces, little doors, and little windows, all filled with colored glass."[74] At the same time, and clearly by the same compulsion, all kinds of in-house "boudoir lighting devices on high feet," lamps, and chandeliers of the most fantastic shapes filled the Indies journal advertisement pages, and thus Indies homes.[75]

There had always been, as we saw in earlier chapters, some kind of nostalgia for the "home back in Holland." Now, we can see, it was a network nostalgia, increasing as the colony passed into its closing years. In its official publication in 1938, the Indies state gas company explained:

> In Dutch towns, the inhabitants live close to each other. In the residential areas the houses touch upon each other, and each building complex can accommodate three or four families. Once a gas pipe or electric cable from the central reaches a street, a relatively few branch connections are sufficient to connect hundreds of consumers.

In the Indies, "on the contrary, the European houses stand far away from each other." Naturally, the company lamented, there was "a relatively high risk of leakage."[76]

Couperus sensed it sooner than most of the others: the modern Dutch in the Indies were entrenched by light.

Dactyloscopy

This was a textbook that Kartini's father, when he went to school as a boy, might have read. Written by Anthonie Cornelis Oudemans, published first in 1873 and then reprinted several times, the textbook opened with: "Once upon a time, in Java, close to the sea, there lived a man who had great knowledge at all the matters of the earth, stars, sun, moon, etc. In the same place, there lived a [native] regent, and he had a son called Abdullah."[77] The usual subjects were dealt with in the textbook. Yet one item got an extra space. Two long chapters, in fact, dealt with *kijker*, a "telescope," "perspective glass," or, as the textbook enunciated it, a "Dutch *kijker*," "Dutch telescope," "Dutch perspective glass."[78]

The wise, white man at the beach instructed the regent's son on how the Dutch *kijker* worked. He aimed it at a distant coconut tree and, after some focusing, the tree was distinctly visible in the perspective glass. Thus, a reality was captured. And one might aim this glass at everything.

> But now, the regent's son looked into the glass, then at his teacher, and said:

> But, sir, can I ask you something? The image of the coconut tree appears turned around, with its crown and leaves pointing downwards. Yet, my father has a

binocular, and also a taller perspective glass, and through them one can see trees not upside-down.[79]

At this moment, we realize how early in the late-colonial age this book was written; how close to the beach and how far still from Java's heart the white man knew he lived. "Your father," the white man answered Abdullah, "has more advanced instruments. That's why."

"Picture no. 2" in the textbook showed the Dutch *kijker*, "telescope," with a coconut tree, indeed, reversed, yet safely within. "Picture no. 1" was a "Map of Java"—even cleaner and more cheerful in its style and message. Straight, abstract lines intersected at points marked by single digits or letters of the alphabet. Almost exactly like this, yet even purer, there were several maps of the heavens farther on in the textbook. In this case, tropical stars, instead of letters and digits, carried Latin names.[80]

A.C. Oudemans, author of the textbook, was a lucky man. He spent the years between 1834 and 1840 as a principal of the Dutch elementary school in Batavia, and he died in quiet retirement "back home" in Holland. He was a father also, blessed with perfect sons: Jean Abram Oudemans was a doctor of medicine, botany, and mycology and professor, eventually, at the University of Amsterdam; Antoine Corneille was a chemist and also the director of the Technical Institute in Delft, the Netherlands;[81] Jean Abraham Chrétien was one of the most prominent Dutch astronomers of his time.[82]

Especially the last of the three sons was almost like his father's textbook. Born in Amsterdam, he received his early education in Batavia. In 1852 in Holland, he defended a dissertation on the determination of the latitude of Leiden,[83] and in 1857 he was appointed "head engineer of the geographical service" in the Indies with the task of mapping the colony by astronomical means.

In the following decades, J.A.C. Oudemans became one of the most respected and best-paid men in the colony: "one needs at least 500 guilders per month here to live decently," he wrote, "and this sum did not take account of capital expenses necessary for setting up a household."[84] He was a white wise man, still very close to the beach:

> Except for Java, no big roads or horses are to be found, the coasts in various places are unapproachable and the inland does not offer anything but marshes. . . . [We] have to drag [ourselves] with chronometers and instruments from one place to the other. For each expedition we depend on the solution to the big question: How shall I get to the places to be defined (geographically)?[85]

Oudemans counted on steamers, the technologically most *à jour* things, and he could become very angry when a ship was not available to him at an appointed place. At the same time, he was fond of using what he strongly believed were native symbols of authority: palanquin and cane.[86]

His, clearly, was a most intense scientific, high-soaring pleasure—to penetrate the dark interior, to measure the islands by means of pure rationality. In his mapping and triangulating of the Indies, Oudemans used the very recently laid submarine transcontinental telegraphic cable, and the signals over the various segments of the line, in the widest range between Perth and Berlin.[87]

Oudemans left for home in Holland in the early 1890s, the huge job, naturally, just beginning. The last thing we hear about Oudemans in connection with the Indies is as cheerful and uplifting as his life, and as his father's textbook: "As early as 1881 he observed that the meter used in Java (which he himself had delivered in the 1850s) was not calibrated directly against the one in Paris; neither had it been corrected for temperature variations." In 1894, Oudemans acquired a new, true, and perfect platinum-iridium meter. This he kept in his new laboratory in Utrecht, Holland, of course. To the colony, he bequeathed a glass copy.[88]

Mapping was a kind of high art in the late-colonial Indies, and increasingly so as the colonial crisis deepened. Each new map edition inevitably called for a literary review. Cultural personages were involved, and the language of the reviews was often the kind one might expect from belles-lettres writers:

> The map is networked by Flamsteed's projection. Thus, our Indies possessions festoon the Equator as the poetical expression of Multatuli had it [the necklace on the Equator]; the most extending points reach to 6° in the north and to 10° in the south. . . . There is no overloading, there is a great clarity and readability, true harmony of tones and colors![89]

By the same token, maps could be criticized as *cacographie*, "cacography," or *ontaarden*, "degenerate":

> Not just the solemnity of the drawing disappeared, but with each following map [of the reviewed collection] an unnatural mixture of colors gets grosser.[90]
>
> I found it in a bad taste to paste up a map with all the colors of the rainbow. I found it child-like to show coffee plantations through the color of coffee, tea plantations through the color of tea. The color of this land is green. Light green are the rice fields, dark green is the rest.[91]

Geometrical beauty, we have seen already, was appreciated—and increasingly so as the Indies' reality grew messier and harder to make into a straight line. In another Indies textbook, published nineteen years after A.C. Oudemans', and this time explicitly for native children, it was explained to the little Javanese, for instance, that their island "has a shape of a very prolonged rectangle, thus for your textbook, it had to be shortened a little."[92] Sumatra, in the same textbook, had "a shape almost the same as the island of Java,

Kelud volcano, measurement after the explosion. (*Nederlandsch Indië
Oud en Nieuw*, 1919–1920)

except that its position lengthwise extends between the north and the south-east."[93]

It took a great war in the West, in 1914–1918, and a great revolution, for the European avant-garde to come up with ideas like "linear cities," or to think about society, as the Dutch leading painter, Piet Mondriaan, did, in terms of "holy mathematics,"[94] all powerful "circles in connection with triangles and squares"; a great part of it "in an attempt . . . to obtain a knowledge . . . about the ascending path, away from matter."[95] The Indies, always a place for pioneers, again seemed to tramp few decades ahead of the West.

Thinking about what a perfect late-colonial Indies map might be—most transparent, beautiful, and powerful—dactyloscopy comes to one's mind. The origins of dactyloscopy are traced to Jan Evangelista Purkině, Czech scholar and educator, who in 1822 under his microscope discovered a breathtaking order of lines on the skin of the tip of a human thumb. Purkině was a hero in the Indies. "He distinguished nine principal categories. Thus he laid the foundation to the present dactyloscopic methods of registration."[96]

In the Indies, the dactyloscopy used became "system Henry," after "chief inspector of the London police, Sir [E.R.] Henry."[97] The "operators" taking the fingerprints ("each operator must work in a neat and orderly manner")[98] became the colony's cartographers par excellence, and also its most intense engineers, dreamers, and policemen.

The word used for fingerprints in Dutch at that time was *vingersporen*, "finger-traces" or "finger-tracks":

> By the government decision dated July 23, 1898, no. 27, and another one dated July 17, 1905, no. 20, offices of judicial anthropometrical identification . . . were established in principal cities: Batavia, Semarang, and Surabaya. . . . In 1901, the anthropometry was supplemented with *photography*. . . . By the decision of the governor-general of March 15, 1917, the Central Office of Dactyloscopy began to work together with the police in Batavia.[99]

The Indies dactyloscopy manual from 1917 named the piece of paper where fingerprints were stored *slip van Simin*, "Simin's tag." Simin, of course, was an exclusively native name. One copy of each "Simin's tag" was to be kept in the central office in Batavia, another copy in the local police dactyloscopic files.[100]

Ernst Mach wrote that "the mind feels relieved, whenever the new and unknown is recognized as a combination of what is known."[101] There must be something to it. Robinson Crusoe, too, felt relieved on his island as long as "stating and squaring everything by reason."[102]

Dutch experts in the Indies spoke about the "dactyloscopic sieve."[103] First, the giant flow of native people between Java and the rest of the Indies, they calculated, might be "strained" dactyloscopically. Medan in East Sumatra, the biggest plantation center in the Indies, and the Javanese towns like Semarang, Surakarta, or Yogyakarta, where most of the plantation labor originated, were the earliest targets for being "dactyloscoped."[104] Some argued that this was progress: with dactyloscopy at last, a modern labor system would be established in the Indies. Instead of the semi-feudal *poenale sanctie*, forced labor system, "came Dactyloscopic Institute . . . all the contract workers began to be registered by the means of fingerprints."[105]

The fingerprints, mapping of fingers, natives, and the colony could fully be appreciated only under a microscope. Dactyloscopy, therefore, was a historical moment not just for the Dutch colonial mapping, but for the Dutch *kijker*, perspective glass, as well. There seemed to be no limit to the new vision:

Of whom will the slips be required?
In the government instruction it is stated: "of persons suspected of a criminal offense and of breaching rules . . . of all suspected individuals in the Netherlands Indies."[106]

The poetics of the maps in the early Indies textbooks, and the poetics of the early Indies triangulation, found its consummation here. The immense archipelago, through which Oudemans junior had dragged his theodolites, the whole colony entering into its last decades, was about to be pinned down by numbers. Here was a digital dream:

With a pair of fingers, we get $2 \times 2 = 4$ variation. . . . With three fingers, the number arises to $2 \times 2 \times 2 = 8$ variations.

So it follows, with ten fingers, there is $2 \times 2 \times 2 \times 2 \times 2 \times 2 \times 2 \times 2 \times 2 \times 2 = 1024$ variations.

Given the fact that there are 1024 possibilities, the slips can be divided in 1024 classes: System *Henry* makes 32 series of these 1024 classes, with 32 numbers for each possibility. . . . Because, on each fingerprint, 20 details can be distinguished, with 10 fingers we arrive at number of combinations $20^{10} = 10,240.000.000.000$.[107]

Here also came forth an amazing possibility for a long-distance map, the best thing any ruler may ever hope for—a touching map, a one-way touching. In 1917, the Indies manual of dactyloscopy explained what might be achieved when the Indies dactyloscopy was solidly installed:

In London, a thief climbed over a gate outfitted with nails at the top. He entered a warehouse and committed a serious crime.

On his way back, as he climbed over the gate again, he got himself caught by one of the nails between his ring and his ring finger. By the weight of his body, the thief fell to the ground, and as he run away, he left behind, on the nail, the ring and, also, a piece of his finger's skin.

At the dactyloscopic central laboratory of Scotland Yard, a print was made of the patch of the finger skin, and it was checked against the other samples in the collection of the central laboratory. The man was arrested, and his wounded finger provided the proof of his guilt beyond any doubt.[108]

This was a dream of utmost modernity. Out of millions of Simin's slips, an Indies *waarschuwing register*, "caution register," would be compiled.[109] "The collection and storing of the slips at the central bureau can be compared with writing a dictionary. Then, if one wishes to find a word, one simply turns pages, and the word has to be there."[110]

Fingerprints were photographed and enlarged.[111] Natives, otherwise moving too fast, would be made to stand still. This was better than photography:

> To be able to photograph somebody quickly, clearly, and at somebody's every possible step, one would have to have special equipment that is rarely available. . . . It is also very difficult to identify even a white man by a photo. To identify a native, often, for a European, is quite impossible. . . . The dactyloscopic identification slips stored according a numerical combination are a solution. . . . If this method, however, is to have universal success, dactyloscopying must be declared compulsory by the government.[112]

Natives, it was the idea, would become as individual as their fingerprints: "not two fingerprints look one like the other."[113] From now on, in the Indies, natives, also, would be as eternal as mapped: "They [the fingerprints] do not change; they remain unchanged since before birth until shortly after death."[114]

The Floodlight

"Two eyes are often not enough to watch them," the Dutch Indies post and telegraph union journal wrote in 1906 about native employees.[115] Whenever two eyes were not enough, or when they themselves appeared vulnerable, glass was tried.

Already at the first big Amsterdam world and colonial exhibition in 1883, a special lighting of the Indies, artfully filtered through glass, was most highly praised:

> We step inside, into the Indies section. What a soft and jolly light! we cry out. Looking up, we notice white cloth arranged in pleats and spread out under the ceiling, over the whole hall. As if the light was falling down through a web of white palm leaves: it is an ivory light. De Heyder & Co. manufactured this cloth without using any impregnation. Even if a tropical rain ever fell here, and should it leak through the glass panes of the roof and the cloth, the rainwater on the floor would leave no smudges whatsoever.
>
> The soft light, by itself, would be enough to justify this whole enterprise.[116]

Maps, not surprisingly, proved to be the most willing matter to be put under the glass: "Maps, models, drawings, photographs, samples . . . graphs

A photographer of the General Deli Immigration Office in Surabaya, 1919. (*Verslag van der Dienst Arbeidsinspectie*, 1920)

. . . collections of insects (no less than a hundred glass bottles with the enemies of a sugar planter)."¹¹⁷

As the colonial times evolved, the trend accelerated. At the "last and greatest"¹¹⁸ colonial exhibition, in Paris in 1931, glassed, cased, and artfully lighted
maps in fact dominated the show. Just before the opening, a fire destroyed
the Dutch Indies pavilion.¹¹⁹ In forty-one days, the pavilion was rebuilt and
celebrated as a "Phoenix pavilion."¹²⁰ As most of the three-dimensional items
had vanished in flames, the new pavilion was filled with "chiefly photographic material . . . sent quickly to Europe through our unsurpassable air
mail service. New entries, of course, of a larger size and greater weight could
not be sent."¹²¹

The memorial book still exudes the pavilion's spirit. The volume opens
with a photo portrait of the Dutch queen; there are plans of the exhibition,
air photos of the Paris fair grounds, and other photographs, maps, and photographs of maps:

> Netherlands Indies compared with Europe. Steamship connections in the Indies.
> Development of the air traffic in the Netherlands Indies.¹²²
>
> Special interest was attracted by "the staircase of abundance," with a large map
> of the Netherlands Indies archipelago as its backdrop, with 50 steps leading
> toward it, each with an image of a human figure carrying a typical product of
> one region that also, by a special symbol, could be located on the map. . . .
>
> On the left side of the hall, there was a map of radio connections of the
> Netherlands Indies with the world. The map was supplemented by a system of
> dioramas.¹²³

There were more photographs of maps than maps themselves. The flames
that destroyed the irreplaceable originals were blamed. Yet, as much as that,
this was an acceleration of a trend. Already in 1883 in Amsterdam, two
groups of natives, one from the West Indies and the other from the East
Indies, were brought to the colonial exhibition and placed so that they might
be easily watched by visitors. A highly respected scholar of the time, Roland
Bonaparte, hired one of the early Amsterdam photographers to portray the
individuals in the two groups. Three photographs of each man, woman, and
child were made; *en face*, a profile from left, and a profile from right.¹²⁴

This was the rudimentary beginnings of the modern gallery of "wanted
natives" and of "wanted colony." At the "last and greatest" Paris exhibition of
1931, in the Dutch pavilion,

> There were 78 heads of folk types from all the parts of the archipelago. The
> heads are by the hand of a well-known Javanese painter Mas Pirngadie, at pres-
> ent on assignment with the Batavia Museum. He painted from photographs, and
> thus these are not character studies, as it would not fit the design, but purely
> ethnological plates.¹²⁵

Youth on a native latrine. (Tillema-archives)

Landscapes were flattened through photographs and maps, and the photographs and maps were placed behind the glass. Faces and peoples were flattened. Watching Bonaparte's photographs and Pirngadie's paintings, one would never believe what Georg Simmel has written: "Within the perceptible world, there is no other structure like the human face which merges with a great variety of shapes and surfaces into an absolute unity of meaning."[126]

Buildings were smothered into mere façades: "sugar cake creations with silly little pavilions . . . triumph of stucco. While [initially], it looked like that iron was winning against stone of the earlier time, now plaster has the word."[127] There was no Nero in the Indies, no burning of Rome. Just exhibition and light diversion. The extraordinary Indonesian journalist and writer of the period, Mas Marco, described the big colonial fair in Semarang in 1914 as "showing off a perfect fake."[128]

H.F. Tillema, the pharmacist in Semarang, was also sometimes considered a founding father of modern Indies photography.[129] Tillema's *Kromoblanda* is filled with photographs, so many of them in fact that it is often considered unreadable.[130] Tillema made photographs, and he also collected photographs by others. He classified and stored the photographs in a "photo central"[131] kind of register, open to newspapers, museums, and anybody willing to pay.[132]

Kromoblanda was meant as a journey. As Tillema traveled, through the lenses of his cameras, the Indies got exposed. The opening series of photographs in *Kromoblanda*—the jungle, Hindu and Buddhist temples, native women with bared breasts, volcanoes—accompany a repeating caption: "—blinded by the glare of majestic tropical nature."[133] As one turns the pages, however, and as Tillema guides us, the scenes become progressively ordered—angular as we have seen—but also focused. The man behind the camera is less blinded. Eventually, the camera gets it all, as it stands and lays.

As on a perfect map, a measured proportion of shade and light adds to the composed beauty of the whole. Landscapes and peoples are snapshot, given no time to move in an unruly manner or on their own.

One is reminded, once more, of the climax in Couperus' novel *The Hidden Force*—a white woman is attacked by a force hidden in the darkness as she takes her evening bath. Almost a full volume of *Kromoblanda* is taken up by photographs of modern bathrooms, shiny and tiled—harmony of straight lines, refuge of cleanliness, and repository of bright and measured light.[134] In dramatic contrast, on the same page as a rule, there are photographs of the messiest latrines of the natives, and also photographs of the natives themselves. In contrast to their murky latrines, natives are well lighted.[135] Some of the natives, in fact, are snapshot in their most private act. As one Tillema's biographer put it, "Tillema archives are a monument of Indies scatography."[136]

Later in his life, in 1930, Tillema published advice on how to "film and photograph in tropical forest."[137] First of all, one has to think of carrying a big white sheet of cloth to be used as a backdrop. The mats of natives might be used too, but they do not always have them. There were other perils, Tillema warned, for which a photographer in the wild had to be ready: "Because the houses of Dayaks are dark, with roofs and walls black from smoke and, thus reflecting no light, one should not be stingy with his powder. I have made dosages with 50 to 75 grams! . . . Then, there is a danger of fire in those easily inflammable huts."[138] Heat and humidity, too, might harm the camera, lenses, and film. There were "myriads of annoying insects," there was "sunlight": "Hands are sticky with sweat. This is not just annoying, but also very dangerous to the film. But, with training, we will make it."[139]

With Tillema, once again, Dutch pioneer, daring colonial, soldier of the empire, or, as he put it, "amateur," were on the move:

> an amateur on a journey in uncivilized regions . . . has no place to rest, and no place to buy anything, a little screw, a nail, a piece of copper or iron wire, a tool. In a word, he lives and works in a photographic vacuum![140]
>
> . . . in a rimboe [Malay for "forest"], one has to walk, and for this one needs shoes of a kind that cannot be bought in the fatherland. So-called "Acheh shoes" [name is from colonial war against Aceh in 1873–1903] are excellent for the purpose, and they are available in many shops in the Indies. . . . Every Indies military man will give you good information about this matter.[141]

Infections and bacteria posed a grave danger to the pioneer/amateur and, at the same time, to the pioneer/amateur's technology:

> all kinds of infection . . . dysentery . . . vaccination . . . must be repeated each 6 months.[142]
>
> . . . in spite of all the precautions, fungi and colonies of other bacteria may develop that can spoil the picture in the end.[143]

While on the mission, one has to take fastidious care of oneself and one's tools: "*Lenses.* One takes lenses of the greatest possible light strength. I used a f2.9 focus 12.5cM. Dallmeijer, which worked well."[144]

When all this is carefully done, the journey makes sense. Subjects may be taken in:

> Soedah! [Malay for "Done at last"] I also noticed that primitive peoples, not unlike ourselves in our childhood, are not accustomed to sitting still; even for a second! Yet we are often forced to work at a very slow pace, with 1/8 second aperture. Remember, whenever light permits, you should take a short-time exposure.[145]

In 1935, H.F. Tillema made one more Indies trip as a camera expert again. This time, he traveled to the deepest interior, to film the tribes of Apo Kaya in Central Borneo.[146] In his report on this adventure, he appeared particularly excited about one item. "It had been of scientific interest," Tillema wrote,

> to discover whether there still remained some old women who knew how to weave the way that was used during their youth. In fact, I was anxious to obtain a filmed record of the whole process. . . . De Rooy [a local Dutch official and Tillema's partner in this] gets on marvelously with those people, and, with tact and sympathy . . . [he made it to] one very old woman, who still knew how to weave. But she was unwilling to give a demonstration for us, because the spirits would not approve.

Tillema and De Rooy quickly decided upon "providing an active force that would neutralize the power claimed by tradition . . . a tin of salt . . . some sewing needles, a piece of thread, a couple of patches of velvet, a few handfuls of tobacco, and some other items like that. . . . Instead of one old woman, three of them turned up."

The weaving and the filming was done; the camera made it. The rest was easy. As if by way of merely an afterthought, "De Rooy purchased the last Dayak loom, and the last ever woven piece of cloth, and he presented both to *Koloniaal Instituut*, the 'Colonial Institute' in Amsterdam." Who could imagine, Tillema wrote, "my delight . . . after . . . the suspense of about a year . . . to see the negative film developed, and, a moment later, to watch, on the screen, the little women . . . intent on their work."[147]

Sometimes, all this appeared like Emilio Marinetti's futurist, avant-garde dream: "We disengage ourselves today . . . [for] chaos [we] substitute . . . mechanical splendor; [we make] the sun relit . . . [we achieve] healthful forgetfulness [through] controlled force . . . [and] speed . . . [and] light."[148] The late-colonial ways of seeing the Indies might remind us of Marinetti, or, perhaps, Frankenstein. The Indies maps and photographs, images through camera, put behind the glass, and artfully illuminated, increasingly acquired a disengaged, cheerful, yet often ghostly life of their own.

Mr. Scheltema, an old Indies hand, visited the celebrated Paris world and colonial exhibition of 1900, and, as he reported, in the Dutch pavilion he was particularly enthusiastic about "A railway map with illuminated scenes of the most picturesque and, from the point of view of construction, most significant sights on the train line between East and West Java."[149]

Had Mr. Scheltema lived long enough to see the "last and biggest" Paris exhibition in 1931, he might (at the time of the deepest world and colonial depression) have enjoyed something even more exciting. In Paris of 1931, "an endless wonderment was aroused," as a contemporary source put it,

> by a show designed by Mr. Utermark, Engineer, and realized by the Technical Service of the Colonial Institute [in Amsterdam]—a map of the world, showing

to the crowds of visitors the weekly KLM air connection between the Netherlands and the Netherlands Indies archipelago. . . .

One after the other, twelve in number altogether, stations on the route light up, and the names of the places flash, a little yellow lamp for an outbound plane, a little blue lamp for a homebound plane, and a little green lamp for their meeting point in Djask . . . between the little lights, moving incessantly, miniature airplanes have an irresistible power of attraction. . . .

Yet another electric map very truthfully demonstrates the difference between a journey by ship and by plane to and from the Netherlands Indies . . . all is done simply by points that automatically light up and get dark again. . . . Tjililitan airfield of Batavia in miniature is, also a part of the complex.[150]

Just across the hall, competing for the visitors' attention, there was

a fourteen-meter long, four-meter wide, and one-and-a-half meter deep hollow, with a relief map of the main island of the Netherlands Indies archipelago on a scale 1:80,000 in the horizontal, and 1:8,000 in the vertical projection.

The aim of the map is, among other things, to point out the tourist qualities of Java; little electric trains move from one part of the map to another, and as they are passing through an especially significant point, one of the fourteen dioramas lights up.[151]

Like Tillema's snapshots, the illuminated maps were impressive by being an entertainment, and not marginal at the same time. In them, the late-colonial paradigm could be expressed with a heightening intensity. Attractive, self-contained, smoothly working, and trivial, the patches of light manifested what the troubled colony more and more intensely wanted to be.

There were, also, "multicolor fountains" and "Edison feasts" in vogue in the Indies since the early twentieth century.[152] The Indies journals witnessed an unfailing popularity of the "electric pictures." Always, a couple of modern buildings in an Indies town, selected with care, were illuminated on a special night—*floodlighted*, as an English word was used to describe it. The modern façades gleamed, as if there were only façades. Sometimes, and this seemed to be preferred, merely the buildings' contours were illuminated. Nothing but clean, sharp lines stood out of the darkness.[153]

Movies appeared in the Indies, according to one source, on December 5, 1900, in *Tanah Abang*, Batavia, as a diversion for higher society. The highlights of the first show, according to the source, were "shots of the Queen Wilhelmina and Duke Hendrik in The Hague, scenes (probably faked) from the Boer War in Transvaal, and a short piece from a fair in Paris."[154]

Soon, enormous movie houses, often "enormous movie bamboo tents,"[155] were built (first in Batavia, especially in its China Town), and then throughout the Indies. During the next four decades, some of the most modern structures in the Indies, including the first air-conditioned buildings, were movie houses.[156]

French, Dutch, Chinese, local-Chinese, and soon American movies with Malay subtitles became wildly popular: "The Jasmine of Java, The Mystery of Borobudur, The Clear Moon, Fatima, Halfhearted (the first Indies jungle film), The Dagger of Mataram, Zoebaida, The Laughing Mask. Tom Mix . . . "[157] This, too, was a serious colonial project. Professor M.W.F. Treub, one of the foremost apologists of the empire, visited a service park, as you will recall, temporary living quarters for native workers in Batavia harbor, in 1923. One of the praiseworthy features that Treub noted was "naturally . . . cinematograph is not missing."[158]

The public in the movies, naturally, sat in the shadows. Many officials soon complained about the natives, who behaved "nonchalantly" in their *kambing*, "goat," section of the movie house, even "when *Wilhelmus* [the Dutch anthem] is played."[159] The natives in the movie house shadows were even reported "whistling their ditties while the image of the Queen is on the screen."[160] Many worried about the native spectator watching Western movie stars in underwear.[161] Some precautions were taken by law, already in 1912, that films showing "murder, robbery, adultery, and matters of that sort . . . can in no case be permitted."[162] The film law of 1916 allowed for banning movies that might be subversive to "public order and morals."[163] Groups of vigilantes were set up, beginning in 1918, to help the police watch the movies.[164]

Actually, not much police seemed to be needed. The light, and even the shadows, in the Indies movie houses were man-made. The electric switch was in a professional hand. These, indeed, are the most impressive, and most troubling, images of the late-colonial period: people floodlighted, and rulers watching the light effect.

H. Maclaine Pont, a prominent architect in the Indies, recalled in 1916 how he had completed a big new modern office building for an electric tramway company, in Tegal, Central Java, and how he celebrated with his Javanese workers:

> Several wajang troupes entertained the festive crowds. Toward the evening, the terrain was electrically illuminated, and an open-air cinematograph began. After a while, a heavy rain came, and it poured down, so that a great deal of the terrain was soon deep under water. . . . Yet . . . between 2,000 and 3,000 spectators remained, standing in the water and watching till the last film reel was at its end.[165]

In the same journal, the same year, an elaborate account was published by C.M. Vissering, of the Batavia colonial exhibition of 1916: "Batavia's biggest square is temporarily changed into a fair, a platform is erected for the *resident* of Batavia, other officials, and their dames . . . below sits a group of native chiefs from the most remote lands of our Indies empire."[166] There were chickens, birds of prey, rabbits, and crocodiles displayed in the square, and "a pack of old apes with indescribably melancholic, deep-set eyes, looking out as if sick with nostalgia for their forest." There were also clowns, orchestras, stalls selling the products of village industries.[167] But the high point came at twilight:

> As soon as the darkness wins over, both of the main entrance gates and the whole front part of the square stands out as it reflects a glow coming from the multiform lights of cinematographs. . . .
>
> An immense crowd sits on the ground in their best plumage of white *baadje* coats, multicolor sarongs, and neat head cloths. . . .
>
> In the thick of the crowd, there are little *warong* stands with oil lamps that glimmer over the multitude, in red, yellow, and orange, whenever the beam of the light from the movies falls in another direction.
>
> In this immense human mass, there is nothing but light, color, and peace; peace first of all, picked up by colors and light, each time as the scene brightens by the sudden foray of light from the big movie theater, as the cinematographic mechanism releases the broad stream of light over the whole square; now, everybody is garish, now carmine red, now deep blue, now fiercely orange. . . .
>
> For us, it is the joy of the show! That look of placid happiness of the thousands; that picture, which repeatedly emerges out of the darkness in the vehement lighting by the cinematograph . . . always the same amazing multicolored still life, and the little *warong* stands' oil flames flickering orange and red.[168]

When the movies are over "The native chiefs and the native workers return to their distant solitary regions; they will live again in their *kampong*, 'village,' work again with their primitive tools, busy with their pure handiwork of peace; for us, Westerners, they leave behind images of a culture of ennobled work."[169]

The Sublime

There was a determination, among the Dutch in the Indies, not to let the natives see their masters watching. Optical tools were important, more and more, as shields. Everything, or so increasingly it was hoped, might become a *kijker*, microscope or telescope. One might get inside, to watch, and to hide.

Trains, besides the things already noted, became optical tools, too. Late-colonial trains, as they were increasingly described, "slithered through the countless turns through the majestic mass of the landscape."[170] As the train sledded on the iron rails, the land along the tracks, and as far as one could see from the train, sledded back with precisely the same smoothness and speed as the train. Often the landscape along the train, logically, was the most modern strip of the colony, made at the same time as the train—with the same design in mind and largely for the train. It was a most easily and most comfortingly watchable modernity. As the fuel for the trains was still coal, and unless a traveler wanted to "leave the train with [one's clothes, face, and hands] black with soot,"[171] he or she was well advised to pull up the window and watch the land along the tracks from behind the glass.

H.P. Berlage, the most influential Dutch architect of the early twentieth century, visited the Indies in 1923. He, as we saw earlier, enjoyed the trip across the seas and through the streets of Batavia, through what he expected and believed was a modern colony on the march. He was a little nervous, however, before he set out further to travel into the colony's interior: "I went to the [Batavia] station very early, anxious about the surprises of a train trip in the Indies." And, indeed. "A Javanese crowd waited for the train . . . in swarms they stormed it."

Yet they, in fact, were coolies, bearers, servants, Berlage realized the next moment: "You can trust them even with your tiniest luggage. . . . My respect and admiration goes to the Javanese servants. . . . Oh! and that so delightful calm and silence of theirs."[172] As the train pulled out of the station, Berlage pulled his window up: "the aspects are fleeting . . . cinematographic . . . palm groves, kampongs, bridges, green sawah rice fields . . . blue and hazy horizon."[173]

As far as Berlage's eyes could see, behind the window glass and framed by the train window's frame, there was *mooi Indië*, the "beautiful Indies," a late-colonial image of the colony known and often more significant than the colony itself. Retiring Dutch officials and planters particularly loved to take the beautiful Indies back home: a souvenir, the colony captured in a little hangable painting, a photograph, a panorama of rice-field terraces, coconut palm trees, buffalo boys, mountains, and touched-up red, carmine, or orange sun.

There were strict rules to the beautiful Indies—no waving of time, no trembling of air, no hidden energies, nothing hinted at in the shadows. The trees, houses, fields, mountains, and people had to be distinctly outlined. The perspective should swell forward, and everything should offer itself to be seen. As in Roland Bonaparte's or Tillema's snapshots, or as in the electric maps of the colony, there was a distinct police quality in the beauty of this craft.

Berlage traveled from Batavia to Bandung: "The trip by train through Preanger is one of the most beautiful in the world."[174] "A wonderful rail network," he wrote, "the most modern transportation, I mean, not merely in horizontal, but also in vertical direction."[175] The Netherlands, compared to this, was so flat a place! Van Sandick, Engineer, the author of *Krakatoa*, in a speech at the new technical college in Bandung three years before Berlage visited Java, spoke about the same thing: "Indies has three dimensions as against the level and low Netherlands, which has only two."[176]

Paysage de Java. Panorama of a beautiful Indies at the Paris Colonial
Exhibition, 1931. (KIT)

This might imply that, in the Indies, a modern Dutch man and woman might feel complete. This also might imply that by the rail, at the rail head in Bandung, a modern Dutch man and woman might reach a space that was high and up, away enough. Berlage wrote that, as the train reached the destination in Bandung, he felt suddenly *Ueber allen Gipfeln*, [Goethe's German for: "O'er all the hilltops"], and watching "in a purple-gray depth the splendid land of Java."[177]

Many went before and after Berlage. For a trip like that, clearly, there was no venue better than the Preanger railway line, with Bandung at its end. As an anonymous Dutch writer put it, more than a decade later, in 1939:

> Bandung . . . Indeed, no Indies town gives a Westerner so sure a feeling that he is on his own ground . . . the segregation in the design . . . all the time a secluded sphere . . . provincial . . . small and un-monumental . . . cheerful and nice . . . historically uninteresting, enclosed . . . among its—indeed beautiful—mountains. . . . The rest of the Indies . . . the heaving tropical land . . . lies far and deep, "in the warmth" . . . reachable and playable as one looks at it from here, as from an airplane.[178]

There was a logic in the airplane being mentioned in the last quote. Just a few years after Berlage's visit, in the same process of smooth "slithering through the mass of landscape," higher and higher, and in the same quest for the third dimension, airplanes became a technology and optical tool as welcome in the Indies as the train—indeed, as technology advanced and crises deepened, more welcome than the train.[179]

When the airplanes appeared, it was as if the Holland of the Golden Age was about to rise in glory again. Beyond the beautiful Indies, with the same qualities (and equally kitschy), as if there was a "beautiful world": "The first day we flew from Amsterdam to Prague, then, second day, to Belgrade."[180] The Dutch planes slithered and soared:

> over the Blue Danube . . .[181] Jupiter temple, Athens, the pyramid of Cheops . . . sphinxes . . . Dead Sea . . . Cashmere Mosque in Baghdad . . . finally, on November 24, we are above Batavia . . . the whole distance done in 127 hours 16 minutes . . . 20 flight days . . . the average speed 120.75 km per hour.[182]

One of the early passengers on the KLM flights wrote "an appreciation" about how, on that kind of trip, one might "obtain a true bird's eye view. From a vantage point of an airplane, one can see the primeval forest of Sumatra and Burma, the blue depth of the Malaccan Archipelago [sic]."[183] It was noticed that, in the last decades of the colonial era, the "Indies public" "became 'air-minded,' and much faster and deeper so than it was, on average, the case with the public in the West."[184]

Like air conditioning, for instance, the airplane was the most wonderful technology of the time and place, technology useful by being and making up the trivial. From an airplane, one could see the Indies volcanoes, for instance, like toys. Seen from a plane, the volcanoes befitted the panorama of the beautiful Indies. Even nature itself seemed captivated by the new invention. Almost precisely at the moment when the first passenger planes started flying in the colony, Krakatoa volcano came to life again, just in time for the early air-minded tourists to see it:

> special "sight-seeing-flights" . . . the most popular are the flights from Bandoeng over the Tangkoeban Prahoe volcano craters, from Batavia over the bay of Tandjong Priok and the islands offshore, and from Soerabaja over Bromo volcano and the Sand Sea. The flights that the KNILM [Indies partner of KLM] organized during the explosion of Krakatoa at the beginning of 1929 were especially unforgettable for all the participants.[185]
>
> . . . toilet is available in each airplane; as the cabin is suspended under the wings, the passengers have an undisturbed view through the large windows made of unbreakable glass.[186]

Through the unbreakable windows of the planes, not only the volcanoes and the rest of the Indies landscape but also the people of the land were made more safely visible: "The Papuans showed a particularly great interest in Dragons, the first airplanes they ever had an opportunity to see."[187]

Soeroso, one of the Indonesian deputies in the *Volksraad*, the advisory assembly of the governor-general, complained in July 1938 that the support by the government for the Indies air traffic was overzealous, and that it made, namely, no sense to the indigenous population: "I was of the opinion [in 1928] that an introduction of the air transport had no significance for the indigenous population. . . . We are now ten years further, but yet not a single *tani* [in Indonesian: peasant] or coolie has made a use of an airplane."[188] It was explained to Soeroso by his Dutch colleagues in the *Volksraad* and by other experts how misguided his logic was. First, there was "a concern for the empire."[189] Second, the indigenous people were sending a great number of postcards, 52 percent of all, indeed by airmail.[190] And, third, the air travel made perfect sense "for all the races" in the colony, "for all nationalities without an exception"—if only one was able to see it as an amusement: "Number of passengers by air between November 1, 1928, and November 1, 1929: Batavia-Bandung: 3953; Bandung-Batavia: 4303; Batavia-Semarang: 997; Semarang-Batavia: 1002; extra and sightseeing flights: 3514; total: 13,769."[191]

Close to one-fourth of all flights in the colony, in 1929 and ever since, were sightseeing flights, and, it was true, in the sightseeing flights, all races were represented. Groups of Dutch tourists and colorful natives were loaded in the planes, just for fun of it. In the amusement argument, the name of dead Kartini (we remember how much she had wanted to fly) figured prominently again. In the publications of the *KNILM* Dutch Indies airlines, throughout the 1930s, in countless advertisements for "observatory vacation air cruises," Kartini schools, founded in her name by compassionate Dutch, became a regular feature:

The photo shows an excursion by Kartini schools in Semarang . . . [before its] sightseeing flight over Semarang.[192]

Great interest by the youth . . . Visit of a Kartini school at the airfield Simongan.[193]

There was Lembang, another thirty minutes or so by car beyond and above Bandung, the end station of Berlage, a hilltop with a magnificent view, and, since the 1920s, with a Dutch observatory. The telescopes of Lembang were as famous, in the Indies, as the Preanger railway and the city of Bandung itself.

The Lembang observatory was built by K.A.R. Boscha, the tea magnate, one of the wealthiest Dutchmen in the colony. Boscha lived on his estate at Malabar, a hill near both Bandung and Lembang.[194] It was Boscha's Malabar house that Professor Treub visited in 1923, as mentioned earlier, excited about the strawberries, caviar, and oysters that were served on ice. Louis Couperus also visited Boscha in his Malabar-Hill house, a year before Treub, and he in his turn noticed, among other things, "a dainty little electric grill . . . electric piano . . . generator by the river . . . seismograph which records any tremor of the ground, which is still volcanic."[195]

Boscha was a son of a Dutch physicist who at one time had served as director of the famous Technical Institute in Delft, Holland. Boscha himself studied civil engineering, but he did not graduate: he was more of an "enthusiastic amateur."[196] He came to the Indies in 1887, and, by the 1910s, he had made enough money to build Lembang observatory and dedicate it "exclusively to science."[197] By the mid-1920s he had already brought some truly amazing machines to the hill:

The 16-cm Secretan refractor donated to the observatory by the grandson of Lie Saay [a wealthy Chinese from Padang, Sumatra] . . . 11-cm Zeiss comet-seeker and . . . the 60-cm double refractor with one barrel for visual work and one for photograph. Around the double refractor . . . rose a round, domed structure of reinforced concrete. . . . The floor beneath the telescope changed elevation by means of an electric motor, to accommodate observers.[198]

Boscha "decided to set up the finest observatory in the Southern Hemisphere."[199] The Dutch skies were obscured most of the year; thus this was yet another effort to compensate for the two-dimensionality of the fatherland. The observatory of Boscha, even before it began to work, was hailed in the colony as "in this land, a monument of the Dutch culture."[200]

The star watching at Lembang was soon as popular as the air cruises and the railway and airplane sightseeing trips. Often they were offered as part of the same package. Every day and night at Lembang, "the same show: young-sters . . . on vacation . . . elders . . . natives, who were attracted by rumors about the great telescopes."[201]

As the colony developed, and as the great depression set in, there was saving in everything, and even in Lembang. Some of the most spectacular technologies, in fact, were never put to work.[202] Yet what seemed to matter was to keep on gazing. As long as one had been star watching, so it appeared, one might survive and not even notice the depression, and, after it, the final destruction. As the Indies fell to the Japanese in 1942, the Dutch director of the Boscha observatory, and Boscha's former protégé at Lembang, indeed, after a short internment,

> returned to the observatory at the request of the Japanese astronomers who took charge in 1943. There he carried out observations of double stars with the tacit approval of the *chargé d'affaires*, Masai Miyadi. Under the Japanese administration, Voûte measured 11,000 pairs, first with the 37-cm telescope and then with the largest one, which the Japanese remounted.[203]

Like the trains and the planes, the late-colonial telescopes were shelters. They allowed a gazer to be absorbed, and to watch the beautiful Indies, beautiful world, and beautiful stars, all on the same plane, distinct and safe, behind the glass, a kaleidoscopic game: "the telescope . . . aimed upon the southern skies . . . beneath . . . the city of Bandoeng . . . with its thousands of visible stars, the vault of heaven."[204]

The late-colonial culture strained to get high. High was as sublime as the hillside trains, air cruises, hill cities, and hill observatories. Highly sublime in the universe, and close-to-perfect optical tools, were the mountain hotels. Hotels, again. In a typical tourist brochure of the time, next to a photo or two, possibly of half-naked hotel maids, preferably Balinese, sworn affidavits often were published, or sometimes facsimiles, with a doctor's signature and government office stamp concerning "the detected number of germs":

Certificate I

Results of the bacteriological chemical microscopical analysis of DRINKING WATER at the BROMO HOTEL in Tosari, taken by Dr. J. Schut in sterilized bottles at Tosari, May 2, 1925. The bacteriological analysis on plates of agar-agar, at a temperature of 30 degrees Celsius: after 2 days 12 germs per cc.; after 3 days 18 germs per cc. The germs were bacilli vulgati, which are harmless. The total number of germs per cubic centimeter is low. The water is clear, of good taste, and it is up to the requirements of good and safe drinking water . . . it can be drunk without being boiled. Soerabaia, May 20, 1925. Signed: Chem.Bact.& Serol. Laboratories of Sugar Industry.

Certificate II

Result of a repeated analysis of Drinking Water on December 30, 1925. Conclusion: The quality of the water is A 1. Signed: The chief of the Government Laboratories.[205]

As the (belief in) optics developed, the close watching and the health of colony were made into one thing. Microphotographs of cerebral malaria were the main highlights of the "First Exhibition of Hygiene" that opened in 1927 in Batavia: "The richness of colors of the sick tissue, and of the affected organ, are a constant attraction, and never fail to make an impression on the crowds."[206]

Like the toy trains at the Paris world and colonial exhibition, and like the colony through the Lembang telescopes, the insides of human bodies moved, ultimately obvious, brimming with man-made light:

Fascinating was the way in which Prof. H. M. Neeb, in his "stand" at the exhibition, demonstrated infection by hookworms: an almost life-size picture showed schematically the blood-vessel and the intestinal system of a human. By pushing a button, one can see a tiny electric light that represents a hookworm larva, moving into the body through a cuticle on the foot. It can be watched as it climbs up through the blood vessels, as it enters the heart, the lungs, the pharynx, the esophagus, the stomach, and, finally, the intestinal canal.[207]

Whatever vigor had been left to the Dutch rulers in the colony of the 1920s and, especially, the 1930s, it was expressed in the intensity of moments like that.

Early in 1933, the Dutch communities in the Indies—precisely half a century after the explosion of Krakatoa—passed through another brief but violent shock. There was a mutiny among the Dutch and Indonesian crew on the *Zeven Provinciën*, "Seven Provinces," a battleship operating in the waters off Java. The commander of Dutch naval aviation in the Indies, P. J. Elias, recalled later in his memoirs how the mutiny was dealt with.

At one point, Captain Elias wrote, the mutinous ship was almost lost to the pursuers' view. However, on the ship, "a young officer pulled it off . . . [he] got into the radio room at the last moment and sent a message . . . to the naval aviation about the position of the ship."[208] Then it was decided that an ultimatum would be issued, and, after the ultimatum passed, the ship would be bombed, first from an airplane by an 800-kilogram bomb, and, should this bomb miss the ship, from a submarine by a torpedo:

between 8.53 and 9.02 the instruments gave their summations . . . on various wavelengths . . . no white flag. . . . Coppers, a young bellwether air-fighter, just before the end of his six-year service shift, grabbed the opportunity. At 1,200-

meter altitude, his Dornier began a foray. Then the ship came into the view-finder of the bombsight. . . . Zero! direct hit . . . exactly on target . . . flames. In a fraction of a second . . . they learned. . . . The mutineers in panic gave themselves up.[209]

This was how the mutiny could be seen, and how powerfully the seeing worked. This was also, perhaps, as far as this story could go. Correct or upside down, the mutineers' ship was gotten into the Dutch perspective glass. Then the mess of the colony "far and deep, 'in the warmth,'" could safely be—as Oudemans or Kant might say—"subsumed under the concept of Quantity."[210] As Captain Elias put it: "Zero! direct hit." They had no chance.

The Mirror

The Dutch watching the Indies makes one think, often, of the concept of "disinterested perception," as the great Dutch avant-garde painter, Piet Mondriaan, for instance, explained it in 1918: "Since perception springs from the universal (within us and outside us), and completely transcends the individual (Schopenhauer's perception), our individual personalities have no more merit than the telescope through which distant worlds are made visible."[211]

The maps of the Indies as the Dutch drew and illuminated them, the photographs, and *kijkers*, the perspective glasses—all this was designed for watching out. Everything was known as far and as well as it was mapped—the exact numbers of prisoners,[212] the costs per prisoner,[213] the number of deserters,[214] the number of strokes with a cane, for what crime,[215] the number of people killed by lightning, tigers, crocodiles, snakes, wild boars, "other animals,"[216] the number of victims of tuberculosis,[217] the number of rat holes even.[218] Yet, and the government itself made no secret about it, all the data made only a very particular sense of knowing. The numbers of accidents were grandly unrelated to the sum of all the necks broken through the Indies, as was the number of tuberculosis sufferers to all the coughing, and the prison statistics—this especially—to the security of the whole, immense, and largely suspect colony. This is the eerie sense one gets all over again of the world that is mounted under glass. The numbers, the maps, and all the other gadgets of the watching technology were applied so that the matter is not touched.

Only the man-made late-colonial light was to pass through the glass. There was no mention of fire. No Nero and no Prometheus in the Indies. Cool light. No smell. No noise either.

When, in fact, the people of the land, the natives, were catching their criminals, still during this late-colonial era, there was, as a rule, a hue and cry, the air was full of sounds, and it trembled. This is a description from a very popular Indonesian novel from the early 1920s, Mohammad Roesli's *Sitti Nurbaya*:

> "We listened and, first, we could hear nothing!" said Arifin, and went on: "Suddenly, *ketuk-ketuk* [a drum] sounded. I was sitting with my parents on the back verandah of our house, and we were just getting ready to eat. The sound alarmed us, *ketuk-ketuk*, it came from a house, clearly, not too far away. . . .
>
> Every now and then, the sound of *ketuk-ketuk* became louder, as if somebody had already been stabbed by a madman. For two hours, we remained in fear, and then at about eleven the sound of *ketuk-ketuk* became quieter. Not very long after it stopped, and everything became calm and silent. At that moment I worried no more, as I knew that the man was already caught."[219]

In *Sitti Nurbaya*, also—a story otherwise just jammed with crime and apprehension—only once is a colonial policeman described as he physically in full view of others enters a native home. On that occasion, supposedly to arrest Sitti Nurbaya's father, the Dutch agent is remarkably clumsy and even almost shy.[220]

"Disinterestedly," indeed, the Dutch wanted to perceive. In a sense they made it. They were seen like that by the Indonesians as well. Immediacy, so uncomfortably kindred, in the tropical land, to the sensations of twilight, and jungle, and blurred edges, was to be avoided—no touching across the line, no warm shadows:

> After being silent for a while, Samsu [Sitti Nurbaya's lover] said slowly: "Don't you think we should run away from here, from the hands of the police?"
>
> "I think, whatever we might do, it will be suspicious," Sitti Nurbaya answered, and she shook her head, "in the end, no doubt, we will fall into the hands of the police anyway. Where do you want to hide yourself? The whole Java is the police."[221]

This was "seeing like a state."[222] The Indies was watched like the river is dammed in Heidegger's story of the Rhine—made good, halfway, as a "standing-reserve," and, halfway, as "an object on call for inspection by a tour group ordered there by the vacation industry."[223]

The Dutch building dams and perfecting optics were surrounded by the brilliance of their own making. The profusion of perspectives, what can make the world worth touching, almost everything beyond the reflection of themselves, was lost. What the Dutch succeeded in building through their careful late-colonial policy of looking turned out to be a hall of mirrors, a maze of flat reflections of themselves. The Dutch glass images of their Indies—to use the terms of Roland Barthes' classic on photography—were unary, without a vacillation, indirection, or fissures, banal and vulgar, not erotic, but pornographic.[224]

The Dutch in the Indies, through the intense watching of the colony, got their Indies all right. Their gaze, as it became so very much technological, became also almost entirely glued to the polished, hard, and unyielding surface of their *kijker*'s lenses. Like the photographs that Barthes described, the Dutch visions of their colony in the end, joined "that class of laminated objects whose two leaves cannot be separated without destroying them both: the windowpane and the landscape."[225]

The world the Dutch got, in the Indies, was the world they photographed, and there was a growing sense of tiredness. What we got, one Dutch author complained, were mostly the "white plastered" objects and not much more than the plaster; the "tasteless sun-blinded" planes, "nothing but white walls with tasteless window blinds . . . white plastered stone pillars . . . scorched by sun . . . "[226] As Walraven put it, "pillars that support nothing."[227]

As the time came for the ultimate test, the Japanese were about to invade the colony, the Dutch tried to search truly deeply where their real identity might be. Some found it, indeed, crystal clear and glass-like—"spotless," as a very prominent author, the second- or third-highest Dutch colonial official of the time, Charles van der Plas, defined it in a series of articles published in 1941. The "Dutchness" in the Indies, van der Plas wrote, stood at present denuded amid its own "dreamed-up world."[228] The Dutch soul in the Indies, van der Plas wrote, was like the good spirit of Shakespeare's *Tempest*, the "light and ethereal" Ariel, "an airy spirit": "Thou who art but air."[229]

Fakkel, "Torch," a liberal Dutch journal that first appeared in November 1941, set an explicit mission for itself to keep up the Dutch spirit and soul in the Indies against the coming and rising deluge. Van der Plas too wrote for the journal. *Fakkel* was very intensely read in the Indies during these very last months of the colony. *Fakkel*, "Torch," indeed, was now expected to illuminate the darkening scene.

Through the little over a year that had been given to *Fakkel* to exist, the journal's emphasis was on folklore, especially Dutch and South African— clear references to the greatness of the Dutch imperial past. Next, poetry was made crucial in the journal, of Willem Brandt typically, a Dutch author with texts like *Tampaksiring* (on an ancient Balinese temple), *Boeddha* (on an ancient Javanese shrine), *Mijn porceleinen Boeddha*, "My Porcelain Buddha," or *Ons Heimwee*, "Our Nostalgia":

> I thought: now, over there, everything is white
> Like daisies in the grass of spring . . .
> Gamelan [Javanese orchestra] sings its sadly fluent song.[230]

Prominent in *Fakkel*, also, were extensive reviews of several exhibitions of ancient and traditional "Hindu" art that, just around this time and with much ado, were opened in Batavia.[231]

Beb Vuyk, a young Dutch woman writer who lived with her husband, a Dutch planter, on the island of Buru in the eastern part of the colony, had a

long story published in *Fakkel*. An ancient petroleum lamp provided a dim
light for the novel. The story was that of a white man who lived resignedly
in a premodern kind of house, and very close to the beach:

> The house leaned low against a hill, fifteen meters above the level of the sea; the
> crowns of the old coconut trees reached almost to the same height. In the eve-
> ning the dark sleepy birds, whose feathers moved in the wind, descended on the
> palms. The white glistening beach marked the edge of the plantation, but, at
> every high tide, it almost disappeared. . . .
>
> The house was large and primitive, with gaba-gaba [sago palm fronds] walls
> and a roof of corrugated iron. . . . At nights, there flickered a little petroleum
> lamp, and the shadows shivered in the corners. . . . Bouvier, who fled the land of
> the living, found a space here, with just a single trail that always, infallibly,
> turned him back toward the shielded house, and the blue bay of dreams.[232]

The artistic and literary avant-garde in Holland, after World War I, conceived a specific notion of a modern house made largely of glass:

> In the April 1920 issue of *De Stijl* [van Doesburg] published a photograph of the
> interior of an American factory, the Bethlehem Shipbuilding Works. He de-
> scribed this in quasi religious terms in the May issue as a new "paradise of glass,
> iron, and light," where machines were "prayers of light and movement in direct
> relationship with space."[233]

As the Dutch Indies glass house was concerned, however, usual dictionary
explanations are more in keeping. If there was a glass house in the Indies,
this was something between, as dictionaries have it, "glassworks, greenhouse,
photographic room, or laboratory," or, in slang, "guard room, or military
prison."

In 1921, at the brand-new late-colonial technical college in Bandung, Professor Clay initiated a student debate society called *Ter Algemeene Ontwikkeling*, "For General Development," abbreviated, "*Tao*."[234] On August 4, 1922,
as the first native student, Sukarno, the future engineer, and still later the
first Indonesian president, gave a speech:

> Only the intellectual and moral qualities could be a measure for us! The human
> intellect is international. One finds it in a Negro with the skin black like ebonite
> wood, as well as in a European, white like silver! . . . Down with the color of
> skin as a criterion! Down with the blood as a standard![235]
>
> We all have to trust that the only way the Indies can be moved forward is
> through the cooperation between the leading personalities and the population.
> And the leading personalities, it always will be you, gentlemen students! The
> cooperation is urgent . . . it rests on your shoulders as a heavy yet wonderful task
> to work together in building the Great Indies for the well-being of all Indies
> residents, for the well-being of all that is Indies.[236]

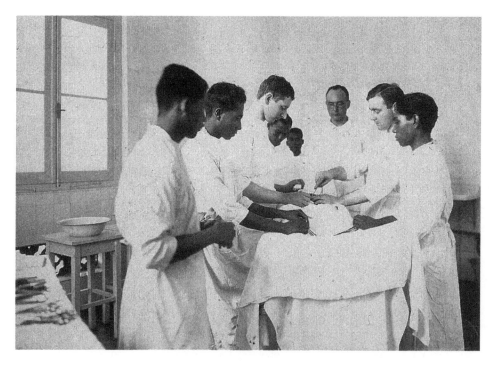

An operation in the Central Civil Hospital "Salemba" in
Batavia, 1925. (KIT)

Young Sukarno gave a good speech, as he was, of course, to give many good speeches in the future. He, however, also gave a perfect opening to his Dutch fellow students at the school. They responded in an instant:

There was a brotherhood preached! Between white and brown! That cannot be. . . . *There was a brotherhood preached between white and brown!!! It is impossible!!!* . . . Just take a Negro, for example, and place him, with an Englishman, let us say, under a glass bell. Best of all, use a glass bell that is airtight and so large that there is just enough space for the two to lay lengthwise, close up next to each other. Then, try to make them be brotherly with each other. Try to make the brotherhood come about under various conditions. Inject some adhesive substance under the glass bell, for example. Something like that. What I am suggesting is merely one of several ways. You may, first, turn to experimental chemistry and physics. Then, you may attempt to do the same experiment on a mass scale.[237]

The pride of every concert hall is in its acoustics. As in a concert hall, anathema to the Dutch glass house in the Indies was a chance of some false reflection, a dull spot in the overall brilliance, a gaze that might evade the strict rules of the total and all-absorbing enlightenment.

The late-colonial glass house had to be light-tight. Everyone approaching the glass house, the circle of light, was supposed, at a safe distance, to be fully revealed and all the revealing, to use Heidegger's words, was to be consumed in ordering. No false reflections! Yet a number of the Dutch in the Indies were noticing exactly that: a gaze that avoided the overall design. With an unease that was increasing, they spoke, like Isabella in Shakespeare's *Measure for Measure*, about a native: "His glassy essence—like an angry ape."[238]

Even glancing through the Indonesian newspapers made it obvious that the native people of the land liked and, indeed were obsessed by, optics; no less than the Dutch, in a very disturbing way "almost" like the Dutch: "There will be portraits taken of all the members of the executive committee of the PTTR [Indonesian Trade Union of Post, Telegraph, Telephone, and Radio], and each member will buy all the photos, because most of us want to recognize our leaders on sight, our fathers, our Gods . . . the fathers of our union."[239] The people of the land, the natives, got excited about cinematography, no less:

Also, while shopping or when on a picnic, some of you, our readers, can now make your own movie!

You can call your own even the newest machine, which is a combination of camera and movie projector. . . . For just Fl 29.50 apiece it will be sent, cash on delivery, on your address . . . With Dallmeijer lenses that are famously perfect.

The lowest price of camera in our stocks is Fl 75 a piece, not including projector; the cheapest projector is Fl 80. . . .

You can make films yourselves, and you can also buy 9 1/2 mm Pathé films, with comedies like Mickey Mouse and others.

And here is a machine that is most easy to operate. Even a little child can make a movie and project it. For the projection, you can use batteries if there is not electricity in your house; that is no problem. If you have electricity at home, we have a transformer for you.[240]

In another sales pitch, in the same Indonesian journal there is a picture: a beam of light is turned upon a masked, well-clothed, and to all appearances European man (with something on his head, even, that looks almost like the Sherlock Holmes cap). The man is about to break into a peaceful and to all appearances, native house:

The Newest Item! The Watcher over Thieves And Burglars.
For only Fl 1.50 it will watch over your peace and property! THE NEWEST PATENT

The usefulness of the WATCHER OVER THIEVES AND BURGLARS was acknowledged by American police. Thieves and burglars, however intelligent they might be, will certainly fail when this WATCHER is at work! It is most easy to use, wherever there are doors and windows. Does not need any electricity or gas. It is safer than building a fence or hiring a guard. It is tested. Price a piece Fl 1.50; 2 pieces Fl 2.75; 3 pieces Fl 4. Will be sent after receiving a money order.[241]

A few pages farther in the Indonesian journal, field glasses were offered, "to watch soccer, tennis, or to use on picnics, price apiece only Fl 0.75."[242]

Indonesian nationalists, "almost" like the Dutch, just a little too late in the period perhaps, and in a more grainy way, fashioned their own *tableaux vivants*, scenes caught in a moment and stilled, photographs made of living people: "Then tableau of unity was performed, and after this other numbers and extra numbers were shown that made everybody present very happy."[243]

Illumination of buildings at night was loved, too, by the native people of the land. It was just a little too difficult, sometimes, to understand—from a technical, rational, and Dutch point of view—what, in fact, was going on here:

Opening the House for the S.R.V. [Indonesian Surakarta Radio Union] Studio August 29, 1936 . . . The opening was accompanied by switching on the lamps, which was done in a very modern way. The lamps were not lighted by pushing a button as, nowadays, is the usual way, but—on the initiative of a member of our Technical Commission, Mr. Tjiong Joe Hok, a technician fresh from America—with the use of a pocket flashlight (*zaklantaarn*) that was connected to an apparatus that, prompted by the flashlight, gave an impulse to all the lamps. In a blink, all the lamps flared up, so that through the entire house of the broadcasting studio everything became bright as at full noon.

The method in which the studio SRV was opened did not differ much from the method in which the recent exhibition in New York, the metropolis of the U.S.A., was opened. The difference was only that, in America, the lighting of lamps was done by the use of the rays of STARS in the sky, while our method was to use the rays of a pocket flashlight.[244]

As the optics of policing permeated the colony, it became increasingly baffling to "watch the natives." One of the Indonesian local post, telephone, and telegraph workers' unions, for instance, among the demands sent to its union central for the upcoming congress, required *"Branch Soekaboemi-Tjiandjoer:* We urgently demand that our trade union establish a *detective-propagandist."*[245] "Why should there be a detective propagandist in our union?" even the editor of the union journal, and a member of the union executive, felt compelled to wonder.[246]

I have not studied the statistics of ophthalmology in the Dutch Indies. Whatever change in the eye health of the colony might have occurred, however, it could not possibly explain one phenomenon: a steady stream, and flood, of advertisements for eyeglasses in the Indonesian papers through the last decades of the colonial era.[247] Willem Walraven, the journalist in the Indies who always seemed able to put things into perspective, wondered in 1939 what could be most ominous to the Dutch and to the Dutch presence and future in the colony, he put this very "optician" item at the top of his list.

The eyeglasses, Walraven pointed out, so much in vogue among the native young men in the colony—and one could check it out easily in every shop, every sidewalk stall—were "almost" eyeglasses, "mostly fake" eyeglasses, "merely make-believe" eyeglasses.[248] Transparently weird, these so popular optical tools, worn so ostensibly, were also—or so it now appeared to Walraven as he watched—trained upon the modern Indies, and, at some moments at least, upon the Dutch themselves. The lenses were "just" panes, "almost like" the windows of the late-colonial trains. Walraven did not say so in so many words, but the feeling was clearly and chillingly there—the time might be imminent when an Indonesian own *kijker*, perspective glass, is about to be invented: "Eyeglasses made of window panes," Walraven exclaimed, "so many specs . . . with ordinary glass, and in fool's gold frames."[249]

FOUR

INDONESIAN DANDY

The next day there was an endless coming and going at the
palace. All the real princesses had to try to put the glass slipper
on; then all the duchesses; then all the ladies of the court; but
not a single one had a small enough foot.
—Cinderella[1]

The Dolls

IN 1893 (ten years after Krakatoa exploded, two years after IJzerman's
expedition started toward Siak), a collection of about 150 dolls was respectfully presented by "the ladies of the Netherlands Indies" to the
Dutch queen. Puppets—in doll houses, in children's rooms, in salons—were
the fashion in those days. The Indies doll collection, as its catalog suggested,
covered the whole recently conquered East Indies Archipelago and more.
Even Aceh, for instance, North Sumatra, not yet completely beaten at that
time, was included. There was, again, a pronounced concept of historic optimism expressed in the gift, and in the dolls as well.

As the catalog of dolls for the Dutch queen suggested, a complete picture
of the Indies native society was meant to be presented through the collection.
Indeed, a large number of ethnic groups and social layers each found a di-
minutive representative in the collection. There were several *hadjis* among
the dolls, Islamic pilgrims to the holiest places, in little all-white garb, their
faces cast to look proud. There was an assumed-to-be complete, regionally
varied hierarchy of native officialdom, carefully carved, painted, and clothed.
A set from Preanger, West Java, contained a *regent adipati*, the princely top
native official, a *raden ajoe adipati*, his wife, *a patih*, his assistant, a *mantri*,
minister, a *djoeroetoelis*, scribe, and on and on, through thirty-five consecutive
ranks, down to an *orang tani*, he-peasant, and a *njai tani*, his wife. Significantly, there were also dolls of native professionals—a *hoofddjaksa*, attorney,
and *raden ajoe hoofddjaksa*, the attorney's wife, an *ondercollecteur*, tax sub-
collector, a *mantri oeloe-oeloe*, surveyor of irrigation, a *mantri goeroe sekolah*,
teacher, a *dokter djawa*, native doctor, a *mantri tjatjar*, official in charge of
vaccination.[2]

The ladies of the Indies, looking back on how the dolls were made and clothed, explained in the catalog that, in the beginning, no native artist could be found to help them who could do the job properly. The natives, the ladies wrote, were perfect in their *wayang*, traditional shadow-puppet fantasies; such puppets they could make. They were, however, hardly capable of producing a true representation of themselves. The ladies had to teach the people of the land. Only as the working hours and days dragged on, they—the ladies and the native experts—slowly moved toward doing it "together."[3]

One is, of course, reminded of one's other reading, and of Foucault in particular.[4] The colony was being miniaturized, inventoried, and surveyed; discipline would come next. There were no dolls of the white ladies. Yet the wholesomeness of the enterprise was striking. The natives, as well as the "ladies of the Indies," appeared to come out of the adventure strikingly in one piece. The Indies appeared to be represented with a complexity and definitiveness comparable to the doll-like representations of the Western societies produced and displayed throughout Europe and North America at the same time. In a way, the dolls and clothes as constructed and displayed were telling the truth as it was believed to be the truth, in the East as well as in the West. "An ordinary native woman (Makassarese)," for instance, was fully defined by naming what she wore:

> Her hair-bun is a usual *simbolong*, and it is decorated with a sort of a flower bouquet, *boenga ni goeba*. Her *baadje* [blouse] is called *badjoe bodo*, and the name of the piece of cloth over her right shoulder is *pasapoe roi-roili*; on it *roi-roili* are fastened. Her dress is decorated with a kind of accessory called *renda*. Her skirt is called *tope*, and the trousers are *tope i lalang*.[5]

Clothes seemed to speak in a full and clear voice. Was it merely because the bodies and souls were made of wood and clay? Was it, as von Kleist wrote in another culture and time, because puppets "have the advantage of being for all practical purposes weightless"?[6]

In the catalog, the language of some sections was Dutch, with technical terms in Malay. Other sections were in Malay, with sometimes full, sometimes summarized, Dutch translation appended. The ladies evidently did not feel that they had to think too hard about this arrangement, or rather this nonarrangement. They did not seem to experience any particular urge to edit, to emphasize a (dis)equilibrium between the two languages used, to keep an order in translation. In language, as in their other relation to the Indies, they did not seem intent to build or dismantle any bridges, as they did not seem to be disturbed by any space in between.

The Modern Times

By 1900, telephone lines were established among Batavia, Cirebon, Tegal, and Pekalongan; and between Semarang and Surakarta.[7] By the 1900s, as we have seen, hundreds of thousands of natives, their bags of rice and, often, goats with them, traveled each month on the new trains of the colony.[8] Reflecting upon the modern times in the Indies, it is difficult not to see what happened to dolls and clothes.

The most remarkable thing about peasant clothes in the Indies, at the time, was how inconspicuous, indeed invisible, they became: rarely were they mentioned in the growing volume of the colony's modern images, in newspapers, books, official reports, posters, and catalogs.[9] While a peasant went on with his or her work in the fields, his or her peculiar dress, and, deeper still, his or her peculiar body movements and manners, were being taken away, piece by piece, in a sadly nonsensual striptease, to be remade, as it soon turned out, into the new attire, body movements, and manners of an anonymous and non-peculiar mass—"the common people," "the folk," *volk* in Dutch or *rakjat* in Malay. The peasant got, if we may use fashion-maker jargon, "practically naked." He or she disappeared in a sense, because this was a time of new costumes dawning upon the Indies.

At the turn of the century, there was an increasing number of men and women in the Indies—peasants, initially, in clothes that lost their color and quaintness, in an increasingly neutral outfit allowing access to the strange new space—leaving their villages, employed on plantations, railways, and the roads, in mines, harbors, and factories. By the 1910s, the native workers already had their own organizations, newspapers, and publishing houses. Occasionally, and sometimes with a passion, they spoke and wrote also about their clothes. But the native workers wanted their clothes, in essence, to make themselves fit into the space as defined by the order of the modern times: to suit the working site, and to shelter their otherwise naked bodies. As soon as the native workers had their clothes, they had little reason to think about the matter any further.

The politics of clothes gained a new intensity elsewhere, in the Indies modern offices and offices' anterooms, among the newly emerging strata of the native subassistants, underclerks, third-rank telegraphists, and sub-substationmasters. There, a real tension developed concerning the clothes: it was a struggle to get an outfit in order to articulate one's own statement about the modern age.

This clearly was a moment when a troubling question arose among the Dutch dominating the colony. If a native became clothed as he or she wanted to, would he or she no longer be a native?[10] Wherein, then, would the native belong? A space to fit in, for such a him or her, could not be found on any recognized map known to the Indies. Among the Dutch in the Indies, a sense of confusing and ominous mix-up emerged and intensified. A new awareness of possible trouble was followed by new images of the colony. These can still be seen in the old photographs of railway stations, bridges, factory courtyards, or office buildings: scenes carefully organized around the Western technologies with immaculately all-white clothed, professional-looking Europeans. With no natives! As the palm trees and other rich tropical flora, they seem to be cut out, or at least retouched and blurred.

Much of this story may be conveyed by evoking a sewing machine or maybe a gramophone. While in Europe, and in the Netherlands itself, the decades around 1900 brought about a great variety of modern technologies employed to change everyday life—democracy was being spread by the trains and the telephones, Amsterdam was being expanded by public housing—in the Netherlands Indies, there was an increasing eagerness to attain modernity by technologies that were unabashedly frivolous, and by machineries that served primarily to produce an appearance and an amusement. The modern Indies, much more than the modern Netherlands, was expected to move to off-the-peg music. Here come the dolls again. Much more than in the West, progress in the colony was identified with fashion.

Tillema's *Kromoblanda: On the Question of Living*[11]—a six-volume opus published between 1915 and 1923 on the Indies asphalt roads, modern bridges, cities, prisons, and bathrooms, classic of the late-colonial Indies— was also, and perhaps most of all, a good-manners book. Tillema suggested ways for *kromo*, "natives," and *blanda*, "Dutch," to live orderly and hygienically together. His ways are perhaps best explained by one of the photographs in *Kromoblanda*. Tillema himself demonstrates in the picture how to teach a native to pour drinking water from a bottle into a glass. His body is poised, the bottle and the glass are held just by the fingertips, clearly not a drop is spilled, and Tillema is in an all-white suit.

Tillema stood with both feet in the modern times. In contrast to the Indies ladies' doll creations described above, for instance, there were already price tags on virtually everything in Tillema's *Kromoblanda*. Tillema was a businessman, and he knew that his bottled mineral water was too expensive for virtually any native to buy. Unlike the ladies' doll collection, Tillema— more modern and less happy—was entangled in translation. The very title of his book, *Kromoblanda*, was an effort at the translation. It was as if to arch over a gap between two languages and two increasingly incongruous worlds. It was made to hide the gap by a gesture, a hyphen. The strangely hybrid term of *Kromo(-)blanda* was, in fact, artfully stitched together to work like a set of snobbish clothes.

Ons Huis in Indië, "Our House in Indies," by Mrs. J.M.T. Catenius-van der Meijden, has already been mentioned, too. It was published in 1908, about the time when Tillema conceived his *Kromoblanda* and when *Boedi Oetomo*, the first modern native organization, was founded in Java. Mrs. Catenius, as we read on the book's title page, was already well known as "the author of 'To the Indies and Back,' 'The New Complete Indies Cookbook,' and 'Spices and Other Ingredients of the Indies Rice Table,' the two last works awarded with a Diploma and a Golden Medal respectively, at The Hague Cook Exhibition in December 1904."[12] Mrs. Catenius, in many ways, was a lady-comrade-in-arms of Tillema. The aim of her *Our House in Indies*, she wrote, was to give the Dutch men and women in the Indies a textbook of manners and a fashion guide. She aimed to tell them how to move, what to wear, and how to believe strongly in how they moved and what they wore. In particular, she directed her book at young Dutch housewives, who were arriving in the Indies in increasing numbers lately, as a part of the modern times, imported housewives filling the Dutch houses in the Indies, pushing away the doll-like and sensual *njais*, the native consorts, common wives, and concubines of the Dutch men.[13]

Mrs. Catenius described, truly, a revolutionary time in the history of the Indies (although she wrote "evolutionary," and we can feel her blushing). Early in the book, "a good lady friend, who had recently returned to the Indies after a sojourn in Holland," is asked by the author about the news. This is not the Indies of "10 à 15 years ago" at all, the lady answers, and the "10 à 15 years ago" becomes a highly unsettling expression reappearing throughout the text.

Mrs. Catenius was a very sensitive woman. Without hesitation, she pinpointed the new communication technologies in the colony as the main agent of the Indies recent (r)evolution. Now, she wrote, the trip from Holland to the Indies took less than five weeks. No more did a traveler to the Indies need what she termed *uitrusting*, the "equipment"—no longer did one have to take all things modern with oneself to the East. To go to the Indies was no longer a deed of a unique adventure; no longer did one have to travel to the Indies as one still might have to travel "to the North Pole" or "into the interior of Africa." Nowadays, one could buy practically everything in the Indies: pans, plates and silver, furniture, and quality underwear. And it was a safe deal; as in Tillema, in Catenius' book there were price tags on everything.

Yet, as Catenius-van der Meijden noted in surprise, close to alarm, even, the "equipment" journeys were still being made. In fact, it seemed, more than before! "Our houses in the Indies," nowadays, the author wrote, sported floors and windows that were almost completely covered with heavy—thus instantly damp and malodorous—carpets and curtains. "Our" rooms, stuffy in the humid tropics anyway, now seemed stuffed even more than before with bric-a-brac from Delft, Brussels, Rome, and Paris. The Batavian lady friend of the author reported in detail, particularly, on hats and boas:

> As far as fashion is concerned, the craze for *de luxe* as it appears at visits, receptions, dinners, picnics, parades is beyond your imagination. Even in an ordinary catalog mail-order shop the display is unbelievable. This is no longer the Indies of 10 à 15 years ago! The wearing of boas, for instance: ladies here wear white,

Newspaper advertisement, ca. 1920. (Archive of the author)

gray, or black *ostrich* boas. A hat for ladies must be decidedly 'reçu.' No longer can someone just walk bareheaded as in the former days. No more just the fur boas. At receptions and parties, gentlemen appear in dressy jackets, ladies in *japons*; thus, no more merely a blouse and a skirt.

Sarong and *kabaja* ("Malay style") in the morning outside one's house is no longer 'reçu.'[14]

Mrs. Catenius raised her eyebrows, but in fact she was a party to the change. Some development, she willingly admitted, was inevitable. There had to be, for instance, new fashion shops in Batavia, and even fashion shops for men, if only because there was a modern gymnasium (school preparatory for the universities) recently opened in the colony, and "our [Dutch] boys," naturally, wanted to wear decent clothes to their classes. Again, with truly an acute sensitivity, the author singled out one item that epitomized the historic change: *pet* (Dutch for a cap), she wrote, became a sort of a new uniform, a piece of cloth, a sign of modernity, worn particularly daringly and elegantly by "our boys," the students of the new Indies.[15]

Like good authors of good looks and good manners guidebooks probably should, Mrs. Catenius had the heart of an engineer. Her book featured a dozen photographs, all of them of modern houses. In the spirit of H.F. Tillema, all of them were, first of all, "hygienic" houses. The space around had to be cleared of trees and bushes, sources of humidity, insects, and darkness. Flowers were to be in flower pots.[16] In contrast to "10 à 15 years ago," modern Dutch men and women retreated from the increasingly uncomfortable and busy outdoor Indies to their back verandahs and adjoining courtyards or gardens, which were also now more often than not fenced or walled in.

The new center was sought there, in the back rooms, and it had to be filled, as we have seen, with "a novelty piece of furniture," an upright piano here, a cozy coffee table there, and, here and there, a few Japanese screens.[17]

As the problems, it seems, grew too many and too complex to keep the order, Catenius' book in its second part was constructed into a series of numbered questions and answers. Stain cleaning, house-doctor problems, and domestic animals, among other things, were dealt with. The emphasis, again, however, was on manners, looks, and clothes.

Like Catenius' modern and hygienic houses, her manners and clothes were to be used to manner up the Dutch in the Indies, and to clothe them in. A carefully mannered retreat was advised from the increasingly exposed social space facing the Europeans. See, for instance, in Catenius' book, question and answer no. 51:

The manner in which to ride a horse.

A particular question was put to me: "Is it against a good norm if a girl sits on a horse like a man?"

Eurasian young man in a tennis costume, Batavia, ca. 1915.
Family album of Mrs. Emely Louisa de Vries-Foltynski. (KITLV)

Very definitely, it is. In this manner, the women of the land (the Indies wives [meaning concubines of Europeans], and the natives) ride.

Recently, an effort had been made to introduce a fashion of ladies riding on men's bicycles, but the attempt failed.[18]

No longer, Madame Catenius warned, and however unnatural it might appear, were the Europeans in the Indies to take their clothes lightly:

> It is known that the natives in the warm climate wear light, airy, and casual clothes. It might seem that someone, acting sensibly, should follow the clothing habits of the natives.
>
> This [however] is contrary to the Western civilization and its rules. A European cannot walk around dressed like a coolie, cannot clothe himself as the natives do; it would not help, even, if he may use the most expensive material for clothing himself that way. . . .
>
> [Merely] when sick, and while inside their houses during the day, many Europeans wear the popular Indies dress (*sarong-kabaja*). When a European, during the day, when every [European] should wear European clothes, is seen undressed, I mean dressed in the Indies way, it signals to everybody that this particular person is *ill*. Then one can attribute the illness, and so it happens most often, either to laziness or to taking matters of life too easily.
>
> Except when in the interior, and far from all [modern] centers of civilization, ladies and gentlemen should never let themselves be spotted in or around their houses in the Indies *negligée*.[19]

The manners, looks, and clothes of natives were also dealt with in Catenius' book. They also, it appeared, had undergone a change from the time of "10 à 15 years ago." In contrast to the Indies ladies' doll collection of 1883, at least, the natives of Mrs. Catenius were no longer made to represent a multiplicity of origins and professions. Now very little diversity was pictured. Instead, as a very modern thing, the standard was sternly set. Only one type of native, Catenius suggested, from now on was to be a "genuine native"; and only this genuine native—the servant—was to be admitted into or near the Dutch modern Indies house.

The servant was a norm, and it was typified primarily by he-servant or she-servant devotion to the modern European house. Most devoted and, thus most genuine was *baboe*, "nanny-servant," as it was she who touched the European babies and children—who had to be admitted into the less sheltered zone in between, among the least civilized, most exposed, and most vulnerable European humans in the Indies.[20]

Devotion, in the best Tillema manner, was defined by Catenius as hygiene. The genuine natives of Mrs. Catenius existed—they were made, indeed—to clean the modern house and to keep themselves clean. Very important, they also had to guard the house and themselves against contamination by those

House servants. (*Gedenboek voor Nederlandsch-Indië ter gelegenheid van het Regeeringsjubileum*)

who might look like them, and who might thus get close to them and to the house—the servant's relatives and the folk from the servant's *kampong*, "native quarters," the untamed crowd, which, like the trees and bushes, uncropped, surrounded the modern Dutch house uncomfortably from all sides. These, indeed, in the good language of Tillema the pharmacist, were called by Mrs. Catenius "parasites of our servants."[21]

Most distinctly, genuine native was defined by clothes, the "cascades of little nothings," "the social power of infinitesimal signs," the technology of "models" that are "left intact," while "copies" are made to be "identical."[22] The design-and-construction element was crucial. Some natives, Catenius warned the inexperienced Dutch housewives, might try to put on some kind of fashion clothes themselves! Such natives, namely, may decide to ask for a job in "our house," and they may dress up for the occasion. One must be on guard, and do not let oneself be misled. These clothes are counterfeits. They are worn by impostors to smuggle themselves into "our house." One has, categorically, to cloth one's native self:

> Those who wish so, and can do it, give their servant a special set of clothes, better said "a livery," which he will wear with pleasure during the day in the house, or when he is taken along on a visit.
>
> The livery may consist of white pantaloons, kaïn, or sarong, a loose white vest (also called *kamisol*) with gray stripes, and lined with red, blue, or yellow cotton. We also should, for the sake of uniformity, give the servant a headcloth in the same style. . . .
>
> Many families give their servants a more dignified dress, for instance a long coat that falls to the knees (*badjo-tóró*), is pulled over the head, and is richly buttoned in front. These *tórós* are usually made of dark-colored cotton or silk, sometimes of yellow twill or drill with red or blue stripes (kaïn raméh).[23]

As Mrs. Catenius emphasized: "Each servant is responsible for the clothes he was given, and, at the same time, the clothes are not his property. This must be explained to the servant at the very moment when he is admitted to service."[24] A process of continuous decontamination, again in the spirit of Tillema, was to be the best way for "our native" to be made conscious of all this and of oneself. "The (servants') clothes," Mrs. Catenius counseled, "are to be turned in every week, and then they are to go to the laundry in bundles, separate from the other household wash."[25]

About the time Tillema was selling his bottled water, and a few years after Mrs. Catenius came out with her book, Louis Couperus, the foremost Dutch novelist at the height of his fame, wrote a text that is not one of his classics yet is of relevance to this particular chapter. In a commercial publication by *De Magazijnen "Nederland,"* "The 'Netherlands' Stores," a chain of fashion shops that was to open a big new branch in The Hague, a five-page essay by

Louis Couperus appeared in 1915 under the title *Meditatie over het Mannelijk Toilet*, "Meditation on Men's Dress."

The *Magazijnen "Nederland"* publication, as well as Couperus' piece, were no doubt advertisements. There were addresses and other information about the new shop, and there was a list of all other branches. The booklet was a catalog—clothes were offered for home, office, culture, sport, golf, and for motor sport most prominently. Remarkable about Couperus' contribution was how often and at the same time how defensively he used the word "modern":

> Is our *modern* men's dress ugly . . . ? . . . Is truly our *modern* men's dress ugly, I mean, absolutely ugly? . . . We have to understand, thus I meditate and reason with a good use of philosophy (and in front of my mirror) on whether our veston, our vest, and our trousers are decidedly harmonious with our *modern* build [*gestalte*], with our truly *modern* being. The folds and the feathers do not suit our existence any more. It is evident that, in our frenzied and feverish lives, we have to lock ourselves neatly out of *modernity*, as into a little box, into our well-fitting pack of perfect styled dress. . . . All the automobiles, trains and tramways, crowded theaters and crowded streets, crowded *cafés*, crowded restaurants; being hastily here, going frantically there, in order to use, to enjoy, to get through it all, to live in three sets of times every minute of our lives—which all cries out for nothing but a "reasonably" charming men's dress, and which is also offered to us by the *modern* times. . . . Our cardigan, our veston, dinner jacket, and overcoat—light for summer, heavy for winter—enclose us safely in the seemly case. Yes, we feel very secure in our *modern* lives, when in our *modern* costume. It is our outfit of armor for our struggle of life, and as it is so closely adapted to our *modern* existence, in its perfect fittingness, it almost becomes a "beauty."[26]

Very significantly, in the same text, Couperus wrote about the "glory of our dandyism."[27]

Couperus' text clearly belonged to the writer who critically helped to shape the consciousness of a generation, in Holland as well as in the Indies. Couperus' feeling for clothes stemmed out of his acute sensitivity to colors and textures, to touches, scents, and vibrations. There was a close connection between Couperus' taste for fashion and his sensing—sooner than most of his contemporaries did—the ominous change in the late-colonial air and the rumbling under the ground of Dutch late-colonial modernity.

Long passages of Couperus' most famous Indies novel, *De Stille Kracht*, "The Hidden Force," published in 1900, the story of a small Dutch community in a provincial town in East Java surrounded by unruly and ominous tropical nature and people, can be read in many ways, including as a grotesque fashion show of sorts.

One of the principal figures, the embodiment of "the hidden force," *haji*, a

Muslim pilgrim without a name or face, appears throughout as merely an all-white garb reflected against the darkness of tropical night, merely clothes.[28] The climax of the novel, as we will recall from chapter 3, a terrifying flash of appearance of "the hidden force," comes when a Dutch woman's naked body is polluted by red stains—it may be blood, it may be betel spittles, certainly something improper and unmodern; you cannot, Mrs. Catenius might say, go out like that. The woman's husband, a high colonial official, crushed, or crumpled, rather, by the hidden force, his marriage and his career destroyed, is pictured at the end of the novel living with a native concubine in a half-Dutch, half-native house.[29] When found out by a Dutch woman friend from a better past, he takes an awkwardly long time to change clothes. When he comes out finally, to greet the friend, we can feel it strongly—Couperus does not have to be explicit: the man's collar is left unbuttoned, and parts and pieces of undergarment protrude from under his jacket and tie.[30]

In yet another part of Couperus' *The Hidden Force*, a fair is described: the wives of the Dutch local officials pose in a "Maduran *proa* fitted out like a Viking ship"; they act out scenes from Arthurian legends or sell Dutch *poffertjes* (pancakes), for exorbitant prices, "dressed like Friesian farmers."[31] The fair is boisterous. It is in fact—and Couperus notes this as if merely in passing—a charity fair for the victims of a recent earthquake, the most terrifying of the hidden forces in the Indies. As in the *tableaux vivants* he liked to organize during his stay in the Indies in 1898 and 1899, in *The Hidden Force* Couperus crowded the stage with daringly fashioned costumes—in a scheme perhaps not just to cover the bodies and make the scene merry, but, as in Mrs. Catenius' houses, to leave no space for fear.

Tillema and Mrs. Catenius, as well as Louis Couperus, sensed and conveyed well, indeed, a historic change. New fashion, new theatricality, and new nationalism emerged at the same time. By the 1910s, an increasing number of modern schools were being opened in the colony. An increasing number of natives, too, attended these schools, and as often as not, out of necessity, they frequented modern fashion shops as well. Not a few of the new emerging native elite daringly wore the *petje*, *peci*, or *pici* in Malay or Indonesian—the Dutch student caps, that Mrs. Catenius, as we have seen, deemed to be so significant.

The "whites" in the modern colony, as they tried increasingly hard not to clothe themselves loosely and to keep themselves distinct from the natives, became flagrant. With increasing fascination the natives were watching the *tableaux vivants* of the whites' own making; they surveyed the whites on the fashion pages of the white newspapers and soon in the white movies—the white fashion-pages in motion. There was new light rising in the Indies. The latest Parisian *mode* was shown on mannequins, in the brightly illuminated shop windows. The *rakjat*, the *volk*, the natives, the "people," might

Gathering of the princes of Surakarta and Yogyakarta. (*Nederlandsch Indië Oud en Nieuw*, 1920–1921)

now, merely by strolling through the streets of Batavia, Surabaya, Semarang, Medan, and other late-colonial cities and towns, watch these exemplary white mannequins, dolls, sometimes, indeed, only in underwear.

Nationalism and the Birth of the Dandy

The first native dandy was conceived, like almost everything in the Indies at the time, by the Dutch themselves, and out of the increasing Dutch drive to engineer the colony. Building up a dandy meant to contort both the body and the clothes.[32] Dandies were also to be sort of "fashion dolls" and "ambassadors of fashion."[33] Tools and machines. As von Kleist remarked, when watching puppets a century before in Germany, one was to be impressed with "those artificial legs made by . . . craftsmen for people who have been unfortunate enough to lose their own limbs."[34]

The newly born native dandies in the Indies were to be those natives who borrowed Dutch clothes to place themselves in the modern colonial society. Thus, clearly, these were not genuine natives. These were, to recall Mrs. Catenius, counterfeits. They were not the standard, thus they should not be fully real. They were yet another Dutch attempt at an overall solution: unreal natives could not be a threat.

Het Indisch Tijdschrift voor Post en Telegraphie, "The Indies Journal for Post and Telegraph," the press organ of the Dutch employees' trade union in the colonial state communication service, wrote in 1912 about the few new natives who made it as far and worked at some low-level post and telegraph jobs. Rather than being "voices of a rumbling volcano," the journal wrote, "they squeeze their necks into high stiff-starched dressy-shirt collars, they drink *jenever* gin . . . [they think of spending their time off] in one of our spas or another recreation spot . . . [they are after] the latest Parisian novelties . . . [they are] caricatures of Europeans."[35] If one asked a Dutchman at that time, as a contemporary remembered, who this new "Indonesian" was, the new breed that clearly did not fit into the Dutch stereotype of the native, the Dutchman would stereotypically answer: "he is a category, wholly apart, an ex-native, somebody with a peculiar skullcap on his head, yellow shoes on his feet, colored eyeglasses on his eyes, and several pens and pencils in his vestpocket."[36] The native dandy was built up, but, somehow, the threat did not go away. A new problem, even, was created. Pushing the ex-natives into the realm of unreal, the Dutch were merely building up yet another specter haunting the orderly Indies.

By the 1910s, the should-be Western-made dolls, in their odd-fitting *fiets-pantaloons*, "knickerbockers," were on the streets, riding trains in increasing numbers, some of them even second-class. There were some of the new breed, Indonesians—almost inevitably the most dandyish—who worked themselves toward the most respectable Dutch academic degrees. These wore the peculiar skullcaps, *petje* or *pici*, that Mrs. Catenius had highlighted a few years before; then, of course, they were worn by Dutch students only. These Indonesians became an all-archipelago type. Like the Indies ladies' dolls from 1893, they were defined, and they wanted to define themselves, by their profession. Like the Indies ladies' doll collection, too, these exotic figures appeared to come from all parts of the colony. ("Modernity," as Baudelaire thought, might indeed derive from "mode.")[37] They began to look like a force.

There had been, on the Dutch side, some reckless driving with the dandy metaphor. There was also some backing off—attempts to blur the dandy.

As the colony moved deeper into the epoch of a disturbing modernity— through the late 1910s and into the 1920s, through the First World War, the postwar depression, and the growth of the Indonesian political movement— modern natives were increasingly portrayed in the Dutch journals and picture books either as "true to themselves," wearing a truly genuine native costume, or as less pure, in kind of late-colonial fatigues. The latter curiously resembled the *pijama-broek*, "pajama trousers," and the *kabaja*, "Malay-style" blouse, the *negligée*—the outfit so intently defined by Mrs. Catenius a decade ago as fitting only for the lazy, sick, and weak.[38]

At the same time, the pressure was growing in the Dutch press, in the colonial government circles, and in the *Volksraad*, the advisory council of the governor-general, for the colonial army officers, even the subalterns, and even while off duty, to wear starched and all-white uniforms, until then used only on ceremonial occasions—no matter the trouble and expense it would cause to keep the uniforms starched and white in the tropics.[39] Keeping the appearance all-white, justifiably so, was discussed as if it were a matter of survival. Perhaps it was. The plan of modern colony was turning into an obsession with neatness.

In 1918, Mas Marco Kartodikromo, Indonesian journalist and writer, was serving one of his prison terms in Batavia, as usual for a press offense. Because he could not go on writing political articles in prison, he tells us, he decided to write a novel. The result was *Student Hidjo*, written in Malay, published instantly in a serialized form in Mas Marco's journal, and then in 1919 reprinted as a book.[40]

Student Hidjo is a text as symptomatic of its time as Mrs. Catenius' *Our House in Indies* or Tillema's *Kromoblanda*. Student Hidjo is an Indonesian young man, and we first meet him as he is about to embark on a journey to Holland to register at the most prestigious of Dutch colleges, the Technical Institute in Delft. Why travel so far? The motivation is a sense of social inferiority on the part of Hidjo's father whenever he meets his fellow-native higher-ups, especially the noble Javanese officials. Having an engineering son might change things.

The hero's name, *Hidjo*, means "Green" in Malay, thus a reader might expect, a suggestion of Islam, the green banner of the Prophet, a steady faith, consistency, and purity. Hidjo's girlfriend, however, is *Biroe*, which means "Blue." (Another girlfriend is *Woengoe*, which is "Violet," and the place where much of the novel happens, home for several principal actors, is called *Djarak*, "Distance.")[41] If this novel is consistent, and it certainly is, then it is so in its consistently drawing conclusions other than a reader is initially convincingly made to expect and in its consistent blurring of outlines and abusing of colors.

Hidjo travels across borders smoothly. The only hindrance on his way to the Netherlands, besides perhaps the submarines on the open sea (the novel was written in 1918), were lustful Dutch women onboard and wherever the ship anchored. Nevertheless, or including this, Hidjo travels nonchalantly. He is jovial. *Plesir*, the Malayized Dutch word *plezier*, "pleasure," as well as a truly breathtaking number of other Dutch words denoting comfort streaming into modernity, and modernity streaming into manners, looks, and fashion, carry the novel forward:

> *Dari mana kamoe?'. . . dari plesir . . . Dag Biroe . . . Heerlijk, kata Raden Adjeng . . . Nee, Lieve . . . Zus . . . Raden Adjeng . . . merasa koerang senang dan berkata: En dan? . . . Kijk! . . . tetapi toch* [Where are you coming from? . . . from pleasure . . . Hallo Biroe—Wonderful, says the princess—No, darling . . . Sister—princess . . . does not feel very happy and says: And what?—Look!—and yet].[42]

"*Nee, Lieve*" (Dutch for "No, darling") is translated, and most probably by Mas Marco himself, as *tidak djantoeng hatikoe* (Malay for "no, my love") and is placed in a footnote. Student Hidjo and his girlfriend, Biroe, could not be caught by their words, their manners, or, as we will see, by their clothes; they cannot be stopped in a certain mood or in a certain place. Watching their elusiveness, one may say that they travel dandy ways and speak dandy language.

Surveying a landscape (almost exclusively a modern landscape), as Hidjo and his friends very often do, they consistently survey *PANORAMA* (often capitalized this way in the text).[43] Student Hidjo eats in *hotels* and *restaurants*. Benedict Anderson defined the early nationalist thought in Indonesia as a time of a dawning light; he pointed out how frequently words like "light" or "dawn" appeared in Indonesian texts of the early twentieth century.[44] Hidjo and Biroe definitely liked the light dimmed.[45]

When in their favorite *restaurant*, under the dimmed and electric light, Biroe and Hidjo talk in muffled voices, of course, and they drink *limonade*, Dutch for "lemonade."[46] Whatever might have happened to the *Kromoblanda* scheme, the hyphen, the abyss between the Dutch and the natives, one is led to feel that these two dandies could do a lot, even drink Tillema's mineral bottled water, if they only wished to.

Hidjo wears trousers, jacket, and tie, *à la mode*, and, no doubt about it, shows off at least two pens in his vest pocket. Biroe flashes "silk yellow *badjoe* [blouse], smart *kain* [skirt], *soebeng berlian* [diamond earrings], and *selop model baroe* [sandals of a new model]."[47] As on the mannequins in the shop windows, flagrantly and without allowing a mistake, there are price tags on everything. Biroe, for instance, and Woengoe, her friend, in one scene wear "silk *badjoe* priced at Fl 200; . . . Solonese *kain* priced at Fl 40."[48]

This is a novel about dolls that work not as it was planned, that are on the verge of starting a life of their own. It is a story of modern natives, as we read through it, penetrating Dutch houses and (Mrs. Catenius would certainly understand the connection) penetrating Dutch women. Hidjo, on his way West, as mentioned, and until the end, is hunted by sex-hungry Dutch women. There seem to be no exclusion zones. The Dutch bodies are not safe, as Dutch hotels are not safe either. When student Hidjo reaches Holland, he is waited upon in a Dutch hotel by a Dutch waiter, a "servant" Mas Marco calls him (in Malay: *boedak*),[49] and, we may safely guess, he wears a livery.

Hidjo's life in Holland, besides his studies, is largely made of theatergoing ("opera Faust, a show much loved by the Dutch," "Lili Green," and others), picnicking, and tramway-riding (tramway lines nos. 7, 2, 8, and 11). Hidjo definitely is not "neat." Dutch women are removed of their clothes in the process, and, on several occasions, they are described as "heavily sweating."[50]

The scenes in *Student Hidjo* move fast: between Holland and Java, between PANORAMA and PANORAMA, and the roles switch even faster. At one of the novel's climaxes, at a traditional, native Javanese gathering in a provincial town in Java, in *kraton*, a native regent's palace, a local high Dutch government official can no longer restrain his years-long urge "to dance like a native." The excited Dutchman begs the regent's son and Hidjo's friend to lend him his genuine native headcloth. He gets the headcloth, puts it on, and dances—like a Dutchman, faking a Javanese—to the roaring laughter of all the natives present.[51]

The novel ends as smoothly as it began. In spite of the submarines and the women, student Hidjo returns from the Netherlands and proposes to a Javanese girl—*by the way*, not to Biroe but to Woengoe (Biroe, as a true she-dandy, smoothly engages herself to a brother of Hidjo's new wife). Another climax comes as Hidjo's friends and sweethearts celebrate their being together in a varsity park in Solo, the ancient princely town of Java. They drive around in a car and honk. They drink *limonade* and watch a historic congress of the *Sarekat Islam*, the first modern Indonesian mass political organization. Everything is possible. Photographs are taken.[52] Long gone, so it seems, is the cute order of the Indies ladies' doll collection. Whoever might try to discipline this becomes dizzy.

Mas Marco liked to be seen in Western clothes, all-white best of all.[53] He liked to tell jokes in his journals, and he numbered the jokes meticulously—

"joke no. 1 . . . joke no. 35."[54] He liked to collect photographs and send postcards, but not like Kartini did. In his journal *Saro Tomo*, in 1915, for instance, he made some money by publishing, selling, and distributing, for 25 cents apiece, portraits of "two Dutch journalists who hate the people of this land more than anyone else."[55]

Mas Marco, as we have mentioned already, paid for his jokes. In 1927 he was exiled to the infamous internment camp in New Guinea at Boven Digoel, and there he died, in 1932, of tuberculosis.

Mas Marco was part of a segment of the Indies culture that, just at that moment—told what to wear and how to talk—cried out, sometimes in irony. Mas Marco's statement of an Indonesian dandy very probably had not been exceptional, but little remained on the record.[56] Thousands of natives, some in knickerbockers, this we know, roared with laughter through the colony, every night, for instance, at *komedie stamboel*, "Istanbul comedy"—a genre revitalized by the early twentieth century through the shock of cinema, a patchwork musical—played and composed by the Indies natives, mixed bloods, Chinese, and few more-than-avant-garde "foreigners" (e.g., Willy Klimanoff alias A. Pedro, "a white Russian born in Penang"),[57] with dark skin, light skin, red skin, and yellow skin Shylock, Hamlet, or Sherlock Holmes, in the most extravagant and grotesque clothes, loud, flagrant, irreverent, dazzling, for all, and played in a language that might easily be Mas Marco's,

> En satoe en satoe en satoe dat is een
> En batoe en batoe en batoe, dat is steen
> En roti en roti en roti, dat is brood
> En mati en mati en mati, dat is dood.

The lyrics (as well as the clothes, the bodies, the blood mixture, and the manners) made a fool of anybody who might suggest a standard to it, a bridge or a translation:

> And one is one is one
> And stone and stone and stone is stone
> And bread and bread and bread is bread
> And dead and dead and dead is dead.[58]

The Death of the Dandy

Diving into things modern, and yet keeping elusive, demanded the courage of a guerrilla fighter. If one did not come up with the right kind of a guise or joke at a given moment, one might die.

In his essay on Paris of the 1850s, Walter Benjamin wrote, using the French term *flâneur* for "dandy":

The operetta is the ironic utopia of the capital [which the *flâneur* inhabits] . . .
Paris [as the *flâneur* enjoys it, is] a submerged city, more submarine than subter-
ranean. . . . [The *flâneur* lives a] "deathly idyll" [ruled by] the law of dialectics
seen at a standstill. . . . The last journey of the flâneur [is toward the] illusion of
novelty . . . the gaze of the allegorist that falls on the city is estranged. It is the
gaze of the *flâneur*, whose mode of life still surrounds the approaching desolation
of city life with a propitiatory luster. The *flâneur* is still on the threshold, of the
city as of the bourgeois class. Neither has yet engulfed him; in neither is he at
home. He seeks refuge in the crowd. . . . In the *flâneur* the intelligentsia pays a
visit to the marketplace.[59]

In Benjamin's view, the *flâneur* was a commentary on a particular period of
Western bourgeois culture. *Flâneur* was temporary, and *flâneur* left, Walter
Benjamin wrote, at the moment when Engels entered. When the solid or-
ganizational and conceptual entities of labor and capital were established,
flâneur had to die; "bohemianism" ended with "professional conspiracy":
"The *Communist Manifesto* puts an end to [the *flâneurs'*] political existence."[60]

The famous Indonesian Generation of 1928, Sukarno, Hatta, Sartono,
Sjahrir, or Amir Sjarifoeddin—names known until now to every Indonesian
pupil—was, arguably, the first group of native professional politicians (or
"conspirators," colonial police might say) in the Netherlands Indies. These
people were also arguably the first generation that dressed in public consistently and completely in a Western way—"completely" meaning trousers and
shoes, jacket, tie, and, ideally, hat and mustache. A famous photograph of
young Sukarno at a Bandung café, sometime shortly before or in 1928, is one
of the countless witnesses to this.[61]

In its early years, in the mid-1920s, the Generation of 1928 was indeed a
colorful, and often dazzling, fast-moving crowd that enjoyed being seen as
connoisseurs of Greek philosophy and the French Revolution as much as of
wayang, the Javanese shadow puppet theater (if they happened to be Javanese), or, and this very much so, of Hollywood. They had much of *flâneurs* in
themselves. However, to use Benjamin's phrase, Engels had by the late 1920s
already entered the Indies.

Indies labor and capital became standardized, however perverted (meaning
colonial) the standard might have been. It was increasingly difficult for a
native to resist the avalanche. The machines of the colonial Indies and of the
Dutch colonial state were producing neatness on a mass scale. In order to
face this power, so it appeared,[62] an Indonesian standard of its own, a standard language, and a standard ethics, a specifically Indonesian kind of neatness, was needed.

The concept of *sini*, truly "us," versus *sana*, truly "them," was conceived
with a sense that time was running out. Time was flowing dangerously,
even for the Indonesians, and had to be dammed. *Sini-sana*, in the late

1920s and especially in the 1930s, became the most fierily propagated tenet by Sukarno and the majority of the Generation of 1928. Everything that could make the contrast between *sini*, "us," and *sana*, "them," sharper, the cleavage line cleaner, the space in between more barren, became good for the struggle for freedom.[63] The Generation of 1928 spoke and wrote correctly—not Mas Marco's and "comedy Istambul" peddler's Malay. Malay became "Indonesian," both languages' vocabularies were identical but the will was different.

Organisasi, "organization," became at the same time to the Generation of 1928, an all-powerful or at least an all-explaining and all-defining word. Concern for clothes remained very strong among the freedom fighters. But no more were they dandies. Not to be a dandy—somebody in between, an enigma—became a battle cry.[64] Sukarno later remembered (and the ghost of Mrs. Catenius reappears with his memories) as one of his big successes at the time: "establishing the *pitji* (*pici*), that black velveteen cap which is my trademark, as our symbol of nationalism . . . I [said], let us hold our heads high bearing this cap as a symbol of Free Indonesia."[65] Sukarno remembered that at the time he also "proposed that all members [of his party] wear a uniform." In this he failed.[66] Facing the colonial mammoth of neatness, the politically awakened Indonesians became anxious that nothing—study clubs, sport unions, political parties, hats, shorts, jackets, and trousers—was left disorderly and uncertain. The fierceness of the struggle bred a neatness of its own kind. Who might blame the fighters that they lost a sense of humor on the way?

At the turn of the 1920s and the 1930s, the Dutch colonial state and the Generation of 1928 clashed. Sukarno and several other top leaders in his movement were arrested, and Sukarno's organization was in tatters. New ways—or manners, one might say—were being desperately sought, but they were difficult to find. In the modern times, it seemed fairly settled how one should make politics, behave, and dress. A new ideological, radical, and very short-lived concept was tried, an Indies mode of Gandhian *swadeshi*. The Indonesian nationalists were to manifest their tenacity in the struggle by wearing clothes made at home.

A historian described a political rally of Sukarno's party, now called *Partai Indonesia (Partindo)*, "Indonesian Party," in the early 1930s:

> All but two of the *Partindo* leaders were dressed in sarongs, which was to be a regular occurrence at subsequent meetings. Sartono [the party leader] spoke at length on the need for Indonesians to wear clothing made in their own country and promised that the promotion of *swadeshi* would be one of the major tasks of *Partindo*. . . . Sartono warned that those who were not dressed in *swadeshi* clothing, or at least did not wear a headcloth made locally, would in future be refused entry to *Partindo* meetings.[67]

It was of course supposed to be deadly serious. Some said that the Indonesian leaders, most of them college educated and very modern in their lifestyle, looked silly.[68] One feels that clothes were becoming more powerful than the bodies and the souls. Each time the leaders were publicly to meet among themselves or with the masses, they changed clothes in haste and in order to appear natural.

In 1933 Sukarno was exiled indefinitely, and he stayed in exile, in fact, until the end of the colonial era. Many prominent members of the Generation of 1928 (including at one point Sukarno himself) renounced their political activity of the past. They entered a more moderate political, better say social, work or simply began to practice some other profession. With most of the radical leaders away or silent, the rest of the era, until the Second World War, was murky, muffled, hopeless, flat, and drab.

Soewarsih Djojopoespito, a woman, a little younger than Sukarno, who used to work as an activist for the Generation of 1928, lived through the gray years with intensity. Through the 1930s, Soewarsih made a living as a schoolteacher, and, by the end of the period, she wrote *Buiten het gareel*, "Out of the Harness," a novel that is sometimes considered the best novel from the pre-independence period written by an Indonesian.[69] In her book, published in 1940 and written in Dutch, Soewarsih expressed the gloom. There were just whispers around, she wrote, some sadness and much nostalgia for the past, when Sukarno was still free. It is an autobiographical novel, and Soewarsih herself appears in it as a modern young woman. Strikingly enough, one may read the novel through and again, very carefully, and still it is not clear how the young woman was clothed[70]—so bad, indeed, the modern times turned out to be. After Mas Marco, Sukarno, and the *swadeshi* attempt to save the style and the struggle, the clothes seemed to lose their voice.

Sutan Sjahrir had been close to Soewarsih. They shared the student culture of the late 1920s, and they became allies in the politics afterward. Sjahrir, in a more complex way, was close to Sukarno, too. He was a leader of the Indonesian emancipation movement at the same time as Sukarno. He was exiled by the Dutch only a few months after Sukarno was, in 1935. He was to became Sukarno's major rival after Indonesia became an independent state. Throughout, their manners and fashion were very different, too.

Sjahrir's biography is full of references to clothes. His sister Rohana was, among many other things, *agen Singer*, the representative for Singer sewing machines in her native town, Kota Gedang, in West Sumatra. Rohana's life dream, when young, was to take part in one of the fashion exhibitions in Europe.[71] At various times in his dramatic life, Sjahrir himself had his own Singer machine, and he worked hard on it. When he was in exile in Banda Neira, an isolated small-island place in eastern Indonesia where he spent the years between 1935 and 1941, he not only adopted four local children, two boys and two girls, at the ages of a few month to mid-teens, but he also patched trousers and even sewed dresses for the girls, "some quite fancy."[72] Sjahrir also subscribed to at least one fashion magazine, and it was sent to him, through the colonial censorship, to his place of exile.

When I wrote few years ago about Sjahrir's early years, I also noted that

> much of the story of Sjahrir's childhood may be told through the way his clothes and costumes changed. Sjahrir got an elementary Islamic education from his father, and one can picture the boy's tiny white-garbed figure at the *jaksa's* [native judicial official] feet. . . . Between the ages of six and twelve, each day after classes in his Dutch-style school, Sjahrir changed from the prescribed trousers and jacket and stepped across the street to attend Koranic courses. And there was a soccer uniform, too, and a dinner jacket, probably, prescribed for the *soires dansantes* at the *Hôtel de Boer* [where Sjahrir made some pocket money playing violin].[73]

The manuscript editor remarked on these few lines: "But, well, what were his ordinary clothes?"

There was something of the student Hidjo in Sjahrir. Like Hidjo, Sjahrir was elusive and thus able to move across the lines. Like Hidjo, Sjahrir went to Holland to study, in 1929, and was said to "blossom" in the West. He enjoyed things Western, like Dutch-young-socialist-student picnics with dancing in *Kijkduin* varsity park near The Hague.[74]

Sjahrir, like student Hidjo, had a capacity for removing himself (and others) of their clothes easily. As a Dutch friend of his from their time in Holland remembered:

> I lost track of Sjahrir for some time. Later I gathered from his stories that he, in search of radical comradeship, had wandered further and further left, coming to rest at last with a handful of anarchists who had managed to keep themselves free of all capitalist taint by avoiding any profitable work, and who survived by sharing everything with each other except for toothbrushes (insofar as there were any) but including contraceptives. He reemerged from this rather quickly and without damage, and afterward his interest in socialism took more practical forms.[75]

This, however, in the Indies as well as in Holland, was already long into, in keeping with Walter Benjamin, the Engels era. When Sjahrir came back from the West, late in 1931, he still deeply impressed his close friends with the way he dressed. According to Soewarsih, "he wore huge, clearly borrowed slippers, a stained sarong, a jacket, and a skullcap."[76] According to the Soewarsih sister, Soewarni, "When I saw [Sjahrir] in such a pitiful state, I had to hold my tears back. The sleeves of his jacket were too long, and it was difficult to guess its original color."[77]

Much earlier, in his school tie, *surau* garb, or soccer sweater, Sjahrir never seemed fully to belong to one particular group. His quick changes in venues,

Advertisement for "Modern Tailor," Batavia. (R.S. Hermani,
Kleermaker Modern [Batavia: Balai Poestaka, 1940])

manners, and clothes dazzled a surveyor. Now, in his too-big skullcap and clearly borrowed coat, he was, in fact, neat and standard. However grotesque, his attire now appeared to fit the ordinary and the genuine. From now on, and for the rest of his life, Sjahrir would be costumed as a political leader, and thus placeable in, and appropriate to, a hierarchy. These, indeed, were the 1930s. In contrast to the *swadeshi* figures of the time, Sjahrir still appeared very much swirl-like, and dandy-like, Mas Marco-like. Contrary to the elusive Mas Marco, however, Sjahrir appeared aloof.

Sjahrir, so it seems, if he was a dandy, was a dandy born after dandy time.

There had been, in the Indies of the 1930s, a group of young men and women who still dressed flagrantly modern. In their own way, however, these people reflected, even more than the *swadeshi* leaders, the plain heroes of Soewarsih, or Sjahrir, the grayness of the time.

Most of the Indonesians of this group were born around the time when Mas Marco wrote his *Student Hidjo*; and so they were about ten years old in 1928. By the late 1930s, they clustered around students of either the Medical School or the Law School, the only two college-level schools in the colony capital before 1940. Practically all of them, even when not students at one of these colleges, had a Dutch-style high school education. Almost as a rule, they came from well-off native families, they were intelligent, and they did many dandyish things.

Unitas Studiosorum Indonesiensis, USI (in Latin), "Union of Indonesian Students," was an organization of which most of the more outstanding in the group were members. The *Liedboek van USI*, "Song Book of USI," from the late 1930s, most probably from 1940, was clearly published to give these young people something to sing when they were together.

The first song in the *Liedboek* was *Gaudeamus*, the Latin international anthem of students. Two songs followed, also anthems of sorts, both in Dutch: *Wij Usianen*, "We the *USI*-members," and *USI Lied*, "Song of USI," from which I quote:

> Are we, the *USI*-members, weak and lazy, and do we complain every day?
> No! . . .
> Does a joyful fire burn in our bodies, and does our blood flow like wine?
> Yes!
> We are *USI*-members, We are students
> Our blood boils
> We do not swim in money
> We live by fun and zest in life.

Other songs in the "Song Book of *USI*" were, for instance,

> *Mutsenlied* [A Hat Song], *Tramp, tramp, tramp, The Boys are marching, Jingle bells, Polly-Wolly-Doodle, O, Die vrouwen* [Oh, these women], *Boeng Boeng Pait,*

Reg-Reg, Drink! Drink! Come Brothers Drink!, Sari Nand, Sipping cider thru a
straw, Who's Emma, Abdul the Bull Ameer, Oh, My Darling Clementine, Groenen-
lied [an initiation song in a student club], *John Brown, Botol Maropat Suhi,* and
Stein Song [A Pitcher Song].[78]

These people let us know a depressingly great deal about themselves
through a caricature. Unlike in the world of *Student Hidjo,* in these people's
world the simplest and most basic colors, sharp and straight lines, and hard-
to-doubt standards seemed to be available. No longer could a headcloth be
easily passed from one head to another. Now, so it appeared, each man and
woman of the group, each movement of their young bodies and of their
group as a whole, each piece of their flagrantly modern clothes, could easily
be fixed into a well-defined space. Despite the noise of their songs, there did
not seem to be much humor.

Instead of elusiveness, these people produced a compartmentalized, and
almost folklore-like, picture, evoking a knowing smile, "ah students."[79] The
manners and clothes of these young Indonesians, however flagrant, did not
dazzle a surveyor. They did not disturb the Indies' mainstream(s), either
Indonesian or Dutch. On the contrary: their songs, manners, and clothes
merely appeared to confirm their status. They were sadly proper.

Theirs appeared, sadly, to be an import culture. Unlike Mas Marco, who
died in exile—and unlike Sukarno or Sjahrir, who had to live in exile until
the end of the Dutch colonial rule—these young people were not given an
opportunity to prove that, in their pranks, they might be serious.

They reflected their time and were certainly sensitive. They distrusted the
emotional and the (temporarily) defeated *sini*-versus-*sana* or the *swadeshi*
nationalism of the slightly older Generation of 1928.[80] Through the dreari-
ness and gloom of the 1930s, these intelligent and well-educated people tried
to keep their souls and bodies, songs and clothes to themselves. If they
wanted to be dandies, however, they, even more so than Sjahrir, arrived after
the dandy time. The more they tried, the more they underscored the image
of how a native nascent, Western-educated, modern elite in the colony might
be expected to behave, to move, and to dress. They were most (or only)
visible in their classrooms and their student cafés, naturally. They might be,
and they were, easily dismissed as merely a "billiard culture."

The Parade

An exalted and high-strung theatricality came to the Indies with the sudden
fall of the Dutch empire. In early 1942, as the Japanese armies entered the
Indies, they brought with them many things. Among them was a new awareness of clothes.

The Japanese were expected, or at least had been presented in the colony before they came, as sort of overdressed and overequipped dolls. In a journal for the native police agents, in November 1940 a long and detailed article was published about how the Japanese would come:

> They will have a pistol model *Schmeiszler* with 24 pieces of ammunition each; some of them will carry machine guns and even small cannons. They will wear "*overalls*" or *badjoe tjelana* [Malay for shirt and trousers in one piece], an outfit like that is often worn by our mechanics and officials. In the pockets of each of the *overalls* will be 'tinned food' (which does not spoil), enough for 14 days. In their helmet (hat), they will have *radio-ontvangers* [Dutch for radio-receivers] for listening to orders from the airplanes that bring them here.[81]

Indeed, the Japanese were not completely unlike the Dutch one historical era before. And, on the whole, like many other things in the forty-two-month Japanese occupation that followed, the Japanese politics of manners and clothes was highly dramatic, but, at the same time, carefully building on the previous period.[82] Theatrical, loud, and flagrant to the extreme, the Japanese era in the Indies (renamed officially Indonesia) also distinguished itself as a high season of standards and neatness.

Among many other things, the Japanese brought about a systematization of clothes in Indonesia, which the Dutch, however much they had tried, had never been able to accomplish. The Japanese as they arrived had very specific rules for their military uniforms, down to regulation tobacco pouches for army engineers, navy officers, and air-force pilots: "army footwear, navy footwear, leggings, gloves and gauntlets, cases and knapsacks, army and navy belts and engineer's equipment, holsters, gasmasks and life jackets, armbands, flasks and mess-tins; field instruments and commemorative towels, bullet-proof vests."[83] The Dutch in their time had aspired for their colonial army subalterns to have their uniforms well ironed, starched, and all-white, even when off duty. The Japanese brought the same idea almost to its absolute perfection. They also radically socialized the almost Kantian idea of fashion. Virtually everything in Japanese Indonesia became a uniform. The Japanese idea of the Indies and empire was a uniform idea, and the uniformity was consistently, efficiently, devotedly, and often brutally fashioned throughout society.

It was often a cruel occupation. Japanese hoarded cloth, and there was, at the end, a famine-like lack of clothing. Yet order and mode was maintained, and even heightened, by the scarcity.

Japanese soldiers wore uniforms, and so did Japanese civil servants. Indonesian civil servants in Japanese employ wore uniforms, as did the Indonesian *Peta* and *Heiho* militia and auxiliary military men, the members of the *Seidenbu* propaganda corps, the teenage Indonesian nationalist unions cooperating with the Japanese, the medium young, the adult, and the elderly Indonesian nationalist associations—each group in its own way. Including all the horrors, this was a massive and well-ordered fashion show. Even being a *romusha*, a corvée laborer, a half-naked, half-dead slave at the bottom, at the railway, harbor, airport, or highway constructions of the Japanese total war, one's muddy and bloody feet, one's eyes, shorts, "volunteer" cap, even one's naked chest (if one happened to be a man) was a uniform. One was standard. And nobody laughed.

Sukarno was picked up by the Japanese from his exile, brought to Java, and, step by step, elevated to the top of the collaborating nationalist political hierarchy. Hatta, Sartono, and other prominent members of the Generation of 1928, as well as almost anybody of some importance in the prewar Indonesian nationalist movement, actively participated in the Japanese "New Order" and showed up, properly clothed, at various levels of the political pyramid.[84]

The engineers, and tailors, of the *sini*-versus-*sana*, "we" versus "them," and also of *swadeshi* campaigns, were present, either physically or in spirit. The specters of Couperus, H.F. Tillema, Mrs. Catenius, and Kartini even, were felt to be around. Only, perhaps, Mas Marco was missing.

As in the early-twentieth-century Dutch colonial photographs of factory sites, railway stations, and modern houses, the unruly was carefully cut out. It was a historic moment when Sukarno, on his way to becoming the first great Indonesian statesman, added a soon-to-be famous set of olive-green, granite-black, and (this especially) snow-white, well-ironed, and starched uniforms to his student-nationalist (Mrs. Catenius-favored) *peci* black skullcap. Sukarno's explanation was logical, and it smelled of the late-colonial past: the people in rags, the people in pain, the people who suffer and may go astray, all need their leaders standard and neat.[85]

Gold-buttoned uniforms of the leader were to come a little later, as was an Indonesia where everybody was expected to move, appear, and dress cognate to Sukarno's standard. No other image of Sukarno was to be available from the Japanese times until the mid-1960s, when he began to lose his power. Only then, dissident, frightening, and liberating photographs began to emerge of Sukarno aging, without a *pici*, bald, in crumpled trousers and an undershirt.

Dressing gaily and behaving bohemian during the Japanese occupation, amid the apocalypse of standard and uniform, made even the billiard generation appear, for a while at least and to some extent, serious and meaningful.

I interviewed some of them later. During the first days after the Japanese arrival, they said they learned, at last, to play bridge truly well, because the Japanese closed the schools and made it impossible to socialize outside one's own house. For others in the group, this was the time to buy really cheap and recent second-hand Western books, as the Dutch, escaping the avalanche, or about to be taken away into the internment camps, were selling everything in bulk.[86]

These special young people, dandies born after dandy times, had listened to the Allied radio, illegally and under the threat of beheading. They were, so they said, eager for news of Japanese defeats, but they searched equally for Artie Shaw, Benny Goodman, and Cab Calloway.[87]

There is a photograph of some of these Indonesians. A small group of young men and a woman sits leisurely in rattan chairs around a coffee table, on a verandah of, evidently, a modern house. There are flowers in a few white flower pots, a well-mowed lawn, and a fence. Everyone in the group is dressed fully European—in all-white, and one almost can smell the men's buttonhole flowers.[88]

Some of these young men and women moved on into the politics of the national revolution when the Japanese were defeated by the Allies on the battle fronts and Indonesian independence was proclaimed on August 17, 1945.

Some of them even appeared to maintain their style. Being often out of place, they provided highly desirable flashes of absurdity to the revolution that followed after August 1945. They helped to disturb the increasingly uniform language, policies, and manners of the political mainstreams in the month that followed, be it the Indonesian mainstream of the Generation of 1928, that soon took over the revolution leadership, or the Dutch main-stream, when the Dutch began to return and tried to push matters in Indo-nesia back to the Indies time. Many of the billiard generation worked on Sjahrir's staff when Sjahrir became the prime minister of the republic in October 1945.

They tried to maintain their style. Mr. Posthuma, a serious middle-aged Dutch negotiator, one of the first to have met the new Indonesian leaders, after many years, and still with noticeable bewilderment, remembered some of Sjahrir's young lieutenants, "amidst the raging battle of revolution," asking such strange questions "about Benedetto Croce," for instance, or "how T. S. Eliot was doing lately."[89]

Making the revolution, however, seemed to be truly a serious business, and these young people increasingly came to believe it. Those who did not were very soon easily dismissed. John Coast, a young Englishman who came to help the revolution, a bit of a bohemian himself, saw some of these young-sters in the offices of the Revolutionary Indonesian Republic, after three years of fighting and negotiations with the British, the Americans, and the Dutch, in 1948. Coast described young men sitting in "steaming offices wearing a suit with collar and tie," and young women in "very short skirts, which showed for the first time the unfortunate broadness of the Javanese female knees. Such costumes had the effect of making this already youthful-looking people appear as overgrown children on the Dutch model."[90]

Dead Indonesian youth with his weapon, a sharpened bamboo spear,
Java, 1947. (Niels Douwes Dekker Collection)

There was also, of course, and very significantly, "the other youth." The boys and girls—painted by Soedjojono[91] and described by Anderson[92]—the young men and women who grew up during the forty-two months of the Japanese occupation, the long-haired and the half-naked children of the revolution. They without a doubt did most of the fighting and dying.

The "other youth" also presented itself by its distinctive fashion—its non-haircut worn loose, *laarzen* boots worn barefoot, samurai swords worn like a stick, *bamboe runtjing*, sharpened bamboo stick, worn like a rifle, headbands worn bloody red, the ammunition belts worn crisscross around a naked chest. Were they also born too late?

There is a long scene in a book that has already been quoted, Robert Musil's *The Man Without Qualities*.[93] The hero takes a stroll through a romantic, unspoiled countryside somewhere near Vienna, Austria. He walks and talks with a woman friend—as educated, brilliant, beautiful, and of society as he himself. There is a growing physical attraction rising between them. How difficult it is nowadays, the man suddenly realizes, to get naked. Being removed of one's clothes, would that not strip one of the signs one carries as a part of this world of radiance and belonging? Getting naked, would that not make one necessarily leave society and descend somewhere into a subhuman world? Would not one disappear to one's society? Can a modern man, in the modern times, make love to a modern woman without keeping his clothes on?

Mas Marco, Tillema, and Catenius, as well as Sukarno, Hatta, Sjahrir, and most of the others, are dead. The dandyish youth of the 1930s, old men and women, now, if they survived, play cards rather than billiards and try to keep remembering the past, sitting under the enlarged photographs of their by now very properly clothed heroes and, often, martyrs. Sukarno, Sjahrir, Hatta in the portraits wear gray suits, mostly, or gray uniforms (also the photographs are fading), and, if any, a very quiet tie.

The survivors of the "other youth," their long hair gone, might still on each August 17 celebrate the anniversary of the Indonesian revolution. They have a place reserved at the head of the official parade, or on the tribune that stands opposite the palace and the president, and they march or watch as well as they can, in shirts and trousers as all-white as possible. Looking at them, one might be struck by an embarrassing idea: are not they scared to get naked?

FIVE

LET US BECOME
RADIO MECHANICS

The Metaphor

I N 1900, twenty-one-year-old Kartini, who liked trains and photographs, and who perhaps best liked to think of herself as a young wild horse, a daughter of the Javanese regent of Japara, a woman who, after her death, became an anointed mother of Indonesian nationalism, wrote to one of her friends and mentors, Henri van Kol, Engineer, a Dutch parliamentarian and a European socialist expert on colonialism: "It would seem as though an invisible telephone cable ran from here to Lali Djiwa and back again."

By "here" Kartini meant her father's palace in Java; "Lali Djiwa" (Javanese for "To Forget One's Soul/Self" or "Utterly Delicious Rapture")[1] was van Kol's residence after he and his wife retired from the Indies, in Prinsenhage in Holland.[2] A few months before she wrote this, to another friend, a Dutch feminist and employee at the Amsterdam post office, Kartini sent a letter about the nearness she felt between herself and her aristocratic father—as if a "telephone cable," she wrote, ran between their hearts.[3] It was the opening of a new century—seventeen years after the Krakatoa explosion and nine years after Dr. IJzerman prospected for the railway in the jungle of Sumatra—and Kartini expressed what appeared to be a distinct mood of the time. Hers, it also appears, was a good case of what Søren Kierkegaard, in another time and culture, defined as "enthusiasm of imagination," a nostalgia for the mythical, a stepping "forth in a new form, that is, as metaphor."[4]

As Kartini was writing, the Netherlands East Indies was—at least so it seemed—on the verge of becoming a part of the modern world. As we have seen there were many modern things in the Indies by that time, and also, already, almost half a century of wire-communication experience. The first telegraph line was laid in 1856 between Weltevreden, a mostly Dutch residential suburb of Batavia, and Buitenzorg, the summer place of the government, about 80 kilometers from Batavia.[5] Forty-four years later, at the turn of the century, when Kartini was writing her letters, half a million long-distance telephone calls were placed through Java every year.[6] In 1900, there were 925 telephone subscribers in Batavia, 371 in Semarang, 568 in Surabaya, 123 in Yogyakarta, and, among other places, 43 in Japara—Kartini's hometown.[7]

In the spring of 1902, another big step forward was made. The first Dutch

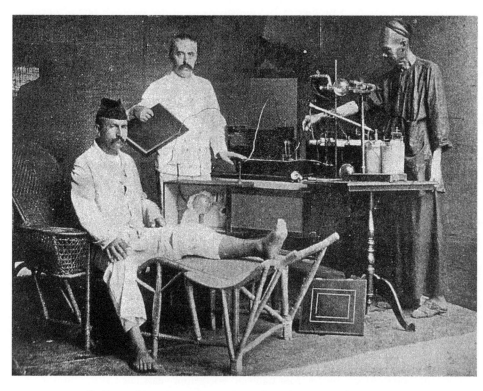

Radiographic examination in the military hospital in Aceh, 1898.
(*Gedenkboek van het Korps, Marechaussee van Atjeh*)

demonstration of the "Marconi Wireless" took place at Villa Jacobson near *Kurhaus*, the "Resort Hotel," in Scheveningen on the Holland coast. The host of the event was Mr. Weiss, a businessman from Batavia. The antenna was turned in the direction of H.M.S. *Evertsen*, a ship laying at anchor off the coast about 80 kilometers away: "After a while, a light bell-like sound was heard, and a little rod, softly tapping, printed dots and bars on a strip of paper unwinding from the receiver. 'That is done by Evertsen!'"[8]

This, clearly, was a time of beginning, in Holland as well as in its colony, and Kartini appeared to enjoy herself immensely. Both her correspondence and her life were crowded—as much as with trains and photographs as with "wires" and "calls" and "cables" that she, her family, and her friends sent and received. Franz Kafka, a Jew in Prague, four years younger than Kartini, caught in a web of a different, Austrian imperial, mail service, complained to his Czech girlfriend that, in his letters, "written kisses don't reach their destination, rather they are drunk on the way by the ghosts." "The telegraph, the telephone, the radiograph," Kafka believed, made things even worse: "these are evidently inventions being made at the moment of crashing. . . . The ghosts won't starve, but we will perish."[9]

Kartini, contrary to Kafka, clearly handled the ghosts with ease; she employed them. Table turning and clairvoyance, ghosts with jobs to do, were a natural and well-functioning part of her communication system. Spiritism, to Kartini, worked faster than trains, and at least as fast as the wire.[10] It was a means, moreover, at the same time traditional and modern: she learned it from her aristocratic Javanese stepmother as well as, and at the same time as, her most *à jour* Dutch friends and especially van Kol, Engineer.[11] The spectral mail, to Kartini, more than anything that could be offered at the time, was international, and it was multilingual: through the medium, Kartini sent and received messages in Javanese, Malay, Arabic, and, most frequently of all, in Dutch, the language of the future.[12] In the case of Kartini, telegraph, and telephone—and as the new century opened—we may talk with Kierkegaard about "pantheism of the imagination."[13]

Kartini died on September 17, 1904, four days after she gave birth to her first son. She did not live to see the Indies wire extending and then turning wireless. We may only speculate about what she might have thought. In 1918, when the great war ended in Europe, and when Kartini's son, Singgih, was fourteen, the first direct wireless communication was realized between the Malabar Hill above Bandung, in West Java, and Blaricum, in Holland.[14] When Singgih was twenty, in 1924, the Indies press agency *ANETA* had been given a concession to receive news from Europe by radio. Since that year, "from 6 A.M. till 1 P.M., *Aneta* broadcast almost without interruption; its radio messages were received in Bandoeng, Djokjakarta, Semarang, Soerabaja, Makassar, Menado, Balikpapan, Bandjermasin, Medan, Palembang,

Radio transmitter in Amsterdam-School style on the North Sea coast
in Meijendel near the Hague used for the connection with the
Netherlands East Indies, 1925. (*Gedenkboek ter herinnering van het
tienjarig bestaan van de Nederlandsche vereeniging voor radiotelegrafie,*
1916–1926)

Padang, and Kotaradja."[15] When Singgih was twenty-five, "a group of [Dutch] enthusiastic amateurs established *Bataviasche Radio Vereeniging*, 'Batavia Radio Association,' and laid a foundation of the radio broadcast in the Netherlands Indies."[16] Two years later, the Dutch queen addressed her subjects in the Indies wirelessly for the first time, speaking from the experimental studio in Eindhoven, South Netherlands.[17] (The same year, the operetta *Hallo!—Batavia* was played at the *Plantage* Theater in Amsterdam with a hit couplet: *Ik sprek met Oost / Ik spreek met West / Mijn apparaat is toch zoo crimineel best*, "I speak with the East / I speak with the West / My apparatus, no doubt, is criminally best.")[18]

In 1933, *Philips Omroep Holland Indië*, or *Phohi*, "Philips Broadcasting Holland Indies," and a year later *Nederlandsch Indische Radio Omroep Mij* or *Nirom*, "Netherlands Indies Radio Broadcasting Co.," from Hilversum and Batavia, began their first regular broadcasts to the colony.[19] When Kartini's son was thirty-eight, in 1941—the last year of the Dutch rule over the Indies—about twenty radio stations in eight cities of Java and Sumatra broadcast from early morning until midnight.[20] By January 1, 1941, there were 101,868 registered listeners throughout the colony.[21]

The Indies were wired, and upgraded to wireless, during the four decades after Kartini's death. Yet, reading carefully the late-colonial Indies magazines, newspapers, novels, poetry, and political statements, one is struck by a decline of the Kartini-like, early-century, mail-and-wire enthusiasm of imagination. Few flashes might be noticed—an otherwise unknown worker-correspondent of a trade union journal *Keras Hati*, "Courage," for instance, signed his report as *'Microfoon!,'* "Microphone!" But, strangely, there is nothing in the text about communication.

It was Nietzsche who said that literal truth was merely dead or fossilized metaphor.[22] Kartini's metaphor of the Indies wire and wireless, in the forty years after her death, indeed appeared to be rapidly dying, or at least fossilizing, into a literal truth.

Wim Wertheim, in the late 1930s, lived with his wife Hetty in the Indies world that Kartini might dream about. He was a Dutchman teaching law in Batavia, and he and his wife were among many other things a prominent radio couple in the Indies before 1942. Hetty, besides writing about the arts for the newspapers, gave regular soprano concerts on the radio. Wim, besides being a college professor, accompanied his wife on piano.[23] They remained in the Indies until the end; in their diaries they described the demise of the era:

> The gasworks stood on fire, there was a heavy black cloud of smoke, the big antenna of Radio Holland lay across the road, harbor installations and ships were in shambles, there was an appalling mess in the offices of the shipping company, with typewriters either on the floor or outside the building in the yard.[24]

The scene was Batavia harbor, the date March 5, 1942, three days before the Dutch capitulation to the Japanese. The antenna fallen across the road was *Archipelago*, the best known radio transmitter of the colonial Indies. Was not Wertheims' entry the literal truth that Nietzsche had in mind? The shock of the Japanese invasion was immense, and, as we know from what followed, it caused amnesia. But how sudden, actually, was the death of the metaphor and of the Kartini-like enthusiasm of imagination? Was it native, or was it Dutch? Was it, indeed and only, the Japanese invasion that brought about the inert vista of the voiceless thing "across the road," in a distanceless landscape? Remembering Kartini—and how uncertain her metaphor (not native, not Dutch, technological most of all) had been—did the Indies radio ever connect over the expanse of the colony? Did the fancy thing called radio ever have a voice in the Indies?

The Thing

When the future of Indies broadcasting was planned in the mid-1920s, the "American experience" was used as a model. The number of listeners should be roughly equal to a quarter of the number of owners of private cars; thus for the whole colony in the first stage, about 10,000 radio owners.[25] Not unlike cars, the radio had become visible in the colony—in advertisements and showrooms, with price tags and technical data prominently attached as the main phenomena of the shiny thing.

Unlike cars, radio sets in the Indies were typically sold in music stores and piano shops.[26] They appeared to the eyes of the Indies public as a kind of sensational new knick-knack, *fantasie meubeltje*, "a novelty piece of furniture"—electric furniture without which, soon, no modern house was thought to be complete. As the design of a car had to suit the image of the fast motion along a modern road and through a modern landscape, so the design of a radio set became, in the Indies, a significant organizing principle of a modern colonial house.

There had always been, as we have seen, a considerable intensity about what a modern colonial house in the Indies should be—an intensity increasing and heating up as the colonial crises outside the modern colonial house deepened. Thus the journal *Alles Electrisch in het Indische Huis*, "All Electric in the Indies House," as we have seen, became popular exactly at the moment when the worldwide economic depression descended upon the Indies and the final stage of the Dutch rule in the Indies set in.[27] This was made as perfect as possible. In the Indies radio, a thing carefully designed and priced, more undoubtedly than in any colonial house, all was electric.

One radio advertisement appearing often in Indies journals during the 1930s deserves a brief mention. In the advertisement, a radio set is presented as a house in a very modern art deco style. Two signs are visible: *TOSCA PHILIPS 119 guilders*, which is the price tag, and *Waar Oost en West elkaar ontmoeten*, "Where East and West meet."[28] There is, on the left, an unmistakably Eastern dancer accompanied by traditional Javanese *gamelan* orchestra of gongs, metallophones, and *rebab* strings. On the right, there is a very Western ensemble: a pianist, a violinist, and a cellist, all busily performing on the house's (which is the radio's) roof. An irksome question arises: why on the roof?

Indeed, Indies radio was essentially, and even desperately, an indoor adventure. This was a universal feature of the technology,[29] yet, in the Indies, it was one of the most striking and most social aspects of the thing. Still at the time of Kartini, at the beginning of the twentieth century, there was a hearty memory of trailblazers, IJzerman-like engineers, and "coolies" cutting their way through the jungle to lay track or a wire. As for the wire, they fastened it, in those early days, on live trees, *kapok* trees best of all, and they built, as their shelters and as the end stations of the wire line, huts of wood, bamboo best of all.[30] The wire opened the old, and so it seemed, previously closed landscape. The wire, so it seemed, built a new land; new maps and journeys seemed simple: "just follow the wire, and you will reach a post office."[31]

Earthquakes, volcano eruptions, floods, elephants, and insects in these early days were the enemies. (A telegraph cable "destroyed by an elephant" was exhibited at the colonial exhibition in Amsterdam, in 1883.)[32] They felled the trees, tore the wire, and leveled the end stations. It is worth of noting that *onlusten*, "riots," as the causes of the wire problems, until the end of the nineteenth century, until the time of Kartini, in the Dutch official telegraph and telephone reports, were extremely rare.[33]

Roughly at the time of Kartini's death, since the early twentieth century, a change began to be visible. Natives, too, now, were reported as felling the trees, cutting the wire, and, even, robbing the post offices. Initially, these were "outcasts," thus merely another element besides the earthquakes, volcanoes, elephants, and insects. Yet, not long afterward, less than two decades after Kartini's death, certainly, wire-cutting natives grew ominous. The purity of a technological axiom appeared in danger of being violated: the natives broke into the post offices and cut the wires with increasing expertise. (A "close watching of line watchers" was advised in a Central Telegraph Office annual report.)[34] Alternative professionals had entered the colonial scene.

The reaction of the colonial regime was techno-logical. As progress and money permitted, live trees were being replaced with iron poles, and the wire-line end stations were moved into buildings made of wood rather than bamboo or, when possible, of brick and concrete. The rush among the wealthy and modern in the Indies, as we have seen already, had generally been toward an all-electric indoors. The introduction of the wireless helped this kind of history to move much faster. At the ideal consummation of this, the end stations in the colony might connect with each other without touching the uncertain ground between them—and without a wire, even.

To a large extent, this became an idea of the colony and empire. In 1934, a group of the most graceful Javanese dancers performed before the queen of the Netherlands and the Indies, Wilhelmina, in one of her audience halls in the Hague, in Holland. The accompaniment to the dancers by *gamelan* was transmitted, wireless, from the palace of the sultan of Yogyakarta, in Central Java. Whatever the distance, there was not supposed to be a single hitch.[35] In the same historic vein, shortly before this event, an ambitious and exciting plan for a "wireless transmission of fingerprints" was announced in the Indies.[36]

Indies radio achieved what the Indies railways, asphalt roads, and even wire telegraph or telephone hardly could. By the late 1930s, a wireless signal from a new system of stations on the main Indies islands, Java and Sumatra, and from the Netherlands, made the entire huge Indonesian archipelago, with a partial exception of New Guinea in the extreme east, open to radio signal.

> Kloengkoeng, a little place in the interior of Bali with three European families. No gas, no electricity, no theater, no movies. But, indeed, radio. . . .[37]
>
> Each evening at half past seven the Wilhelmus anthem sounds on our gallery. . . . First, it is a metronome, *tik, tik, tik, tik, tik, tik*, etc. Then, the Wilhelmus. We all sing with the choir.[38]

The reception was not always perfect, but a new ideology built up fast: as far as the signal could be heard, so far there was the Netherlands East Indies. This was "our greatest task," the *Phohi* newsletter wrote in 1935, "to make the Netherlands and the Indies into one, through the ether."[39]

Indies radio became yet another tool to define a modern colonial space. In the Netherlands itself, through the 1920s and 1930s, according to law, radio broadcast time was distributed in proportionate sections among the major Dutch political parties. After a short pause of hesitation, this was rejected in the Indies. With not so distant memories of the outdoor adventures and perils of the Indies communication system, and with the alternative professionals, the wire-cutting natives, uncomfortably felt to be around, the policy-making radio men in the colony decided it imperative to be categorical. As in a perfect all-electric house, everything in the Indies radio was to be made so well as to be easily switchable on and switchable off.

No politics, propaganda of religion, or advertising was to be admitted to the colonial ether.[40] In 1933, for instance, a lecture by a Catholic speaker was banned by the state broadcast *Indië Programma Commissie*, "Indies Program Commission," with the explanation that "the community in the Indies is more sensitive than is usual elsewhere; even Europeans become oversensitive in the Indies climate. The natives, in their turn, are oversensitive by nature."[41]

At least as late as the mid-1930s, the modern inhabitants of the Indies knew about the great political and social potential of radio as it was being discovered and realized elsewhere. A radio journal in the colony wrote in 1934, for instance, about how radio was being introduced into Italian class-rooms. A detailed article described the first day at an Italian school and an opening radio oration by the Italian minister of education on "*Duce* and Children."[42] It had also been well known in the Indies how the cheap *Volks-radios*, "People's Radios," powerfully beamed the government's message into every "new German family."[43]

Apparently, however, the Indies was not a place for that kind of wireless power. The cause of it, perhaps, was not entirely the fact that the late-colonial Indies government was controlled by a democratic regime in Holland. Not entirely for this reason, so it seems, the Indies was not a place for a Nazi-like communication system that functioned, as it was described else-where, through false echoes of a "disjunctive instaneity," as a "thickly dissem-inated network of near misses" and "traceless politics of denunciation."[44] There had not been much fertile soil in the Indies for thick culture of a "chthonic voice" working "from under a mask."[45] Whatever the principle might be, the Indies radio was clearly still too much of a shiny-little-furni-ture thing ever to attain that particular perfection.

In fact, it might be a question of principle. This was, no doubt, a modern technology; transmitters and receivers in the Indies, as everywhere else, were measured by watts and hertzes. The more advanced the communication sys-tem became in the colony, principally, the more Euclidean it was—the live trees, the defining points of the carefully mapped network turned iron, the wireless lines moved off the ground and thus became straighter. The Indies radio signal, on principle, was supposed not to disturb anything along the line, and neither should it take in any echoes from the colonial space it passed through—false or not.

Before the time of radio, it took the Surabaya train about a day to reach Batavia along the north coast of Java. As a rule, even the biggest newspapers in the Indies, traveling by train, remained essentially local newspapers. Even newspaper rumors remained local rumors, and politics remained local poli-tics. Indies radio, born at the time of economic depression and mounting colonial crises, anxiously worked to use its extraordinary new technical prin-ciples of power to bypass the troubled locality. It had been remarkably suc-cessful. The straight radio lines cut up the Indies landscape without building a new one, and the radio stations and radio receivers at the lines' end stood out flagrant, and surreally irrelevant at the same time.

On second thought, a new landscape was being built by the wireless. At the moment one switched on one's radio set in one's house in a colonial city, provincial town, or plantation in the jungle and heard the more-or-less clean signal from afar, amid the multiple-voice colony, one felt like a stranger. The solution offered was to make oneself even more so—to move even closer to the radio set, and to make the house even more all-electric, and, if there were any, to close the windows. This was the crux of late-colonial architecture, too. Like the straight and off-ground lines of the wireless, so the soundproof walls of the house and the wooden finish of the radio set were expected to enclose, keep away, or, best of all, absorb without a trace all the anxiety there was.[46]

It was the surface of the thing that became charged. The Indies radio studios exuded power, artificiality, and temporality at the same time. Radio "booths" made a most perfect attraction at the Indies colonial fairs throughout these years.[47]

The Malabar Transmitter in the mountains of West Java above Bandung (near the Boscha telescope, plantations, and electric house) was a natural focal point and the most powerful symbol of Indies radio. It featured, among other things, the tallest antenna in Southeast Asia. In a Malabar Transmitter tourist guidebook published in 1929, photographs show the big complex but, even more so, the views in the opposite direction, panoramas of the distance below, the colonial landscape as a tropical paradise. The pictures of the studios present soundproof rooms with walls and tables covered with traditional Javanese batik cloth. There are indoor palms in flowerpots. Very few pictures of the period come as close as these to the culture of beautiful Indies and of *tempo doeloe*—a vigorously sentimental and escapist nostalgia for the never truly existing "time past" and time lost.[48]

The falsity, of course, was in the very tranquillity of the scene. Even as Indies radio was born, colonial peace was already no more than a matter of façade and polish. There had been an open war since 1931 in East Asia, and soon it broke out in Africa and Europe as well. By the mid-1930s, Japanese invasion southward and into the Netherlands East Indies became a real threat. On the eve of the war, the Indies radio studios—more artificial and temporary than ever—were suddenly decreed important by the colonial government. The unpredictable enemy, if given access, might broadcast something from the studios!

This (possible) brush with reality made the Indies radio studios into rather embarrassed strategic points. "We cannot tell you, of course, all what is going on," the journal of the "Batavia Radio Association" told its readers-listeners in June 1940; however:

> to those who take their stroll along our studio regularly, the measures taken will be no secret.
>
> Given the changed circumstances, also, we are sorry to have to put an end to the hospitality, which we gladly offered to our members in the past.[49]

It was meant to mean that nothing had fundamentally changed. Indeed, expecting the enemy and the apocalyptic end of the colonial era made the

Indies radio, in a way, more what it had been all the time. Neither the radio's power nor its artificiality and temporality were to be touched. On the contrary—the appealing superficiality of the thing was made even more visible and central.

With the war and the end approaching, it was as if all the goings on of the Indies radio were yet more strenuously pushed indoors. In the spring of 1940, as, in Europe, the Netherlands was invaded by the Nazi armies, *Radio Bode*, "Radio Messenger," a prominent radio journal in Batavia, reported on the Indies radio reaction:

> On May 10, immediately after the invasion of the Netherlands became a *fait accompli*, the Board [of the studio] expressed a wish to withhold all German music, and it gave an order to pull all the German gramophone records from our discotheque and, in order to avoid a possible mistake, to lock them in a special cabinet.[50]

As the Netherlands capitulated to the Nazis a few days later, and as the Dutch queen left for her exile in London, *Radio Bode* reported again:

> we have decided to put an end to the German music once and for all, and to destroy all the available German gramophone records.
>
> There was no need to ask our staff members twice. They took out all the gramophone records, close to 2,000, and they let the military guard [at the studio] vent their rage; as a consequence, the records ended in thousands of pieces, as rubbish, which they truly deserved.[51]

The latest model of a radio specially designed for the Netherlands East Indies had been advertised as late—and on a date as notable as—December 7, 1941:

614 VN.

A new 6-lamp battery receiver with all the advantages of an alternating-current apparatus. A complete world receiver. An apparatus for the jungle. The price: 296 guilders.[52]

The more the Indies house grew all-electric, the more it appeared as if standing deep in the jungle. A radio movement developing since long before the war, *autoactiviteit*, "do-it-yourself," was proof that this was the case. Listeners in the Indies showed an increasing passion for building their radios themselves. Easy-to-assemble segments (like arms, legs, and heads of dolls) were provided by Philips, direct from the Netherlands.

The deep affection for the thing, so the advertisements were putting it, was to make one feel self-sufficient and self-contained—in the jungle, at home, in the colony. A long article in *Indische Luistergids*, "Indies Listener's Guide," sketched for the Dutch colony radio owners what might be achieved:

Antenna romantic

Among the listeners living in the picturesque town of Bánská Bystrica, Czechoslovakia, one may discover a remarkable attitude toward how to build antennas. Because there is no industry in the region, the radio reception is clear, disturbances are low, and the Budapest radio signal comes through strongly. It is not necessary to build truly big antennas in order to achieve good reception. The inhabitants of Bánská Bystrica, however, clearly believe that an antenna is good not just for radio reception. Besides the ordinary T or L antenna, they have invented countless original forms. Many hang cylinder-like or globe-like objects made of copper plates and pans on separate isolators and connect them to their radio sets. Others use various wire spirals so that one can see truly imposing metal constructions erected on dozens of roofs, and in most extraordinary shapes. . . . One feels a hidden struggle going on under the surface over who may invent the most original antenna of all.[53]

It may be that the Indies do-it-yourself radio lovers tapped into the spirit of the late-colonial culture deeper than most of the rest of the population. In their houses, they listened to the codes of the electric scratches from very far away and searched for another man or woman defined not by belonging to a certain place, or by having a certain mood or opinion, but exclusively by the fact that he or she also had built the thing.

Radio amateurs were in vogue everywhere in the world in the 1930s. In the Indies their agility was amazing. Through the second half of the 1930s, *Orgaan van de Ned.Ind.Radio Amateurs Vereeniging*, "Journal of the Association of Netherlands Indies Radio Amateurs," never sold less than 5,000 copies a month—about one-fifth of *Alles Electrisch in het Indische Huis*, it is true. But very few other Indies journals of the era, professional or popular, cultural or political, Dutch or Indonesian, could match this.

It was not completely by accident that *Indisch Spiritistisch Tijdschrift*, "Indies Spiritist Journal," grew as popular at the same pace and at the same time as the journal of the Indies radio amateurs. The table turning did not wane in the modern decades after Kartini. It even seemed to assume a new strength.[54] Unlike Kartini in 1900, however, the Indies spiritists of that very late time seemed to be excited—less about engaging in dialogue, two-way communication, than about listening from seclusion or about watching and peeping as the technology advanced:

T. "Hello, Hello."

O. "Hello. Good morning, Mr. T."

T. "I am here to see you."

O. "That is a very strange remark as you are so far from here."

T. "Yet I can see you."

O "But you cannot see through the wire, can you?"

T. "Oh yes, I can do it very well. I can see that you are wearing a brown suit. I can also see . . . a silver medal pinned on your vest. You won the medal in one or another kind of sport." (Red.: All correct)[55]

The voyeurism of sound and light seemed to promise the spiritists of the late-colonial Indies what even the Indies radio amateurs might only dream about: how they might enjoy a life without moving out of their houses. The spiritists were the perfectionists of the late-colonial culture, and their language was a perfect case of language as asphalt. They were thoroughly, even obsessively, materialistic. Reading reports on their seances—ghosts called, ghosts caught, ghosts photographed—one is overwhelmed by the richness of the technical jargon and the amount of detail, types of hardware, kinds of wood for the table to be turned or tapped (also models of camera and types of film to snap the specter). The language helped the process mightily. Like the Indies radio amateurs, and more so, the Indies spiritists appeared increasingly to live by "equipment" only, by "apparatuses," "tools," "instruments," the segments, the thing.[56]

On May 12, 1940, two days after the Netherlands was invaded, an advertisement appeared in Batavia, in *Radio Bode*, introducing another new Philips radio receiver for the Indies house: "PHILIPS, the greatest radio industry in the world, is announcing *1940 Super 3 Faust* type 419X, *price 320 guilders*."[57] As time got even shorter, and as the Japanese armies neared the Indies, so the Philips radio campaign intensified: "*PHILIPS FAUST Super 3 . . . See our special 'Faust' etagere and ask for a demonstration*."[58] The radio jargon, once more, got it right. What happened next in the Indies was, indeed, a Faustian tragedy.

On Sunday morning, March 8, 1942, the day of the Dutch capitulation to the Japanese, Hetty Wertheim, in her and Wim's house in Batavia, thinking how best to arrange everything for the evacuation and trying to keep herself and her two children calm, was startled by a sudden noise. She turned around toward where it was coming from, the window of the house across her yard: "from very close, I heard *Wilhelmus* [the Dutch anthem] on the radio of the neighbors, which was turned on. I was about to go there, but then I noticed the man and the woman, both standing at attention facing the radio."[59] The Indies modern colonial house was just about to be destroyed, its privacy broken in. Facing the radio, in an erect position, eyes to the front, arms to the sides, and heels together, as the last thing there was, seemed logical. Yet, and even more disturbingly than at the beginning of this chapter, the questions remain: what did Wertheims' neighbors "at attention" think they were facing? And what did they believe they heard?

The Voice

Be not afeard, the isle is full of noises,
Sounds, and sweet airs, that give delight and hurt not.
Sometimes a thousand twangling instruments
Will hum about mine ears; and sometime voices. . . .
—Ch. van der Plas, the governor of East Java, in 1941 quoting *The Tempest*[60]

G.A. van Bovene was a big man of the Indies' new culture and radio throughout the 1930s and until the outbreak of war. He became especially well known in 1938 by his live radio reportage from the opening *KNILM*, "Netherlands East Indies Airways," flight from Batavia to Sydney. It was the first time that an Indies radio microphone worked on a plane.[61] In 1941, at the end of the period, van Bovene published an instantly popular book, *Nieuws! Een boek over pers, film en radio*, "News! A book about press, film, and radio."[62] In the book, he summed up his personal experience, and, in a sense, the history of Indies radio as well.

There was a distinct mood in van Bovene's book that could not be explained merely by the fact that the book was written in 1940, and that the war was near and appeared inevitable. The ambiguous feeling about the future, so striking in the book, ran more broadly. "What we can see above the roofs," van Bovene wrote about the modern Indies of the present, "is not yet a forest of antennas. These are still the poles to hang high the cages with *perkoetoet* [the songbirds of Java]. . . . But the times are changing. Some *perkoetoet* poles already serve the radio."[63]

Van Bovene did not say whether the little songbirds still sang on the poles that "served the radio." Instead, he talked at length about the "island mentality" of the colony, about radio waves already reaching every corner of the Indies, and about the fact that "the fastest of all modern ways of communication," radio, was also "the most fleeting." Not always, van Bovene added, could one "find fulfillment in the speed."[64] Every newsman, he wrote, looked, at moments at least, for a tranquil place, for *stilte*, "silence, stillness." "I am," he wrote, "looking for a place of stillness every day."[65]

Not surprisingly for the Indies in 1940, a large part of van Bovene's book dealt with the trip he had just made to Japan, and a very large segment of it was about Japanese radio. Japanese radio, van Bovene wrote, was full of news: there was a fifteen- to twenty-minute news program four times a day.[66] There were gymnastics on Japanese radio daily, and "millions of listeners" exercised.[67] In commercials, according to van Bovene, Japanese radio was "a great success in the battle against unemployment."[68] Japanese radio sets were made in Japan, cheap and pocket-size. Not furniture, that is: one could easily carry them wherever and whenever one might decide to go.[69]

Van Bovene's story, and its implicit Indies sadness, peaked as he took his readers from Japan across the Pacific Ocean to the West Coast of the United States. There in a house of a friend, on the Montecito Hill above Los Angeles, with a truly American powerful machine of a radio, he searched for, caught, and with the highest excitement listened to indistinct electric scratches from one of the anonymous radio amateurs somewhere in the faraway Indies.[70]

The important, perhaps crucial, issue in van Bovene's story was silence or, rather, the lack of it. In this the Indies Dutchman and elite journalist of the colonial establishment came very close to an author who, hypothetically, could not be more different: Muhammad Dimyati, an ardent Muslim, a native, a popular-novel writer, and an Indonesian nationalist.

The book by Dimyati, *Dibalik Tabir Gelombang Radio*, "Behind Screen and Waves of Radio," was published only a few months before van Bovene's *Nieuws*.[71] The story of the book moves between Bukittinggi, a provincial town in West Sumatra, and Batavia, the place of radio. The novel opens as a young Indonesian man, Amir, returns to Bukittinggi, his native town. He has dropped out of a modern high school in Batavia because of his infatuation with radio.[72] Afterward for some time he drifted around Batavia, much of the time sitting on juries for popular radio-song competitions.

"Amir!" a servant calls Amir, who sits in his room back home: "Calm and quiet [*hening soenji*]. 'Amir!!' There is no answer. 'Amir!' Quiet [*soenji*] again. 'Amir!!!'"[73] Amir's father does not yet know about his son's failing. He plans to marry Amir to a local girl according to old customs. At that moment, alas, a letter arrives: *Miss Anna*, a popular radio singer from Batavia, wants Amir to join her.

Amir has met Anna only fleetingly before. But he has heard her often on the radio. This is what counts now. He leaves at once. Even the image of his mother's grave suddenly appearing as the train passes the Bukittinggi cemetery cannot stop him. "'Radio, again Radio,' Amir thinks. 'If it was not for Radio, I would not be like this.'"[74]

With Bukittinggi left behind, in Batavia, Amir and Anna buy, on an installment plan, a radio set, and they also rent a room. Anna signs a good contract with *Nirom*, "Netherlands Indies Radio Broadcast," the main Indies radio station.[75] Every night she is out in front of the microphone, or so she says. Amir stays at home, keeps on turning the knob on his "expensive and beautiful" radio, nights and days,[76] as he searches for a song he likes, and for Anna's voice in particular. It takes some turning before he realizes that something goes wrong.

Things have gone wrong indeed. Anna has switched back to her former lover, a fellow radio performer. Amir finds out at last, he becomes increasingly desperate, and, at the terrifying climax of the story, in a fit of indoor fury, he smashes the "expensive and beautiful" radio against the wall. "It sounded like broken crockery."[77]

Almost all the rest is Amir's aimless wandering through the streets of

Anna and Amir. (Muhammad Dimyati, *Dibalik Tabir
Gelombang Radio*)

modern Batavia. Amir is empty and down. At one moment, a couple of prostitutes make fun of him. At another moment, Amir hears a radio again through an open window of a house he passes. But it is apparently a Dutch house, and the sound carries no meaning to Amir—it passes over, it is indeed a broadcast from Europe of a soccer match between Holland and Belgium.[78]

Things could not get worse. In a senseless brawl with Anna's new lover as they accidentally meet at night in a local park, Amir is hit hard, and his head smashes down on the concrete—not unlike the radio set, of course. The old father at home, in Bukittinggi, with his remaining children around, receives the news as he looks over the crime column in Batavia papers. There is a deep sadness as the Amir-less family sits close together. The rest is calmness: *hening soenji*, the quietness, returns.

Van Bovene seemed to be aware and, deep inside, tired of the empty noise of Indies radio. Dimyati seemed to suggest that personal and social integrities remained, or might recover, as the radio noise in the Indies gets dim.

Soewarsih Djojopoespito's important novel *Buiten het Gareel*, "Out of the Harness," was published in 1940, chronologically between the books of van Bovene and Dimyati. Its author, as mentioned, was a young Javanese woman, activist, and personal friend of some of the top Indonesian nationalist leaders of the late 1920s and the early 1930s. In the novel, written after most of these leaders were either exiled or otherwise suppressed, amid the general depression of the 1930s and under the threat of coming war, Soewarsih searched for a space in which to gather strength and to survive the difficult time in one piece.

Radio was mentioned only once in the novel, but it clearly played a role:

> The sound of radio died down; it was quiet on the street. Now and then, the light of a bicycle lamp flickered in the dark, and vanished. . . .
>
> Soedarmo and Soelastri faced each other in silence. Outside, a night-watchman clattered with his bamboo stick. A rose-apple fell from a tree, and it thudded on the tin roof. An owl cried out, and a shiver of the cry passed through the stillness. . . . "Memories do come," she said in a whisper . . . "inspiration . . . memories . . . memories . . . "[79]

There is another text by Soewarsih relevant to this, written roughly at the same time and published in the Indies liberal Dutch journal, *Kritiek en Opbouw*, "Criticism and Construction." In this text too, radio is mentioned, and, as in the radio book by van Bovene, there also are *perkoetoet*, the songbirds:

> My mother speaks another language than do I. Yet, I speak my mother tongue. The same words in her mouth obtain another color and another quiver than they get in my mouth. . . .

> She makes love to her sentences' undulations as they reveal her thought in a flowing rhythm of symbols. She is fond of an adorning detail; she delights in cadences of words. . . . [Hers are] the trembling sounds of *perkoetoet* songbirds in the dark crowns of trees. . . .
>
> I live in the world of bombs and radars, apparatuses of radio and of television.[80]

Derrida was quoted: "If voice says nothing, that doesn't mean that it doesn't name. Or at least, it doesn't mean that it doesn't fray a path for naming."[81] *Stilte* or *hening soenji*—the silence as invoked by van Bovene, Dimyati, or Soewarsih—might be or might become the "naming" or, at least, the "fraying a path for naming," as much and as powerfully as a voice. This was perhaps why in the Indies, and on Indies radio, silence, that kind of silence, was categorically not to be given.

In its January 1934 issue, Batavia *Indische Luistergids*, "Indies Listener Guide," let its readers know about the views of the Italian avant-garde writer, futurist Emilio Marinetti, on radio in general and the language of the radio in particular:

According to Marinetti, the present radio is still too realistic, and it completely lacks in originality. The radio must become: 1. free of all traditions: any attempt to resort to tradition is grotesque; 2. a new genre of art to begin, where theater, cinema, and literature left off; 3. a way to broaden space; 4. a spectacle that is no longer visible and framed, and that becomes universal and cosmic; 5. a way of hearing, amplifying, and putting on the air the vibration of unanimated things; 6. as we can listen to the songs of the forests and the seas, at present, should we, in the future, hear and enjoy the ripples coming out of a diamond or a flower; 7. a new organic reality of wireless consciousness; 8. detached words, a repetition of verbs ad infinitum; 9. a sketch and geometric construction of silence; 10. [an endeavor] independent of listeners, whose demands never had but a baneful influence, obscuring the radio.[82]

Marinetti's views were placed by the Indies radio journal among "curiosities from abroad." "Particularly, as the last point of Mr. Marinetti is concerned," was the journal's good-natured and only comment, "of course, we have very different views."[83]

"How noisy everything grows," a journalist, Karl Kraus, complained in Europe at about the time Marinetti composed his radio principles.[84] What Karl Kraus was referring to was deafening news, one-way communication, sound as floodlight, avalanches of information that buried under its mass the possible meaning of a message. Uninterrupted flow of a quasi-language with a collapsed structure, a smooth noise, a sound without a pause and articulation. No silence.

In this sense, Indies radio was truly noisy. Its signal connected the stucco studios with the all-electric houses as far as there were any, not taking in false echoes and without faltering. If there was a halting, it was easy to explain it away in purely technical terms. Amid the depressed colony, Indies radio managed to keep to its smoothness as a technological plaything and a fake. The ultimate Indies radio voice, in fact, might be the metallic *sirenegeloei*, "wail of sirens"—simulated raid-alert signals broadcast regularly by Indies radio from 1940 on as a test. When the reality happened, and the enemy's planes appeared in the Indies skies, Indies radio went off the air.

In fact, there had been some kind of structure to the noise. In contrast to so many other things in this tropical colony in crisis, the radio indeed was highly and efficiently organized. Each program page of each radio journal by its very layout, and one may compare this with a newspaper page, prevented any thought of disorder. There were morning, noon, and evening news on the radio, minutes of gymnastics, children's, housewives,' and do-it-yourself programs, occasional causeries, moments of prayer, all punctuated and tied together by day of week and exact time signals repeated each half hour at least.

Only exceptionally was there an *event* on the Indies radio, something that would not fit the program page as prepared at least a week in advance. Only rarely did Indies radio broadcast from outside the studio. Indies radio might offer, so to speak, only what could be found in the house. In this sense, gramophone records were crucially important. They determined the structure more than anything else.

In 1939, 7 percent of *Nirom*, the main Indies radio broadcast, was made up of news, 4 percent of church services, 5.5 percent of lectures, 2 percent of gymnastics, and 0.5 percent of original radio plays. The rest, 81 percent, was music, about three-fourths of it from gramophone records.[85] These were "78s." They played about three minutes each. Then a little word had to be put in, to make time for placing another record on the gramophone.

If merely out of technological necessity, the decisive part of the Indies radio broadcast, and the broadcast's message, was cut in three-minute segments. Not fragments, not flashes of inspiration, as they repeated regularly, and, truly, mechanically. Whatever might be on the air tended to sound like an encore.

In 1939, also, more than four-fifths of the music broadcast by the Indies radio was placed, by the radio officials themselves, in the categories of either "light classics" or "light music and cabaret."[86] Through the 1930s, this structure of the broadcast—poetics of the noise, grammar of the fake—remained impressively consistent. Even at the moment when everything in the colony was tumbling down, until the very last days of the Netherlands Indies, this culture persisted unchanged. On the biggest (and a sort of last) date of the era, December 7, 1941, the *Nirom* Indies radio, for instance, put on the air:

Our story of the week: LEAVE ME FREE *by Marian Sims:*

Until the moment when she heard Teddy's voice on the telephone, Letty could not understand that the gaping space in her life had slowly closed. . . .

Song of the week: I, yi, yi, yi, yi
I, yi, yi, yi, yi, I like you very much,
I, yi, yi, yi, yi, I think you are great,
Why, why, why is it when I feel your touch,
My heart starts to beat, to beat so much.[87]

Images by an artist who signed himself as Jetses, an immensely popular Indies illustrator of textbooks and books for boys and girls through the 1920s and 1930s, may help to explain all this from yet another angle. In one of the most often reproduced images, a jolly old man sits in a rattan easy chair, surrounded by all the imaginable properties of the colonial private Indies house. He listens to a *tempo doeloe*, "time past," a jolly gramophone with a huge horn: "A stay with the Uncle in the hills? . . . the funnel mouth of His Masters Voice blared . . . even the garden behind the house laughed."[88]

As we observe the size and shape of the thing—and as we recall what kind of radio sets Philips was putting on the Indies market at the time, and what kind of programs the Indies radio was putting on the air—the jolly old man in Jetses' picture, in the same way, with the same face, to the same effect, might have been listening to the Indies radio. The scene is filled with jolly things. There is no place for Marinetti, and, clearly, not for silence. What we can see can best be summed up in a Dutch word as popular in the Indies of the 1930s as Jetses himself: *gezelligheid*, translated in dictionaries as "1. sociability => companionableness, 2. coziness => snugness."[89]

But now comes the difficult part. The late Japanese scholar Kenji Tsuchiya (of whom I think so often, writing this chapter) once noted that it was radio that helped to spread *kroncong*, the popular Hawaiian-like songs of the Indies, among the wider circles of the indigenous population. (*Kroncong* was a public music, too. *Kroncong* musicians—indigenous, Eurasian, Chinese, a mixture, indeed very much like that of the *komedie stamboel*[90]—wandered through the streets "singing serenades." *Kroncong* competitions were held in parks and public places. Gramophone records by local companies made the street music modern.)[91] The radio, Tsuchiya wrote, helped the *kroncong* to become the dominant Indonesian national, indeed nationalist, music.[92] Tsuchiya was no doubt correct. His remark, however, gives rise to a disquieting thought. How different was the *kroncong*, broadcast by Indies radio, from other Indies radio culture? And how different was the Indonesian modern and radio-minded nationalism?

Some affinities between *kroncong* and other Indies radio forms are, in fact, readily evident. The *kroncong* songs, like operettas, popular studio-polished pieces of jazz music, or "light classic," were easy, and indeed "willing," to be fixed into the "78" gramophone-record mold. They did not have much inner resistance against being condensed or cut into the three-minute pieces. Their structure appeared modern in the Indies late-colonial radio sense—they resembled machines, or puppets, easy to disassemble into a final product, petite enough to go safely on the air.

The language of *kroncong* was a combination of Portuguese, Dutch, and less often English words, mixed with the base of Malay and later Indonesian. This linguistic whole, the force keeping it together, was sometimes referred to as *strand-of kazerne-Maleisch*, "beach or barracks language,"[93] or an "atavism of barracks language."[94] Indeed, through the colonial centuries, the language used in *kroncong* songs was a language of the colonial army, civil service, and, more broadly, the colonial explaining modernity to those of the colonized who could not speak Dutch, the real colonial language, well. *Kroncong* language was a part of lingua franca, a language existing loosely on the fringes of colonial society and facing the foreign. It was this quality in particular that made the language kindred to Indies radio even before Indies radio was born.

Indies radio did not seem to face too difficult obstacles in editing the *kroncong* language into a perfection of its own style. *Kroncong*'s long colonial memory offered itself to be edited into a *tempo doeloe*, "time past," nostalgia; *kroncong*'s underlying sense of openness offered itself to be flattened into a subjunctive sentimentality of (kind of) idle troubadours; *kroncong*'s memory of barracks offered itself to be made into a mood of unmoored neatness. The *kroncong* Malay—and thus embryonic Indonesian language of nationalism—offered itself to be mellowed, to melt smoothly on the Indies radio, into the metallic unison of the radio-ordered late-colonial noise, and civilization. Or so it seemed.

It is unsettling that there initially might be something in common with the dandyish language of Mas Marco. The lyrics of *kroncong* by the 1930s seemed almost perfectly suited to the specific expeditiousness of Indies radio. This, for instance, was a favorite—and favorite radio—*kroncong* of the late 1930s:

> *Kindje tabé*
> *Jij gaat niet mee*
> *Anne Marie.*

> Baby [in Dutch], hello [in Malay or Indonesian]
> Don't go out with me [in Dutch]
> Anne Marie [?][95]

Kartini, too, comes to mind again—her telegraph, telephone, and spiritist communication with Javanese, Malays or Indonesians, Arabs, and Dutch, live and dead, mortals and ghosts. Kartini's media were hers to decipher all the different languages she was so excited about. By the late 1930s, over the noise of the radio, it might become a difficult question to answer: why should there be any pleasure in translation?

To sum it up, there was an impressive power contained in this kind of culture. Virtually everything was made possible. Bach, of course, was sung in

German on Indies Dutch radio. But when the Nazi armies invaded the Netherlands, as Wim Wertheim later recalled, the arias of the great German composer instantly were switched to be sung in French: "The famous aria from the St. Matthew Passion, *Blute nur*, Hetty now had to sing as *Saigne à flots*."[96]

The Closed Circuits

In November 1936, radio became, and not for the first time, an issue in the Indies colonial advisory and semi-elected *Volksraad*, the "People's Council." One of the Indonesian deputies, Soetardjo Kartohadikoesoemo, arguing about the low living standards of the natives as compared with the Europeans, demanded that the monthly contributions by subscribers to the Indies radio be adjusted. As a surprise to many, including perhaps Soetardjo himself, the government of the colony responded quickly with a radical initiative. An autonomous "Eastern radio" was to be established in the Indies. In March 1937, *Perikatan Perhimpoenan Radio Ketimoeran*, or *PPRK*, the "Federation of Eastern Radio Societies," indeed came into being.[97]

There had been some Eastern radio activity before. The first native studio in the Indies had already been established in 1932, with twenty members, in Surakarta, Central Java, *Solosche Radio Vereeniging, SRV*, "Surakarta Radio Society." In 1933, *SRV* established branches in Batavia, Bandung, and Semarang.[98] Another Eastern studio emerged in Surabaya, the port city of East Java, and two additional stations were opened in the palaces of the two highest Javanese aristocratic families, one in Surakarta, and the other in nearby Yogyakarta.[99]

In 1935 there were officially 8,469 European subscribers to the radio in the Indies, 1,500 natives, and 1,500 foreign orientals, mostly Chinese and Arabs. By 1938, the number of radio subscribers in the Indies had grown to 31,857 Europeans, 12,238 natives, and 9,468 foreign orientals.[100] There had been some Euroasians listening to the Eastern radio as well. In 1938, for instance, Soetardjo estimated that his Federation of Eastern Radio Societies organized "about 6,000 Eastern listeners, of whom 1,300 were Chinese, 200 Europeans, and 160 Arabs."[101]

Soetardjo, who became the decisive figure of the Federation of Eastern Radio Societies, was—like Kartini's father, for instance—a high Javanese aristocrat in service of the colonial government. He believed in only a very gradual and cautious change of the Indies system.[102] He also had no choice but to leave his Eastern radio under the financial and technical supervision of the government radio. Legally too, his Federation remained the Indies government radio's Eastern branch.[103]

Paradoxically, however, it was because of Soetardjo's political timidity and his radio's technical and financial impotence that the Eastern radio, from the beginning, sounded so very independent. The Eastern radio, in fact, was plugged in by the colonial authorities as just another channel. *Loquere, ut videam te,* "speak, so that I may see you," in a new form. As it should be, to suit the spirit of the time, two sets of radio stations were established, parallel, each with its own set of wavelengths. There was no controversy, no crashing, not even a debate. Signals of the Eastern and European radios passed each other in the ether of the Indies, taking in no false echoes, as good radio signals indeed properly should.

In contrast to newspapers and to society at large, there had been little attempt at crossing swords or voices, either in dialogue or in confrontation, between the two sets of the radio system. The Eastern radio was tuned to Eastern listeners, whatever their definition was: its mission was, first of all, exclusive. One can almost hear a sigh of relief on the European side of Indies radio as it was freed (to listen and watch, and) to keep on broadcasting more unaffected than ever.

When the Japanese armies began to move, and the European radio decided in the emergency to send word at last across the line, now in the moment of truth the principle of the past years was merely confirmed.

A large part of this last-moment European radio broadcast across the line for the Eastern listeners was fully made up of either air-raid directions (how a listener should fall flat when a bomb sounded) or explanations of evacuation plans in case of a possible invasion. In September 1941, a major Indonesian moderate nationalist weekly in Batavia, *Nationale Commentaren,* "National Commentaries," gave a good account of this hurried talking across the line—or, rather, talking over the other: "Hundreds of village leaders, who were made to listen . . . the first evening in Dutch, and the second in Javanese, were left with an unshakable conviction that the enemy ships, planes, and tanks were already at the beach. . . . Among some groups the exodus has already begun, mostly toward the hills."[104]

Starting in late 1940, a new, active, and innovative Dutch director of the colonial economy department, Dr. H.J. van Mook, went on the air repeatedly in Batavia to address the Eastern listeners. At the same time, Dr. Charles van der Plas, a close colleague of van Mook, the governor of East Java and known as an "expert on native matters," launched a series of half-hour weekly radio talks, some of them actually in Malay or Indonesian. The *Nationale Commentaren,* again, reported on the phenomenon. Commenting on van der Plas' radio speech of October 2, 1941, for instance, the journal wrote:

the speech was given in Indonesian . . . what a nice language it was, we thought, what literary depth, and what a patriotic virtuousness. . . . The speech had a very long, indeed a breathtaking, title: "The entire population of this land can work for the progress of this land and its population." The talking went on about many things: about the contacts between the native civil servants and the middle class, which should be advanced; about marriage ethics in West Java; about the gossip predicting the end of the world; about the world war, which is the work of Satan; about the Indonesian militia; and about superstitions. . . . [We wondered] how many Indonesians, after attentively listening to the radio speech by van der Plas, jumped up from their chairs, and cried out: "Yeah, true, all this we must do!"??[105]

What truly could be heard across the line, the journal summed up, writing about another speech by van der Plas in the same series, was a "microphone acrobatics."[106]

This was how the parallelism of the Eastern and European radios in the Indies worked. Heideggerian "thinging" in late-colonial style. Instead of one shiny thing, there were two. According, once again, to *Nationale Commentaren*:

We have it from a good source that His Excellency van der Plas recently happened to be directed to the Eastern studio of the *Nirom* Indies radio, which is located near the W.C., and—as, unlike in the [Western] studio *Nirom I*, there is no air conditioning—it is truly warm. His Excellency spoke to his fellow countrymen [*landgenoten*], all the people of the Indies, and before long was sweating heavily and had to ask for an additional glass of water. Since that event, van der Plas has never again spoken from the studio of the second class. The *Nirom* Indies station ought to build a connector between its studio of the first class and the Eastern radio studio, so that Mr. van der Plas may be spared our climate. Mr. van Mook, too, when he addresses his Eastern listeners, always speaks from the studio of the first class. It is, we think, a good thing to do, because Mr. van Mook may get cold were he to return directly from the second-class studio to his overcooled house. We sincerely hope that *Nirom* Indies radio can do something, and that it may deal with this truly deplorable situation swiftly.[107]

They were like two planets moving along their own orbits, in a very confined universe. If merely in order to stay on track, it was logical and imperative that each of them, the Eastern as much as the European radio, work strenuously on building up and shielding their own wholeness or, at least, the appearance of it.

There were minutes of indoor gymnastics on the Eastern as well as the European radio, there was a program for children, cooking, news, and time signals. In the *Volksraad* assembly, Soetardjo argued often and vehemently that Indonesians could make their radio wholly by themselves: "We have civil engineers and other technicians, we have lawyers, and we have other academicians, like experts in the field of religion and Eastern culture."[108]

The lower the technical state of Eastern radio was compared to that of the European network, the more anxious the Eastern studios appeared to be about how shiny their thing was. Even more than Indies radio as a whole, the Eastern radio strikes one as "microphone acrobatic," and jolly proud in its temporality and artificiality.[109]

However vast the gap was between the quality of the inventories of the European and Eastern Indies radios, there was a common poetics of the inventories. In May 1938, for example, a prominent radio journal published a list of equipment and other property of the second largest Eastern studio in the colony, Bandung *Vereeniging voor Oostersche Radio Luisteraars* or *VORL*, the "Association for Eastern Radio Listeners":

> Transmitter 1 piece; Service table with modulation 1 piece; Gramophone motors 2 pieces; Pic-up 2 pieces; Cupboard for gramophone-records 1 piece; Desks 2 pieces; Chair 1 piece; Tabouret 1 piece; Loudspeaker 1 piece (defective 1 piece); Microphones 2 pieces (defective 1 piece); Wall clock 2 pieces (defective 1 piece).

The largest part of the rest, clearly, the trope of the inventory, was made of gramophone records. Altogether 1,014 were listed, 705 of them "breakable" and 309 "unbreakable."[110]

According to its leading man, Soetardjo Kartohadikoesoemo, Eastern radio's apogee in defining itself came on January 11, 1941. On the evening of that day, the Federation of Eastern Radio Societies organized and broadcast its first public concert from the fashionable *Stadsschouwburg*, "City Theater" in Batavia, and indeed of Eastern music. "Thanks be to God!" the Eastern radio official journal wrote to welcome the event.[111]

Ali Boediardjo was a young, promising, Indonesian law candidate at the time, a student of Wertheim and a friend of Soewarsih Djojopoespito, whose texts about radio and silence are quoted earlier. Ali Boediardjo saw the concert and wrote about it the same night, it seems, in *Kritiek en Opbouw*:

> The concert deserves to sink into oblivion. . . . The evening was devoted ("devoted," of course, is ridiculous in this context)—to *kroncong* of all possible things, and it was called "Indonesian People's Concert." . . . [We saw] a *show*. . . . The members of the band *Lief Java* [in Dutch: "Sweet Java"] wore Javanese costumes! *Miss Soeami* (who might as well be *Miss Ijem*) curtseyed, *Miss Ijem* (who might as well be *Miss Roekijah*) was daintily cheerful, the members of the Radio Federation ensemble, *Botol Kosong* [Malay-Indonesian for "Empty Bottle"], all in *dinner jackets*, frolicked in the background.[112]

It is tempting to call this a dissident view of Indonesian radio. But what Ali Boediardjo suggested instead of *kroncong* was—remember Marinetti, van Bovene, Dimyati, or Soewarsih—not silence. However angry Ali Boediardjo might have been, his radio solution was on the level of the accepted. Against the *kroncong* noise, he resorted to tradition. Next time, he wrote, instead of *kroncong*, the subtle, soft, and ancient Javanese *gamelan* should go on the air.[113]

Soetardjo, compared with Ali Boediardjo, sounded robust. In his speech opening the same "Indonesian People's Concert," he made it clear that, in his view, nothing but *kroncong* could make Eastern radio connect with and resound through the whole huge and awakening land:

> this is our first "Eastern People's Concert" . . . we begin today by presenting what usually is called *kroncong*, a music that is loved by truly the whole Eastern nation, throughout all our islands. After the *"kroncong* evening," some other time, performances of *gamelan* of Java and Sunda may come, and music from Minangkabau, Batak, Ambon, and songs by the Chinese, the Arabs, and so forth, and so forth.[114]

This is, again, essentially a technological story. Thinking of the possibility of dissent, or even resistance, by the Eastern radio, what was the center of the radio? Initially—and this was also why aristocrat Soetardjo could be so big a radio man—*kraton*, palaces of the Javanese nobility, were the natural places where an Eastern studio might find a shelter from which it might broadcast, and where its identity might be defined. The most ancient Javanese palaces in Surakarta and Yogyakarta, as we have seen, were the locations of some of the first Eastern radio studios. Some of the early shows of Soetardjo's Federation still took place and were broadcast from the *pendopo*, "audience hall," of Soetardjo's most *kraton*-like private residence. Gradually, however, democracy—a sort of Indies radio democracy—was setting in.

Under the protection of Soetardjo, a new, radio-minded, and manifestly young and democratic journal, *Soeara Timoer*, "The Voice of the East," was launched in January 1941. On the journal's editorial board and among its advisers one could find names that, in a few years, after August 1945, would come to be identified with the independent Indonesian state: Oetojo Ramelan would become the first Indonesian ambassador to Australia; Hamid Algadrie would serve in several ministerial positions, as would Maria Ullfah; Soebardjo would be the first Indonesian minister of foreign affairs; Sjafroeddin would in 1948 stand at the head of the Indonesian emergency government fighting the Dutch. "*A.S.*," signed under a long article in the second issue of *Soeara Timoer*, might stand for yet another soon-to-be famous, and tragic, name of the Indonesian revolution—Amir Sjarifoeddin, who in 1948 was executed by Sukarno's republic as a suspected leader of a left-wing rebellion. The article by *A.S.* in *Soeara Timoer* had the title "From Palace to Microphone":

> Don't worry. I am not about to write about Their Highnesses Regents or Their Highnesses Regents' Wives who might, perhaps, descend from their palaces down to the studios and decide to work as announcers. No! This is not my idea. I want to write about *art*. Art which, in the past, we were able to hear and see only in the palaces, and which, today, we can easily enjoy at home, stretched out, sitting at our coffee table at leisure—tuned to a broadcast of a *wajang wong*

[Javanese dance drama] or sweet, first-rate *gamelan*. (Note by the editor of *Soeara Timoer*: We can listen to *gamelan* now in our own houses, in our easy chairs, and not just some *gamelan* from some palaces—but even from the most royal *kraton*s, from their innermost shrines.)

All this is now possible, thanks to the technology that gave us radio receiver sets, or—as we simply call it in our everyday language—the radio.

At the former times, in Java as well as in Madura, the center of art was the palace.

Indeed, not merely of art, but also of culture in general, and of manners, too, and of all the matters that were formerly connected, first of all, with the world of nobles. . . . The regents were seen by the people as an all-generating elite. . . . Their orders came as if they were from heaven, and they had to be obeyed. . . . Some people worry now that this center wanes and disappears. . . . This is a *critical* time, and some people feel like fish in a pond that is drying up. But there is a rain, which came suddenly, as if the heavens had indeed opened up again; not one but several studios and radio transmitters are working already. In this age of ours, a microphone becomes the center . . . *microphone!*[115]

The philosophy of the Indies radio seemed to hold to the same principle in its Eastern as well as its European part. With the radio microphone as the center, in a house, and in a chair, one might hear—and bear—the outside world humming. With this philosophy, there was to be no increased tension on Indies radio, no trembling, no shaking, not much of a voice, even as the old order in the colony rusted and crumbled down. Just the volume of the noise was to be turned up proportionately.

As the Japanese armies were really near, on February 5, 1942, the liberal Dutch journal *Kritiek en Opbouw*, writing in Bandung about Indies radio, tried a bit of irony:

Actually, we in *Kritiek en Opbouw* cannot find any cause for complaint about Indies radio. They do more than we think is possible, and they do it before we might ever wish them to do so. . . . We know everything about how Alfred van Sprang, Bert Garthof, and all these little what-are-their-names people are doing, how important their jobs are, how interesting they themselves are, how they stand, walk, smoke, and stammer. . . . We know how is it lately with cigarettes for Anita, and next week she may tell us how big her debts are.[116]

There was, of course, no next issue of *Kritiek en Opbouw*,[117] and there were merely three more weeks left for Anita and her friends to put their trivial stories on the air. Irony had always been weak in the late-colonial Indies, and, as it did this time, it mostly backfired. Indies radio kept working, as the jolly survival kit, to the very end.

The last program pages of the Eastern radio journals, like the programs of

Microphone. Title page of *Soeara Timoer*, March 27, 1941.

the "European" broadcast, remained true to the culture. For the evening of January 17, 1942, for instance—the Japanese were already in Malaya, and deep in the eastern parts of the Indies—the Eastern Indies stations sent out to their listeners "19.20 Songs of Peking, 19.30 News; 20.00 Songs; 20.05 His Excellency H. J. van Mook speaking; 20.20 From our gramophone records; 20.45-5.45 *Wajang Koelit* [Javanese puppet shadow theater]."[118] It was a moment of hollow laughter. On the last pages of the Eastern radio journals, at the edge of the apocalypse, there were announcements of the last radio speeches and portraits of Van Mook, van der Plas, and other pillars of the colonial rule, surrounded by portraits of the *kroncong* Indies pop stars, *Mister Darwin* and *Miss Annah*, for instance.[119]

The Mechanics

Søren Kierkegaard also said about metaphor that at a certain point in time "metaphor overwhelms the individual—he loses his freedom, or rather he sinks into a state in which he does not have reality. . . . This shows how the mythical, too, may affirm itself in an isolated individual. The prototype of this must, of course, have affirmed itself in the development of nations."[120] And further: "The traditional is like the lullaby that also constitutes an element in the dream; yet it is authentically mythical precisely in the moment when the spirit wanders away and no one knows whence it comes or whither it goes."[121]

Kartini, in the Netherlands Indies at the turn of the century, appeared at the crucial moment of transition. She was able to imagine the authentically mythical because, at that moment, the spirit of modernity indeed wandered, and nobody knew whence it came or whither it went. Forty years after Kartini, in the Indies of the 1930s and early 1940s, the journey of the spirit was mapped: Dutch colonial thought, as well as Indonesian nationalist thought, it appeared, did not have reality and was overwhelmed by metaphor. The mythical affirmed itself in isolated individuals, isolated groups, and isolated nations.

Through the myths of what was often not much more than a dead thing in a distanceless landscape, now, at the end of the era, the wildest happenings were believed possible. True *happenings* indeed, as they were not supposed to really change or even disturb the affirmed state of the world.

In an Indonesian novel by A. Damhoeri, *Zender Nirom*, "Transmitter Nirom," published in 1940, that kind of happening is described: Indonesians, people attracted by the sound of *kroncong* music, entering the thing:

> The longer he sang, the more sonorous and sweet became his voice. The people on the main street, as they heard their popular radio singer, suddenly stopped. Those who happened to be riding bicycles climbed down. Those who were driv-

ing an automobile got out of the car. . . . For half an hour Tengkoe Jazid sang. Without pausing Tengkoe Salman played on his violin. The sound of the violin and the song was like a powerful magnet; so much so that the public gathered the courage and moved toward the staircase. The two performers were unaware that there were already people inside. "Hey, who gave you permission?" Tengkoe Jazid said. "Now, the performance is over, gentlemen; another night it may be repeated." The public dispersed.[122]

In another happening in a thriller-mystery series that was immensely popular just before the Japanese invasion, the Indonesian hero, "detective Soebrata," makes himself all-powerful through building himself an all-powerful radio. The thing could receive even on "waves as short as 5.55 meters"!

> And now Aida [the detective's wife] had an unusual job to do in the house—to listen on the very special radio receiver made by Soebrata to all sorts of programs from abroad and home. What they were looking for was the broadcast of a secret message that was to come neither from nor to the government. Aida and Soebrata carefully learned all sorts of codes, and then Aida tried the possible keys. It was truly exhausting work, and frustrating, too, because all the ordinary news had to be listened to before being discarded as useless.[123]

Overwhelmed by metaphor, and suggesting a happening, an influential Indonesian nationalist journal *Keboedajaän dan Masjarakat*, "Culture and Society," wrote in an editorial in September 1939:

> We all know that, for the Indonesian people, it is impossible to find work in the cities, especially in the administration. . . . To improve the fate of our people . . . it may help to establish a kind of Depression Institute. . . . Let us come together, for instance, in a radio course, let us become radio mechanics. In a course like that, high-school graduates can sit in the same classroom with the pupils of elementary schools. . . . In high schools they may learn about electricity, but does a high-school graduate know what volt and ampere truly mean? . . . To become a radio mechanic, it is enough to know just a little bit about logarithms and trigonometry. To handle a *real mechanism*, one does not need to know much more than differential equations. Clearly in the classroom to become radio mechanics we can all sit together. . . . Students in the radio course will learn to build an *audio frequency amplifier, valve voltmeter,* and *cathode ray oscillograph.* . . . As they become good enough to connect a wire, check a *voltmeter* and *ammeter,* and as they learn how to prompt a spark out of *terminals* and *condensers* . . . they, without a moment of hesitation, will come out in the world.[124]

Indies radio grew out of, and further inspired, a potent late-colonial paradigm. Indies radio spread the notion of power as largely mechanical, and it helped to loosen the ties between sound and voice, between word and deed. The future of the Indies eventually might turn a different way. As Indies radio wanted it, however, everything sounded as if it had jolly well been put in place—sparks controlled, microphone in the center—and long before the war.

Except for television! Yet it was to come, too. This was at least how the Indonesian radical nationalist paper, *Soeloeh Indonesia Moeda*, "Torch of Young Indonesia," close to Sukarno, a long time before the actual victory of the Indonesian nationalism, envisaged the revolution coming: "Precisely! Oh, how do I yearn in my heart for that moment, one day in the future, when my revolutionary nationalist fellow countrymen, just quietly, with the help of the most modern means, will watch on television how the last imperialist rolls up his colonial mat—"[125]

EPILOGUE

ONLY THE DEAF
CAN HEAR WELL

University Team: Where did you get your writing paper from?
Pramoedya Ananta Toer: I have eight chickens.[1]

Sjahrir Recalled

On August 16, 1969, Pramoedya Ananta Toer—an Indonesian who wanted very much to become an engineer as a boy and apparently never will, who did not get the Nobel Prize in literature but may still—after four years at Salemba prison in Jakarta and one month in the transport camp at Nusa Kambangan, was put on a ship and sent into exile on the island of Buru 1,500 kilometers farther east. To people with a memory in modern Indonesia, the journey should be familiar. Thirty-four years before Pramoedya, another Indonesian, an intellectual and later statesman, Sutan Sjahrir, long before independence, in the late-colonial period, was shipped to his exile along the same route. From Jakarta, then still Batavia, Sjahrir's ship also traveled to Surabaya and then eastward between Sulawesi and Lesser Sundas to the Banda Sea. Both men, as their ships sailed and as all distances appeared to increase, thought intensely about continuing to write letters home. By what might just be a whim of history, or even of editors, the letters by the two men when later published, Sjahrir's in 1945 and Pramoedya's in 1988, perhaps inappropriately for letters, were given the same title, although one in Dutch and the other in Indonesian: *overpeinzingen* or *permenungan*, "meditations," "reflections," "musing," or "day-dreaming."[2]

In his writing—or thinking about writing—from the ship and then from Buru, Pramoedya mentions Sjahrir several times, always, however, as if merely in passing.[3] He acknowledges Sjahrir as a great figure of twentieth-century Indonesian history, incomparably greater than those of the current regime—the "New Order" of President and General Soeharto—to whom he has to listen so often as they talk about greatness. Nevertheless, whenever Sjahrir is mentioned, Pramoedya's tone is cold.

As Pramoedya recalls on Buru, Sjahrir, in contrast to other great men of modern Indonesia, never became a part of Pramoedya's family lore. When Pramoedya was a boy, his mother did not talk to him about Sjahrir:

Sukarno in front of a painting of Indonesian revolutionary youth by
Soedjojono, 1949. Photography by Henri Cartier-Bresson.
(MAGNUM PHOTOS)

She could tell hundreds of stories about the medical doctors, the doctors of law, and the civil engineers who were also active in the national movement. But she did not give much about "candidates," the people who studied but did not get their decrees; we never heard about them. She was full of stories about Imam Sujudi, Soetomo, Satiman Wirjosandojo, Soekarno, Sartono, Tjipto Mangoenkoesoemo. But she never brought up candidate Soetan Sjahrir or candidate Hatta.[4]

Pramoedya recalls that he first read Sjahrir's published letters, "the meditations," in 1961, when by another coincidence he was locked (by the "Old Order" of President Sukarno) in the same prison at Cipinang near Jakarta where Sjahrir's exile correspondence had started. Pramoedya does not recall being very excited by what he read. Day dreaming indeed, it seemed. And not befitting a patriotic intellectual in colonial Indonesia. "He," Pramoedya writes about Sjahrir, "so longed to return, return to Holland."[5]

Sjahrir was twenty-six when he was sent into exile, and Pramoedya was forty-three when he went. Yet the differences between the two men did not seem to lie in their ages. Rather, the time Sjahrir lived in appears younger. Like Pramoedya, Sjahrir was made aware that freedom was being taken from him for an undefined number of years: there was for Sjahrir as for Pramoedya no trial, and thus no time limit to their exiles. The experience, however, of the Dutch leaving the country in 1942, the Japanese occupying it, the Indonesians revolting in 1945, and all that followed throughout the 1940s, 1950s, and 1960s, was still hidden to Sjahrir by the wall of an unknown future. Like the innocent East Europeans before 1989, as long as the Wall loomed, Sjahrir could believe that there might be a better world on the other side.

The letters by Sjahrir from the mid-1930s, on his way to exile and from his internments on Boven Digoel and Banda Neira, at the most unexpected moments and describing the most unlikely places, could still exude a sense of an almost boyish adventure. However dreadful it was, Sjahrir's exile might still be believed to be some kind of beginning.

Whatever was ahead, Sjahrir's ship sailed fast, his seas looked blue, and his horizons seemed rather wide.[6] As Pramoedya travels to his exile in the late 1960s, and all that had been the future for Sjahrir—the war, the revolution—is known, his ship lingers. This is another "thinging." Everything is sticky or has sharp edges. The ship's deck is crammed with technology that makes it hard to move. There are the same "phosphor flickers" following the ship that Sjahrir observed and described in amazement.[7] As far as Pramoedya is concerned, however, "the old and rusting iron" of the ship is all around, the noise of the "weary" machines, the "W.C." and the "hygiene" that do not work.[8]

As far as the eye can see, death is everywhere: the sea, the latrines of the ship that incessantly squeaks and gasps for breath, bullets, bayonets, orders, roll calls, badges, pistols, rifles, commando knives.

And the ship radio is incessantly on, though nobody listens, the *kroncong* jingles popular songs that make one nauseous, a sermon by a preacher who wishes us good luck in a new life. . . .

What can we do in the cage of a ship like this; we all think of death. And the squealing *kroncong* songs force themselves upon our thoughts. *Kroncong* still had some power before independence, it still had some vitality—a vitality of a nation that was not yet free. As the Revolution erupted and passed, *kroncong* remained just a kind of narcissism, a posy of empty words, a culture of masturbation. Equal to the culture of great speeches and of puppet shadow theater.[9]

Along the ship, the silver skipjack fish of the eastern sea perform their acrobatics—now as then. Sjahrir was greatly taken with them. As Pramoedya observes it, the skipjacks have no choice but to keep on moving, otherwise their blood clots, they stop dead, and they sink into the darkness of the sea.[10]

Sjahrir was young by his historical time. In the malarial internment camp in New Guinea, one day he canoed upstream in the Digoel River, still inside the restricted area where inmates were permitted to go. He visited the Kaya-Kayas, the savage, stark naked, and black cannibals. He swapped some tobacco for sago, did not part with his hatchet however badly the wild men wanted it, and made it back downstream to his camp, like a Jules Verne hero, just before sunset. Thus, at least, Sjahrir wrote about it to his wife and her son and daughter back home.

Pramoedya also writes, as the ship reaches its destination, from the camp to his son, a little boy who, we may expect, is eager, like Sjahrir's children were, for a little trapper story of his own: "You might think of the stories from the American prairies, about Schultz' book *My life as an Indian*, the autobiography of Buffalo Bill, or the movie westerns. But here, for sure, there are no Indians, and not a single horse."[11]

This, Pramoedya writes, is a "heartless prairie."[12] Some hunting, true, is going on. In a letter to another child of his at home, his teenage daughter Rita, Pramoedya explains:

> Your dad never goes hunting. Once I went with someone who was looking for a deer, but we came upon an enormous wild boar, and instantly I was up a tree. Another friend of mine from Cengkareng whose name is Matuyani likes to catch crocodiles. . . . In the water, a crocodile cannot do much. . . . There are no monkeys here, but there is an enormous variety of cockatoos.[13]

The hunting story goes on. Yet another of Pramoedya's friends in the camp, an academician in his former life, taught himself how to catch insects and larvae for meat. Three other friends of Pramoedya on Buru, hunting too, came upon a piece of raw flesh hanging from a tree. They had a rare feast and only afterward found out what they had eaten. Does Rita know what was it?

The placenta of an indigenous baby. . . .
Now you can see, Rita, what people do to keep themselves up, to remain alive, and healthy, and strong.[14]

This is yet another chapter in the history of the Indies, and Indonesian technology, and technology of communication in particular. The letters by Sjahrir, in the late-colonial Netherlands Indies, after being read by a censor, traveled by air mail once a month, and later once a fortnight as the service improved. The letters by Pramoedya, with few exceptions, were never expected to reach their destination, and they rarely did. Most of the time, in fact, they could not even be sent.[15] They were being thought out by Pramoedya, they stuck silently in his head, or they moved in circles around him as myths about his letters whispered among the others in the camp.[16] Describing his letters from Buru, in the independent state of Indonesia of the 1970s, Pramoedya wrote with much precision: "as paper kites" they flew.[17]

Memories of Holland

Sometimes, suddenly, against our wishing it, Pramoedya Ananta Toer—the writer of the revolution, an Indonesian to the root of his soul, and an exile— reminds us of the radio amateurs of the late-colonial Netherlands East Indies of the previous chapter. They were, let us recall, modern and apprehensive, Dutch or Westerners mostly, but not necessarily. They sat in their houses, in the colony sliding into disaster, pressed themselves closer to their radio sets, strained to catch a signal from afar, Holland most preferably, the West certainly—the main source of their modernity and their anxiety as well.[18]

The radio played as Pramoedya was shipped to Buru. Radio is mentioned in Pramoedya's letters from the camp often. Often, too, an image of Holland far away appears, connected with radio. At the end of a long letter from Buru to his son, as he tries to make a sense of what happened to Indonesia during his lifetime, Pramoedya recalls the shock and trauma he felt when, shortly after the Indonesian revolution, some time early in the 1950s, he heard a program on Radio Holland from Hilversum "that mused about what was going on with Indonesia, and was called 'Indonesia is on the Moon.'"[19]

Dutch, Pramoedya remembers in the camp, "was not the language of our family."[20] He explains what he means by that:

Later, I found out that we had quite a few books in our house by Eurasian colonial Indies writers, in pretty yellow covers and in Dutch. But [my mother]

Removal of the Dutch official portraits from the governor-general's palace in Batavia-Jakarta, 1949. Photography by Henri Cartier-Bresson. (MAGNUM PHOTOS)

never read to us from them. Neither did she ever read from [Javanese] *Pancatantra.* . . .

My mother liked to sing. But her repertoire was quite narrow. Most rarely did she sing Javanese songs. Probably as a child she had more often heard Malay than Javanese songs for children. Her favorite was Malay *Sang Bangau*, "The heron." Next, of course, the Dutch songs that she had learned in school followed. But when [Dutch song] *Als de orchideeën bloeien*, "When the orchids bloom," became so popular, strangely, she never sang it. She also very much liked political songs of the time, and she liked to sing the songs that my father wrote, mostly in Dutch.[21]

Dutch seems to be remembered with an ambiguity, like indistinct but constant buzzing in the air, a background sound that is heard only when the family music, or the true family language, Javanese or Malay, gets weak or silent. Like a radio, indeed, that is permanently on, but nobody pays particular attention most of the time. "When the orchids bloom," by the way, was one of the biggest late-colonial radio hits of the time. Some of the songs written by Pramoedya's father and sung by his mother also got on the Eastern section of prewar Netherlands Indies radio.

There is light and darkness in the memories. A dimmed light (and one may recall Mas Marco and the Indonesian dandy of the late-colonial period).[22] Holland, too, is remembered by Pramoedya on Buru as movie stars' faces. In a painful letter to his daughter from Buru about how he met her mother, his first wife, in 1945, and why the marriage broke down so dismally, Pramoedya writes: "When I was a boy, I often looked at the pictures in Dutch magazines. The American and European movie stars always had such big eyes, not like the half slant-eyes of my people. . . . Since childhood, big eyes have had an esthetic effect on me."[23]

Holland recalled, in Pramoedya's letters from Buru, has the distinct quality of a technological project. As nearness, an unmediated sense of other humans and things, appears to weaken and about to vanish from Pramoedya's life, Holland is ready to be switched on. In the early 1950s, Pramoedya writes to his daughter in the letter just mentioned:

I worked day and night, and more and more often went with your mother to the movies. . . . Like mad, I just kept on working, like a volcano discharging lava. At the time, M. Balfas [a fellow writer] said to me when we met accidentally: it is writing no more, what you do, Pram, it is ordinary shit.[24]

. . . Truly, I became a bread writer. . . . No more could I connect myself with the dedication that Han [G. J. Resink] wrote for me in his poem *Het koraaldier*, "The Coral Polyp": "To Pramoedya, the writer of youth." I was nothing, just a bread writer, a writer who never has a break.[25]

Holland was switched on at the moment of human failure:

> Via Han, from *Sticusa* [Dutch "Foundation for Cultural Cooperation"], an invitation came for me to spend a year visiting to the Netherlands.
> Truly a honor. However, again, a cause for doubt. My Dutch was bad. I could neither speak nor write it; only read. Whenever I was to face a European, my inferiority complex became even stronger than usual. However, I had to think of the advice my mother used to give me: go to school in Europe if you ever can. I accepted. . . . You'd better take your wife and children with you, Han told me. Yes, I thought. Not everybody is so lucky that he can visit Europe.[26]

With his wife and two children, Pramoedya boarded a ship in July 1952. He felt, he remembers it twenty years later on Buru, "as on vacation": "For the first time I traveled abroad. And we had a second-class berth."[27] In another letter from Buru many years after this happened, he reminds his daughter that she was not sick once through the whole twenty-six-day trip.[28]

The year Pramoedya spent in Holland is recalled in the camp on Buru two decades later smoothly. Mechanically. Photogenically, Pramoedya recalls the thinness of his life, the sudden easiness of his writing, the easiness of his marriage, as all of it was fast dissolving.

Pramoedya remembers himself feeling clumsy among Dutch "big shots" while in the Netherlands. He could not understand how other Indonesian writers and artists who were on the *Sticusa* invitation with him could socialize so closely with men who, only a few years earlier during the revolution, on the other side of the battle front, might have been responsible for the death of their and his Indonesian friends. Comfortingly enough, however, coldly, Holland throughout was unreal.

Pramoedya, if he still needed it, learns about disinterested watching. He recalls an invitation to a symposium on modern Indonesian literature where scholars from Holland, England, Germany, and Australia read their papers. Looking back at the event from the camp on Buru, "the symposium was truly important to me. . . . They debated Indonesian literature that was not theirs, not as a sport but because they were interested in it as intellectuals."[29]

He seems to recall himself in Holland, even, as being able to find some love without touching. In a letter from Buru, he writes to his daughter:

> I used to go just by myself and sit on a bench made of concrete in the park. Once a young woman came around, and she sat down besides me. . . . She appeared to be a cultured person. . . . In detail, she spoke about French literature of which I did not know anything at all since I read Victor Hugo, Zola, and Balzac in school in Surabaya as a boy. The little I knew about the new literature was a couple of thin little books by André Gide. . . .
> We also talked about my complex. I do not know whether she studied psy-

chology or not, but, in any case, I have learned from her that everybody has a complex, only some people are aware of it and others are not. . . .

Your mother was not affected by this friendship at all, and she did not have to know about it. At home, I lived happily with my wife and children. All day I could listen to good music from all over Europe on the radio. Left to myself, I worked without pause on my fine, flat mini-Olivetti. . . . I am telling you this to show you that there was not the slightest discord between your mother and me. Exactly as in Jakarta, we often went to movies, concerts, and for walks with your eldest sister, and with Etty in a stroller.[30]

In the same letter, Pramoedya recalls the climax of his Dutch affair and of the visit to Holland almost tenderly, as an upbeat carnality of milk and yogurt:

The contact with my friend, the young woman, gave me back the self-confidence that I had almost believed was lost forever. It made some sense, now, to study. . . .

I did not mind the fall with its bristly winds, smell of rotting leaves, bare branches, and falling temperature. I became healthier and healthier. Daily I drank one liter of milk and three-quarters liter of yogurt. In the morning, no more blood was running from my mouth and nose. I felt physically and mentally like a bird. And I could work up to twenty hours *non-stop*—something unthinkable in my own country. I gained twenty-eight kilos.

Snow, in December 1953, was not yet falling.[31]

Pramoedya's wife and his children had returned to Indonesia earlier that fall. He followed them by plane at the beginning of 1954. It was, and Pramoedya repeats it, like a return from a vacation. He also divorced his wife very soon afterward:

The Netherlands let me see the beauty of an organized society, where everyone's service was appreciated on its merit, and where everyone had a right to attain one's means of existence. . . . Your mother began to find fault with the family house in Blora. . . . No longer could I work like before. No longer could I be merely a bread writer.[32]

The image of the late-colonial Indies radio amateur reappears, however inappropriate it seems, so long after the independence and in a man like this. The anxious modernity. The modern awareness that unplugging will bring a soundless loneliness in one's own house:

In Jakarta, during the couple of months I was away [in Holland], nothing changed. The dirt, the torment right in front of you, the baseness of human feelings everywhere around you. I have written one piece about our neighborhood, which was no more than a few hundred meters from the [presidential] palace. But there was no response except for some stones thrown at our house by the fellow community dwellers, who felt that the locale had been offended.

. . . inflation . . . as long as the Indonesian bourgeoisie remains unproductive and uncreative, in contrast to that in Europe, Japan, or America. . . .

Indonesia, indeed, was no America, and no Europe. Here, I stood alone.[33]

Time in Three Dimensions

There is as much on forgetting in Pramoedya's letters from Buru as there is on remembering. It took him months, for instance, he writes, before he was able in the camp to recall all the names of his children. One reason might be, he offers as explanation, that he was hit on head with a rifle butt by one of the soldiers who arrested him on October 13, 1965.

Pramoedya remembers that, sometimes in the period when Sukarno was the president of Indonesia, in the 1950s or in the 1960s, he, as a writer, served on the board of a pencil factory. He also remembers some of his other official or semi-official functions in the regime at the time. He does not recall in the camp when asked whether, in one of these functions, he spoke against a fellow writer, Mochtar Lubis, a dissident at the time, being let out of prison. Pramoedya's significance, however, does not stem from his being an innocent victim.

At one point during his exile on Buru, as he contemplates a historical novel he once wanted to write, Pramoedya suddenly realizes that the name of the most famous fifteenth-century treaty dividing the world between Portugal and Spain had slipped out of his memory. Friends on Buru all try hard to help him out, and after a while, at last, one of them, a younger man and fresher from school, comes up with something that sounds like *Tor de Sila*. It is evident, Pramoedya comments, that the young man actually constructed the treaty's name in his memory out of an resonating echo of Pramoedya's own family name, *Toer*, and that of the name of the current Indonesian state philosophy, Panca *Sila*.[34]

Memory is faced with anxiety on Buru, and Pramoedya is aware how much is at stake. Memory, not rarely, is the only foundation on which the camp's present and its prisoners' integrity depend. In one letter from Buru, Pramoedya thinks about the possibility of untying himself from time. May time, as it flows from the origins in the past, be constructed, engineered, and thus taken apart? Time "with its three dimensions," Pramoedya writes: the past, the present, and the future. This may indeed be the culmination of the century-long efforts by modern Indonesians to landscape their history: "*There's a happy land somewhere* . . . universal symbol of the future."[35]

Long before Pramoedya and his fellow prisoners arrived, time on Buru was seen as moving in a peculiar way. Since deep in the late-colonial period also, Buru appeared to be a place fated to serve as exile to the modern humans in the Indies. As if time on the island in fact was always broken into three segments, the past, the present, and the future, adhering only loosely to each other, and each of them to the Buru airs. As far back as 1858, a Dutch report described Buru as a "remote corner long forgotten among the public." At the same time, the report deemed Buru to be, in the future, a "place to relieve the fatherland of overpopulation, and to empty its prisons as much as possible."[36]

The airs of Buru, better than any other place, might help Pramoedya to lose, to construct, and to break the flow of time. Pramoedya, an engineer of post-colonial Indonesia, in fact, screws his time together, and there is nothing he likes more than to bring a little screw, a nail, a piece of copper wire from the outside. The tribes of Buru, the "aborigines" of the island, Pramoedya writes, for instance, in one of his long and detailed letters from the camp, may easily be the lost descendants of the glorious Javanese fourteenth-century empire of Majapahit. True, they do not know any Old Javanese or Middle Javanese, but,

> as you observe their clothes, Majapahit pops into your mind. They wear batik headdresses with reddish brown as the prevailing color. . . .
>
> When you look at the way they wear their headdresses, it is as the soldiers of Majapahit, in their own time, might have worn them. When they move around Buru in small groups, they proceed in a single file. This may be, also of course, because there are no wider roads here, only footpaths.[37]

Any suggestion of time's uncontrolled flow could be perilous on Buru, much more than anywhere else. After Majapahit, Pramoedya writes, came colonialism, after the revolution came General Soeharto. Pramoedya, on Buru, plans to write "a novel about a Period of the National Awakening—as a period and not as an event."[38] When he was a little boy, Pramoedya now recalls, more than anything he liked to read the old Javanese time calculations of "lucky" and "unlucky" days as they were published in the popular almanacs of *Balai Pustaka*, "House of Books."[39] This was how it was pleasant to be. This is how it still should be, if possible: time in timetables, tabulated time, time broken into columns and rows of signs, time in groups and sub-groups, "periods, not events," time in multiple dimensions, disassembled time.

Pramoedya, writing on Buru, like the generation of modern Indonesians before him, still believes that history began in "pre-logical time,"[40] and that it moved "along its path toward the time of slavery, the feudal time, the colonial time, and the present time, my own time."[41] Pramoedya still believes that history is logical, and that Indonesia moves, logically, with the rest of the modern world. In the sixteenth century, he writes, the world colonialism was born out of the conquest of the Spice Islands in Eastern Indonesia. It was thus "not accidental" that three centuries later the Indonesian revolution, again in the same spot of the world, opened the logically next-to-come, global era of decolonization.[42] "Inevitably" as well, in the postwar period in Bandung in 1955, it was Indonesia, again, that inspired the new historic Asian-African "brotherhood."[43]

Early in the twentieth century, in Kartini's times,[44] and maybe as late as in Sjahrir's time, the logic of history might still promise an Indonesian future built as simply as a modern bridge or house. In the early 1970s, however, Pramoedya wonders about the way history is wound: "like a rope whose beginning is tangled into its end, and the circle is closed."[45] What is the function of the revolution, in the logical construct of history as he learned it, and as he still clings to it?

> I don't know. . . . The process cannot be judged with any meaningful result until there is a written history of it. And I have not yet encountered any true [Indonesian] work on the Indonesian Revolution.
>
> Even as far as the question is concerned, whether the Revolution is finished or not, historians remain silent, and their mouths are full merely of teeth.[46]

Describing a night-long *wayang*, puppet-shadow theater performance he saw on Buru performed by a prisoner puppeteer and with prisoners as the audience, Pramoedya thinks—logically—about a "vicious circle":

> A vicious circle, out of a point of departure, and back to the point of departure. The blood we all have bathed in did not make us any younger. We became an old nation, and we are keeping on the move in vicious circles: from an open door of colonialism to an open door of post-colonialism, a blood bath after a blood bath?[47]
>
> Those who came after [the past Javanese rebels of] Arok, and those who came after those who came after them, did not carry on; they just were returning to the old beginnings, bent just upon going back, reentering the vicious circle. And what now? Whom should I ask? What, indeed, is so wrong about keeping on moving with the ghosts of our ancestors in the vicious circles? It goes on like that for centuries, doesn't it?[48]
>
> And bodies are thrown into the air by just a sleight of hand by a noble and delicate knight . . . men are changed into animals, and animals are changed into men. There is no evolution. And no revolution is heard about. . . . As the sun rises again, the gods, the priests, and the knights [the puppets of the shadow theater] return to the perspective of their true forms. The only ones who have no perspective are we, all the rest, who watched the play.[49]

Bacteria

"And we have to ask again, Et," Pramoedya writes from Buru to his little child at the end of another letter almost entirely devoted to history, "whether we may ever break from these ever bloody vicious circles? And whether we ever may fly free *to the unknown*, as the saying goes?"[50] This is, let me repeat, just another chapter on building modern Indies, and Indonesia. The constructions of all those who preceded Pramoedya, and his own construction, close upon him. The perilous flow of time is feared, dammed, divided into little streams, and, as it had been so often before, subjected to a bird's-eye perspective. It must be anchored too, tied to a secure point. This is why Pramoedya in his Buru letters returns so often to the Indonesian revolution:

> Whatever the other circumstances might be: whoever did not pass the test of the national revolution will fail in all other tests that may follow; whatever he might do otherwise, he will be only pretending to be living; the intensity of his life vanishes, all that a man might achieve for himself, all the qualities he possibly might have given to his nation, vanishes.[51]

To Pramoedya, locked on Buru, the Indonesian revolution of 1945 is the absolute. It is, indeed, what may be beyond all the constructs of history. It may be the "flying *to the unknown*." It may be the moment when time stops:

> from colonial men to free men . . . from nothing to something. . . . As she-liberator of tens of millions of people, the Indonesian Revolution does not lack anything when compared to the other revolutions; it is greater, yes, endlessly greater than [for instance] the industrial revolution.
>
> It is often said that the industrial revolution was the first revolution in production, when manpower was replaced by machines. The time in which we live, we also hear, is the time of the second revolution of production in which computers take the place of human brains. However, although a man may indeed be lost without his computer-delivered information and without his machine-defined station in life, we must still stick to the conviction that man is more primal, and that one can raise oneself above the conditions of his subsistence. . . . The revolutions of production, therefore, mean totally nothing when compared with the Indonesian Revolution. Only he who is blinded by the earthly brilliance can be more impressed by the other revolutions. . . . How beautiful is the Indonesian Revolution, the mother of all virtues, this Pradnya Paramita.[52]

As if it was still necessary at this point, Pramoedya explains that, in fact, Pradnya Paramita is the name of his mother; that to his memory, and in his view of the world, these two are identical: his mother is the Revolution, and the Revolution is his mother.[53] This is the anchor! Or at least for a moment, it is comforting. If nothing else, it permits Pramoedya, in another letter to his child, to stay with the revolution as one can stay, so he writes, with a "story" or a "fable."[54]

This is, indeed, like building a modern house amid a jungle. Time on the outside swirls as Pramoedya's mother is recalled. The past, the present, and the future lose their sharp edges, and even their shape, as Pramoedya's mother is placed in the center:

I found out what her name was only after she died, when I saw it carved on her wooden memorial. . . .

. . . My mother always gave the impression that she was tired, while my grandmother was always fresh. . . . On my grandmother's lips there always played a little smile, and she looked at the world with bright eyes. My mother had a melancholy in her eyes, and she observed the things around her with an evident uncertainty. My grandmother was different. She looked younger than my mother.[55]

The mother, in fact, is never remembered by Pramoedya in the camp as telling him or her other children "stories" or "fables." She "read" to them, and, as she had never learned the Javanese script, the most natural source of stories and fables for Javanese kids could never be tapped. Pradnya Paramita read to her children books mostly in Malay and Dutch.

Indeed, she is sometimes recalled by Pramoedya as a child rather than a mother: as she fell asleep, for instance, often before he, the boy, did, with an open book on her lap. She died young, of tuberculosis, much younger than Pramoedya was when he wrote this. She, if anyone, was immediacy to Pramoedya. The thing to connect with. But how to touch her then, and now, when recalling her on Buru? "I knew," Pramoedya writes about his mother's death,

> that when someone died of tuberculosis, bacteria came out of the mouth and nose. People said that, if you place an omelet over the face of the dead, the tiny holes in the omelet would swallow everything.
>
> No, my mother could not murder me with the bacteria. I threw my arms around her cold body and kissed her on her forehead.[56]

This is the point where memory and senseless flow of time might be supposed to come to rest. Pramoedya remembers that he was out of the house when his mother died. He remembers running back to the house a few moments too late. He asked the old woman who watched over his mother's death whether there was any message left for him by her, before she died. "No message."

This is the latest chapter on the Indies and Indonesian technology, and the technology of connecting and of escaping most of all. Pramoedya's father, he recalls on Buru, wrote songs that Pramoedya's mother sang. Many of them were composed in *kroncong* style, and some became, as mentioned earlier, late-colonial Eastern radio hits.

Pramoedya recalls his father's aloofness from the family as it increased over time. These particular recollections are like a description of a space being laid bare, progressively, piece by piece. There is no suggestion, even, of a kiss. Initially, Pramoedya recalls, his father was at home most of the time, and the house, also a school at the same time, had been crowded, and it bustled with working things:

first, there was the printing done on a Godir machine: forms, textbooks for courses. Then they found Godir too slow. A new Roneo printing machine appeared in the house, and a Gestetner stencil machine with big boxes full of paper. Besides, there were seven standard typewriters acquired second-hand, but still in good shape. . . . Father hired a typist from Semarang to give typing lessons. As I estimated later, when I myself became a typist, the man was able to type almost three hundred characters in a minute. . . .

Zinc bins for the Godir machine stood all around, and some were used to store water. . . .

The printing and stencil machines ran constantly each afternoon. In the storage of the house, there were two sets of *pelog* and *slendro* gamelan in rows along the walls, and they could be borrowed freely at any time. On the verandah of the house, there were some iron bits and pieces in one corner, spare parts waiting for building a studio for training pupils as smiths and carpenters.

With the great depression of the 1930s, however, and with other misfortunes,

with the end of Partindo [pro-Sukarno nationalist party], with the Wild Schools Ordinance [strict government control of national schools], the machines vanished one after another from our house, I did not know where. Not a single typewriter remained. The printing and stencil machines disappeared also, leaving behind just two spools, and soon nobody knew what they were for. The boxes for paper were empty. Not even a bicycle remained behind the house. . . .

. . . My father now was gambling all the time in the houses of his friends. . . . He remained away from the family more and more. Whenever he happened to be staying a few hours at home, he sat, sunk in his thoughts, then, quickly, he put his clothes on and vanished again. . . .

Nobody dared to say anything to my father. His eyes took all your courage away. He almost never opened his mouth at home, except when he asked for a glowing cinder from the kitchen when his cigarette lighter ran out of gas.[57]

Pramoedya, as eagerly as any Indies engineer we have met until now, searches for a center and an axis. The mother is in the shadows, and, ultimately, she does not leave a message anyway. The father's eyes, and an occasional gleam from the hot cinder, gas cigarette lighter, or the cigarette's end, is all that is visible. There is a distance, stiffness, and lack of voice in the memory.

The Splendid Radio

The dream of the boy Pramoedya, thirty years ago, is recalled on Buru as a "plan":

I saved; everything went into savings. I nourished a plan about which nobody knew. Not even my mother.

An uncle who had just returned from New Caledonia had, among his things, a book about electrotechnics. I leafed through it, and it fascinated me. I went deeper. The formulas, however, became very complicated. I could not understand them, and neither did I ask anybody.

With my savings I bought liters of sulfuric acid, some copper, and contacts, and just by myself I built a few batteries. As the contacts began to spark, I was so happy that I clapped my hands and applauded the thing. We had never had electricity at home. I was the first to bring it into our house. . . . In a little while my savings were gone, and I could go no further.[58]

Pramoedya had finished with his local elementary and lower-middle school by 1940.[59] He wanted to study further, and, indeed, he chose electrotechnics. But this was clearly beyond the family means, and, besides, Pramoedya's father did not think his son was a promising enough student. Looking back, Pramoedya admits that his father might have been right. Everything pointed to Pramoedya never making it as, for instance, Sukarno or Anwari did, two prominent national leaders and the favorites of his mother, both of whom became engineers.[60]

Pramoedya's mother was already ill at the time, and she had only about two more years to live. By the summer of 1940, however, she unexpectedly made some money of her own. With her husband mostly away, Pramoedya recalls,

She asked me: "To which school do you want to go? And where?"
. . . I decided: a technical course, divided in three terms, each only six months long. . . . And there was a connection with electrotechnics: it was the Radio Technical School in Surabaya.[61]

He got a wristwatch from his mother, a pair of leather Bata shoes, and two gold rings in case of emergency. On Semarang-Joana Steam Tram (the tram on which Kartini used to travel four decades earlier) the fifteen-year-old Pramoedya left for the city of Surabaya to become a radio mechanic.[62]

On Buru, he remembers himself in Surabaya as touching the radio very lightly. Yet he was focused on communication. In his boarding house, he writes, he lived "with no student but clerks and workers": "They placed me in a back room, and there, on the top shelf of a wardrobe, there was a whole little collection of books on black magic."[63]

Pramoedya also read in the back room "a book on sex by some Swedish scholar," a thick book, he recalls, and he read it several times. Besides:

One truly captivating book there was *De laatste stuiptrekkingen*, "The agony of death," an account of the Boer War in South Africa. . . . From a housemate I borrowed a Javanese book in Latin script by Pak Poeh, which had a great influence on my mental development. . . . I also read, in that room, cycles of novels by Emile Zola and Balzac. . . . In short, everything . . . Handwritten porno. Little detective books.[64]

In the radio school, as he remembers, he was not particularly interested in actual repairing:

> The school did not attract me very much any more, yet I still got good grades in electrotechnics and in radio theory. The practicals were torment, notwithstanding the fact that I could walk around in a white laboratory overall. The source of all the trouble here was my fear that, by one mistake or another, I might break a radio set. If this should happen—I saw my mother before me. How dreadful it would be for her to have to come out with the money for the thing. . . .
>
> At the term examinations . . . in technical drawing, my assignment was to make a sketch of a television, at that time not generally known, and I was not seriously worried about the result of this particular test.[65]

This is how it is remembered. Pramoedya, with his wristwatch, a pair of Bata shoes, and two gold rings, came from a house where cavities and silence gaped after the father withdrew or was not speaking, and the machines were removed or disconnected. His mother sent him a new bicycle to Surabaya, but she could not do much more. And here is the familiar equation. With too few strings attached to people and to things, the dreams or plans of the Pramoedya boy focus, rather, on the amazing wireless: "yet another six months, and I might acquire a *marconist*, 'wireless telegrapher,' diploma. Then, I could go and work on a ship, a radio station, or in a news agency."[66]

When the final examinations at the Surabaya radio school, with all their ceremony, were to take place, Pramoedya recalls on Buru, the month was December 1941:

> When we were all sitting in one row of seats in the big laboratory, the most powerful radio receiver in the school was switched on and tuned to the transmitter in Batavia. A signal came in, very hard but distinct and not too coarse, given the fact that at the time there as yet were no hi-fi systems. The Dutch news reported that Japanese war machines had without warning attacked Pearl Harbor in Hawaii. America declared war on Japan; England, too. Then, the Netherlands-Indies declaration was read in which, also, war was declared on Japan.
>
> As one man, the students shot out of the classroom, jumped on their bikes, and went home. Myself, too.[67]

Diplomas of the graduating class of the Surabaya radio school were lost somewhere in the colonial office in Bandung, where they were supposed to be stamped. Pramoedya waited a few weeks to see if there would be a graduation, and then went home to Blora. He tried to get a job at the only radio shop in town, but there were as yet so few radios in the little town that even the shop's owner was not kept fully employed.

It was at this moment, in May 1942, that Pramoedya's mother died.[68] He waited, again, for a few more weeks and then decided that it made no sense to stay any longer.[69] An uncle of his lived in Batavia, now renamed Jakarta, and he was willing to help the young man out. Pramoedya recalls on Buru that he took some spare radio parts with him. He still might, sometime, build a good radio, at least for himself: "I had a rectifier with me [in Blora], and some other radio segments. . . . I had planned to build a radio, but because there were no additional spare radio parts available, I took everything with me to Jakarta."[70] As far as he was concerned, or as far as he recalls on Buru, the rectifier and the other things were never put together. "Since then," he writes on Buru, "I have cut all my ties with the radio trade."[71]

The Mouth of Karundeng

In August 1942, with his "best outfit, in a white shirt, and long white trousers," truly "a good looker," Pramoedya came up for an interview at the official Japanese news agency, *Domei*, in Jakarta. The interview was conducted in English. The uncle, with whom Pramoedya stayed in Jakarta, knew English: "and I have learned the language in just two weeks from my uncle. . . . I answered just with yes or no. . . . From behind the wall there could be heard a rattle of at least a half a dozen typewriters."[72]

Pramoedya's coded language, *"yes* or *no,"* it is recalled, carried the day. The sound of working machines once more filled the space, and Pramoedya found a new life in typing:

> First I had to retype an Indonesian text, and then an English text. At that time, I could type two hundred characters a minute, and I was hired.
>
> . . . I had my own desk and a typewriter. . . . My first assignment was to type short news reports on stencil paper. . . .
>
> . . . reporting fast became a part of my life . . . Adam Malik . . . P. F. Dahler . . . Djawoto [all were working there] while, in administration, there was Nur Nasution and the brothers Lubis. . . .[73]
>
> . . . With daily training, I improved so much in just two months that I could type two hundred and eighty characters a minute, and thus was probably the fastest in the office.[74]

Pramoedya recalls on Buru that, while in the Domei news agency, he read Spengler's *Decline of the West* and Ortega's *Revolt of the Masses*. There was war all around. But he does not seem to recall having a particularly troubled mind. Rather, he seems to remember a sense of almost a liberation, or a detachment, or a liberation in a detachment: "Nobody gave a damn about the Pacific War; it merely filled the pages of the bulletins that I typed."[75]

Deeper than Kartini, more somber than Mas Marco, Pramoedya recalls himself living in technology. The rhythm of typing and the principle of the typewriter appear to organize Pramoedya's recollections of the time. Number of words per minute appears to matter at least as much as the words themselves. Technique of message is recalled with a greater liveliness than the message itself:

> Then Adam Malik had me do something in the office that was completely new: start chronicles. First, I had to put together a chronicle of the China-Japan War. . . . I had to arrange everything in folders. This did not last long, and I was moved into a section of the archives. The archives were not kept in accordance with the decimal classification as I learned it later. Anyway, this was my first encounter with inventories. Each day, I sifted through newspapers, divided the articles according to the categories that the agency used, clipped the articles out, and pasted the clippings in files, hundreds of them, that were placed to be ready for use on a large and high rack.[76]

History could be made liable to coding. Occurrences of history could be clipped off and pasted in. Time could be ordered. In his Buru camp memories of the Japanese news agency editorial room, Pramoedya's staying and moving makes sense:

> Against the wall, among the other bookcases, and not looking any different from them, stood a wooden cabinet with glass doors. Its shelves were filled with rows of volumes: *Winkler Prins Encyclopedie*, *Encyclopaedia Britannica*, two volumes of *Who's Who* published in the United States, three English documentation folders, and a complete set of *Adat Recht*, "[Indonesian] Customary Law," bound in green cloth.[77]

Pramoedya recalls how intensely he worked and schemed to secure at least one volume at a time of the *Winkler Prins Encyclopedie*, to get one of these volumes to his desk from the adjoining room, which was their permanent place:

> I could not read the *Britannica*. The too-wide spacing and also the big round letters were not truly attractive. In the volumes of the "Customary Law" I read often and, thus, picked up something about the rules of life.
> Reading in these big volumes I felt myself bigger, too. The world seemed less narrow-minded than before, and, what was more important, nicer.[78]

This is how Pramoedya lets us know the path that, as we all know, led him to become a writer. The cavities appeared, now, that they might be filled. The world appeared that it might be structured, however only into the thinly spread order of the news room and of the intermittently ticking typewriters.

Pramoedya kissed his dead mother on her forehead. He did not dare to kiss her mouth; this, he knew, might be lethal. He went on, lightly touching all the matters of his life as they followed. There was some message. Yet, a suggestion of another kind of touching, the mouth, an idea of mouth, kept coming back. "Beginning in 1944 . . . Domei looked for two employees to register for a year-long course in stenography. . . . Adam Malik let me know that I had been chosen and should get ready for the course."[79]

The setting for the course when it started in the middle of 1944, as Pramoedya remembers, was impressive:

> It was a very fine place, where the school was located. . . . All the doors and windows were decorated with black and dark-red corduroy curtains. All the furniture was equally pretty, everything was beautiful, including the W.C. This was because the school was allocated a small section in the building of the former [Netherlands Indies governor-general advisory council] *Volksraad.*[80]

This was the first time in his life that Pramoedya, a son of devotedly nationalist parents, met the most famous leaders of the Indonesian struggle for freedom face to face:

> On the first day [of the stenography school] the Head of *Chuoo Sangi-in* [Japanese occupation authority] appeared with his staff, and with Soekarno, Engineer. . . . The first lecture to us students was given by Soekarno, and he spoke on the subject of politics.[81]

Other lectures followed by other almost equally famous Indonesian figures like Hatta, Soekardjo Wirjopranoto, and Maroeto Nitimihardjo. Yet all this seems to come out in Pramoedya's memory on Buru as if it were secondary, a mere backdrop to the real thing:

> For the rest, there were the stenography lessons by Karundeng. . . .
>
> Karundeng was a tremendous teacher, and I will remain grateful to him my whole life. . . . He taught a blind system of stenography.
>
> Karundeng unlocked for me the beauty of the Indonesian language. As he read out a text, all the stenographers held his lips in their eyes. The lips opened, then closed, they were pressed together, they gaped, they spouted, they exposed his yellow teeth, but nobody noticed. The voice that came out of his mouth formed clear and flawless words . . . not a single letter was suppressed, not a single comma, full stop, or a question mark was clouded. Everything stood in its place.[82]

All the great living personalities of the Indonesian nation paraded before Pramoedya's eyes at the sessions of the Japanese-approved *Panitia Persiapan Kemerdekaan Indonesia*, "Committee for the Preparation of Indonesian Independence." The Indonesian revolution was fast approaching. Yet Pramoedya's memory on Buru, in search of meaning, spins and whirls around Karundeng's mouth:

> The first practical in the school was an unforgettable sensation.
>
> . . . in the plenary hall of the *Chuoo Sang-in* building . . . [where the meetings of the Committee for the Preparation of Indonesian Independence were

taking place] on a high podium, there was a special chair for Soekarno, Engineer, the chairman of the gathering.

. . . The practical was done according to the blind system; it lasted ten minutes, and then the next student came to replace you. . . . Two months later the students began to practice as assistant stenographers at the *Chuoo Sangi-in* building. . . . In these practical lessons I learned that the average Indonesian speaker does not go faster than a hundred and seventy-five letter-groups a minute.[83]

Mohammad Yamin, a well-known Indonesian historian and flamboyant public figure, asked Pramoedya at the time to transcribe a series of his lectures on the nineteenth-century warrior against the Dutch, Javanese Prince Diponegoro. The book that came out of this became, in the postwar period, one of the most influential texts on the Indonesian national history. In Buru camp, Pramoedya remembers the event, however, this way: "The four lectures dealt with Diponegoro and lasted in total eight hours. . . . Fortunately the speaker, though he sputtered and sounded at first like he was speaking fairly fast, never went beyond a hundred and sixty letter-groups a minute."[84]

Something should always be left wanting. Words, if utterly words, might discharge from the mouth like bacteria and kill. Touching, if utterly touching, might draw blood. The mouth of Karundeng, in all its appeal, and in passion that bared his yellow teeth, still feels to us—and should, probably—like the cold forehead of the dead Pramoedya's mother. And, as we look closely at Karundeng's mouth as remembered by Pramoedya, it is made very much like a machine.

How may the revolution, then, be remembered?

On August 17, 1945, Sukarno and Hatta declared the independence of the Indonesian Republic. On September 18, 1945, at a large demonstration in the main square of Jakarta—formerly *Koningsplein*, "King's Square," now *Ikeda*, later *Medan Merdeka*, "Freedom Square"—the leaders of the Indonesian revolution met face to face with the Indonesian people for the first time. Japanese troops, asked to keep order until the victorious Allies, the British and the Dutch, arrived, stood alert in positions around the square and watched. In Pramoedya's Buru memory:

On high platforms, kind of watchtowers, stood groups of armed Japanese soldiers. The loudspeakers that were announcing the arrival of the president [Sukarno] and the vice-president [Hatta] could not be heard over the cheering and shouting. Finally the escort got in. And when the president stepped forward, everything stopped quiet. From the platform I heard him say softly: just go calmly home.[85]

Pramoedya seems to be intent on this particular detail. There were loudspeakers in the square, but Sukarno's voice could be heard unaided. The crowds listened, as Pramoedya believes he remembers, to a "soft voice."[86]

Commandant of the Command for Restoring Security and Order,
General Soemitro. (David Jenkins, *Suharto and His Generals*)

The people left Ikeda square. The streets were crammed with the Japanese military, technology, actually, that made it hard to move. Everything had a hard surface and sharp edges: the Japanese panzers, the armed Japanese soldiers. We can go on in describing the scene almost verbatim, from how Pramoedya described the ship that, in 1969, was carrying him to the camp: "bullets, bayonets, orders, roll calls, badges, pistols, rifles, commando knives . . . radio . . . however nobody listens."[87]

The streams of people leaving Ikeda square pushed into the narrowing space left between the Japanese hardware and the soldiers. The people were not afraid. Some of them already had sharpened bamboo spears, but this is clearly not the point Pramoedya wants to make. The people dared to get close to the Japanese, who were edgy with their vehicles, swords, and rifles. Utterly close! In the crowd, as Pramoedya recalls it in Buru camp, a young Indonesian man with a naked hand grabbed at a Japanese. Naturally, he could get hold of nothing but the soldier's bayonet. Some fingers were lost.[88]

Sportsmen-Dandies-Jokers-Engineers

Thirty years after his mother died, and more than quarter of a century after the revolution—three-quarters of a century after Kartini, and forty years after Mas Marco died—in the seventh year of his exile, Pramoedya writes to his child from Buru: "With my common sense, I understand the situation I am in less and less."[89] Other approaches are tried. Little around Pramoedya, for sure, as he lives in the camp, is described straight. There is a very distinct carnival quality emphasized about the place.

The Indonesian attorney general, for instance, as portrayed, was witty. When asked at a news conference in 1969 whether it was possible for Pramoedya to write on Buru, he said: "He can write, of course, but he has no pen and paper."[90]

There was, as we read Pramoedya, no end to that kind of laughter. The cage of the island Pramoedya was locked in was natural. In one passage by Pramoedya, Buru "forms a natural prison, as it is located in a valley with barriers all around made of forest and bush, an unbroken mountain chain in the North, West, and South, and the sea in the East."[91]

The prison camp, and more and more so, in fact, as time went on, was worth a visit: "as officers arrive at one of the cultural evenings, one can hear marching songs being played, among others the ones composed by Nurjaslan in the Salemba prison, such as 'March of the Resettlement/ Buru' and 'The Units of Buru are Building Up.'"[92]

The military guards at the camp, as well as the supervising army officers arriving on inspection tours from the province capital and Jakarta, as often as not, almost as a rule in fact, were dandies:

Camp commandant has been replaced by Lt. Col. A. S. Rangkuti, who, according to what people have said, acted in movies.[93]

The first Commandant of Unit III, Lieutenant Eddy Tuswara, a former officer of the Cakrabirawa Division . . . likes to walk around in his swimming suit to show the attractive shape of his thighs; his waist is always adorned with all sorts of hand firearms. He is a soccer fan, and a punctual administrator.[94]

Captain Sudjoso is a devotee of *kroncong* music, and, above all, he likes cultural evenings in the camp . . . a slow and swaying step [is] typical for him.[95]

A tall and slim officer approached me: "You have forgotten me?" . . . [It was] Wing Wirjawan. He was a Commandant of Pattimura: "Yes," he explained, "I had an operation in Holland. Now I am slim. They took out twenty-eight kilos of fat."[96]

The prisoners, to look at Buru from a slightly different angle, were good sports. Pramoedya repeatedly writes about push-ups and also "jogging":

> I wake up at five or half past four. While still in bed I do some quick exercise, including *push-ups*. Then I boil water for my group. . . . After going to the toilet, I walk for a quarter of an hour or do some other sport: weightlifting or taisho, and, at the end, I run about 1 to 2 kilometers; not a fast run, rather kakiashi or *jogging*. After this, I gather wood for the kitchen.[97]

The push-ups were a favorite punishment with some of the guards, but they could not destroy Pramoedya because he had been in training since the Salemba prison, at least. A member of the Indonesian Research University Team that visited Buru late in 1973 to study prisoners was curious and a little dumbfounded about all this. Pramoedya explains patiently: sport is sport.

> P.A.T. [Pramoedya Ananta Toer]: Because I had a feeling in the Salemba prison that I would soon have to work hard, I began to practice sport.
> University Team (a woman member): Isn't work a kind of sport, too?
> P.A.T.: No, it is something different.[98]

There were dandies and sportsmen on Buru, and there was (to follow the themes of this book) some optics, too. The truth on Buru, we get the impression, was behind a looking glass. "When I stand in front of my sliver of a mirror," Pramoedya writes to his daughter from the camp,

> how completely white my hair has become. . . . Defective though the mirror is, small, dull, and cracked, it gives a picture that is clear: on the other side is death! . . . Chairil Anwar[99] wrote in a poem: "once, to mean something / then, to die." How romantic.
> . . . my hernia is swollen. . . . Sports and exercise, however, keep up one's belief in oneself. . . . Somebody who witnessed it first hand told me that, in

Glodok prison [in Jakarta], the Japanese officers, convicted war criminals, just minutes before they went up to their gallows, were jogging. I forgot which year it was. Maybe 1947. Whether there is hope or not: all the treasures of life are in the body you have.[100]

Living year after year amid the camp guards, who are also dandies, Pramoedya writes often about clothes. On his arrival on Buru: "From the Salemba prison, we were permitted to carry only two sets of clothes."[101] From his mother, he writes, "I have learned how to work with needle and thread. She taught me also to tailor my own clothes. Since then, I have often made my own trousers and shirts."[102]

Pramoedya writes with an affection about his trousers. If one looks for a good and concise history of late-colonial Indies and post-colonial Indonesia, here it is, zipped up:

> It is sixteen years already that the piece of clothing I am so proud of has served me faithfully. Initially, I did not feel so warm about it, as a Chinese tailor had made the legs of the trousers three centimeters too wide. Today, be the cut judged correct or not, the stuff of the trousers is my pride. I mend them with the strongest thread I could find—I unraveled it out from an old umbrella.[103]

Haircuts, naturally, are touched upon by Pramoedya, and even a beauty salon appears. In one of his letters from Buru, Pramoedya mentions Mulyoso, a fellow internee, who has just been murdered in the camp: "People say that, three days before he died, he received a letter from his family. There was a photograph of his wife in the letter. He did not recognize his wife as he had known her in the past. It appears that she had, just before the photograph was taken, visited a beauty salon."[104]

At moments—and it was Pramoedya who coined the term in modern Indonesia—the camp on Buru appears like a glass house.[105] The dreams and plans of the engineers of the Indies and Indonesia, modern Europeans and modern Indonesians, are consumed, at last, at its (carnival sort of) perfection. Now on Buru, indeed as Pramoedya describes it, everything becomes transparent, and there are no false reflections:

> at the end, there is just death, from whatever angle one may look; even when one tries just to peep at the world from a creepy corner. With or without lenses, whatever the instrument of looking, and of whatever the stuff it is made of.
> . . . In all sorts of fashions, and at all moments death appears.[106]

Maybe there is hope in perceiving the incongruity of life as it is. Pramoedya sits on Buru, in the glass house, and he listens to a gramophone:

> I feel the twilight, and my life as it drifts through the space without clear contours and filled with indistinguishable shapes, like a leech trying to suck where there are no veins. The life-appeal returns, over and over again, and it sounds

like a gramophone record, His Master's Voice: thirty years more, Pram, thirty years more.[107]

As is natural in a modern place, Buru was also full of engineers. The *Appendix 4* that Pramoedya compiled in the camp and later published in one volume with his letters deserves to be quoted extensively:

Agriculture area and road network built by political prisoners
During the ten years when the political prisoners were isolated on the island of Buru, they cleared and prepared for cultivation 3,532.5881 [sic] hectares of wet and dry rice fields, and they built 175 km of main and secondary roads, including the communications through the newly irrigated areas. This was accomplished solely by the manpower of the prisoners themselves, without a budgetary aid from the outside, from the Indonesian state Five-Year Development Plan or any other source. In addition, several barracks, a mosque, a church, a school, a marketplace, a wharf, and other structures were built.[108]

The number of hectares cleared by the prisoners, of wet rice fields, dry rice fields, and fields in total, is tabulated by Pramoedya and set in columns.[109] There is a table of thirty-nine *Distances between the units of the camp on the island of Buru* "1. Mako-Unit I 1.435 km; Mako-Unit II 1.227 km . . . 21. Unit I-Unit XV 7.402 km . . . 24 Unit XV Unit XIV 7.725 km . . . 34. Unit IV-Unit XV 9.238 km."[110] In *Road network of the camp on the island of Buru*, which was compiled by Pramoedya and published next, Pramoedya lists distances on the road from Mako, the center, to the twenty-six other destinations of the Buru camp system.[111]

Let us not take the academic decrees as seriously as Pramoedya's mother did, for a moment. Pramoedya on Buru may also be the last in the long line of the modern Indies and Indonesia engineers. After Dr. IJzerman, prospector of the Sumatra railway; Sukarno and Anwari, nationalist leaders-engineers; comes Pramoedya:

The upgrading of the roads of the indigenous people, where there were any, was our first task. The roads had to be widened up to 3 meters, bridges built or strengthened, canals dug; there were countless meandering to be handled, and large areas under water because of the overall low altitude of the place.[112]
. . . often there was not even a footpath of the indigenous people that we might use as a guideline.[113]
Theodolite was just a fantasy. There was not even a compass. Pencils and drawing paper as well were not easy to get.[114]
We began with improving the roads and bridges in December 1969.[115]

The authorities referred to Buru as a "project." They "planned" and they "regulated" a "transmigration" of the prisoners to Buru, and they often spoke about "building up" the island. This may be the burden of this book. Amid the guards, who were also engineers, Pramoedya was designing a special project of his own.

Investing much time and energy, in the camp, Pramoedya compiled, and then published as the *Appendix 2* to his letters, a *List of friends who died on Buru.*[116] In long, straight columns, each of the Buru dead is listed and tabulated; under a catalog number, there is a name, a prison photograph number, the date of birth, the date of death or missing, the religion, marital status, and number of children, the education, home address, and note on the cause of death, when it could be ascertained; in cases of suicide it is specified, whether it was by poison, an overdose of *Endrin, Diazenon, Thiodan,* or "by rope."[117] There is a commentary by Pramoedya on the project as a whole:

> We all had a similar experience. First, in the period of 1965–1970, we were arrested by the authorities of the New Order without any official document issued. . . . Second, our families were never officially informed about where we were being put in prison. Third, our families were never officially informed that we were transported to Buru . . . the New Order and its apparatus also never informed our families whether we died, and, if we did, whether we died from exhaustion, a work accident, murder, or an illness.
>
> . . . I have asked all the units of the camp on the island of Buru to make notes for me about the friends who died. The list that I compiled is not based on any [information from] the authorities of the New Order. . . .
>
> This list is made primarily in order to let the families know about the political prisoners who died in this place of exile. The list is also made so it becomes a monument to the island of Buru in particular, and to a certain period of the history of mankind in general.
>
> [When they began to be released eventually, and] before the last group was about to leave the island of Buru and return to Java, the political prisoners made some effort so that graves of their friends do not disappear with passing time or by evil hand. They chiseled names and addresses of the dead into the slabs of concrete, and they placed the slabs on the graves. . . . There is a constant danger that some measure may be taken by the apparatus [of the state] against what they might deem a "heroization" of the deceased. Yet, some of the families may still find the graves of their dead, as they will come upon the slabs of concrete.[118]

Throughout the twentieth-century history of the Indies, and Indonesia, concrete, as we have seen, was a substance of great technological importance. Many of the most visible and prestigious structures of the late-colonial and post-colonial country were built of concrete, and their concrete-ness was proudly declared. Professor Roosseno, Engineer, friend of Sukarno, and, through the independence period, builder of the most visible architectural symbols of the Republic, had affectionately been known as *Bapak Beton Indonesia*, the "Father of Indonesian Concrete."[119] Concrete, and reinforced concrete, was a powerful symbol of modern Indies and modern Indonesia. It was natural that Pramoedya, too, decided to use concrete to cover the patches of the modern landscape of his motherland, that he, in the 1970s, still believed it possible to save.

Ear Culture

> One evening, the camp commandant called me to himself
> and ordered me to write [texts for] some comics. It was
> simple, and I might have been finished with it quickly. But I
> was troubled. They pushed me into a corner. Already long
> before my arrest, I belonged to those who were against comics
> . . . these *ready-made* stories cannot inspire children with
> any power of fantasy. . . . The children will grow into
> just technicians. . . .
>
> As someone, however, who was stripped of his rights of a
> citizen, I could do nothing but to listen.[120]

Surrounded by the guards in the camp who also are technicians, forced to write ready-made stories and having but to listen, Pramoedya thinks very much about "ear culture": "ear culture for those who are below, and still more below." It is part of a modern culture, technological culture, radio culture, without any doubt: "so many words disseminated through the ether, all with the same content: such a long file of *must*s and *don't*s, and obligations, and obligations, and obligations."[121]

"I too belong," Pramoedya writes from Buru, "to the ear culture."[122] Sometimes, it even looks as if Pramoedya can imagine his body as a peculiar science-fiction machine equipped with nothing but ears: "And when all that happens, what can you do, you, a creature with nothing but a pair of ears?"[123]

Eyes can see a design, and a perspective. The eyes of the prisoners can see their prison merely as it was made for them:

> Surely not in vain have they chosen this place for us. However far your eyes might see, there is the enclosure of the mountains; that is all that is visible. The only crack in this huge dish of a land, the only opening to the outside, the valley of river Apu, is fully covered with watchtowers, rifles, and machine guns.[124]

Lips, too, and tongues, and vocal cords as well, may easily be taken from the prisoners, and they may even be used against them: "Most of the political prisoners speak only as much as they have to, and most of the time they remain silent."[125]

Of all the senses that may still remain his, Pramoedya most intensely thinks about ears and hearing:

The shouting, the "hey" that is now heard everywhere, is flat, but it is shrieked out in a high-pitched voice. It might be of some profit—and in our culture all is valued by profit—if this telling us what to do, and all the brainwashing, is tuned down to the normal room temperature, let us say, to twenty degrees Celsius on the thermometer of power.[126]

This, of course, may be the way a radio expert would think. Tuning down the volume of sound, one might be able to make even the dumbest (radio) thing soften. Radio knob again, is taken for a gadget of freedom. To hear softer and to hear less is what Pramoedya in the camp, with the ether filthy with words, increasingly hopes to achieve.

At one point of their exile on Buru, the prisoners were informed by the camp authorities that they were not prisoners actually—captured fighters, souls of resistance—but "settlers," thus a part of the New Order Indonesian government's long-term, social-engineering program to open new lands and to build up for the future. "Just at that time, one day," writes Pramoedya, "I lost my hearing entirely. Degeneration. I was deaf. I was scared, and I began to panic. Will I never again hear the voices of my children?"[127]

From that moment on, hearing becomes a crucial topic of Pramoedya's letters. When he later publishes them, he places, above the *Appendix 4: Agriculture area and road network built by political prisoners* and *Appendix 2: List of friends who died on Buru*, an equally extensive and evidently an item no less important:

Appendix 1: Data about my deafness

1. Since the beginning of the nineteen-sixties, I was less able to understand words spoken by a voice in the lower frequencies, especially those by loudspeakers.

2. 13 October 1965—My left ear was struck hard with an iron butt of a Sten gun, a bone was probably broken . . . 1971—Repeatedly, for periods of several hours, hearing in the left ear disappeared completely. . . .

4. 9 October 1978—Both my ears swelled, the canal of the left ear was completely closed. In the right ear canal, there was still a narrow crevice. The hearing in both ears was at the level of about 5% of normal. . . .

5. Present: capability to hear voices in the low frequencies did not improve. The volume of hearing is steady: about 25–30% of normal.

. . . Dated Buru, January 30, 1979

N.B.—The percentage of the capacity of my hearing was measured by myself. It may be not fully accurate.

—Doctor was not consulted.[128]

Deafness becomes a known quality of Pramoedya in the camp and, soon, beyond the island. No more can Pramoedya hear well. By now, questioning Pramoedya can often make some people uncertain, notwithstanding the fact that Pramoedya's answers are, in a sense, eager, innocent, and logical.

When a member of the already mentioned University Team decides to ask Pramoedya, "Where did you get your writing paper from?" Pramoedya answers: "I have eight chickens."[129] Of course, one gets eggs from chickens on Buru, and one may get writing paper in exchange for the eggs. An offbeat turn of speech, a hint of a possibility of a deliberate disconnection, a joke, a flash of resistance, a freedom, may pass in an instant. On the same occasion, Mochtar Lubis, a writer who came from Jakarta on the same plane with the Research University Team, asks Pramoedya:

Mochtar Lubis: Pram, have you found God here?
P.A.T.: Which one? . . .[130]
M.L.: I can see your hearing is not good. Which ear is weak?
P.A.T.: Both.
M.L.: I may send a *hearing aid* to you.
P.A.T.: Thank you. I do not need it.[131]

In the glass house of Buru, amid the perfect machinery of watching, as we read Pramoedya's letters, there might be a place to hide and, even, to keep on talking. Pramoedya jots this down in a letter about some of the worst moments of his deafness: "Fever . . . As the fever disappears a little, I hear just my voice, and the sound of my own heart."[132] There are hours and days on Buru when Pramoedya feels that he might have lost his hearing completely and forever. In a letter about these gruesome moments, Pramoedya uses one word otherwise rare and extremely precious in his loneliness—a "dialogue": "Like a baby, I have no other means of communication but my voice: to scream, to moan, to sigh, to whine. When even that means of communication is seized upon and taken away from me, ah yes, who may seize upon and take away my right to have a dialogue with myself?"[133]

The Happy End

In the middle of 1973, Pramoedya was ordered to walk from his camp to one of the other points in the Buru prison system, "Unit IV Savanajaya." Along the Buru roads projected by the camp authorities and built by the prisoners, after a day of walking, he reached the place on bloody feet: "On October 9, a helicopter arrived from Namlea and landed at about 10 in the morning on the Savanajaya soccer field. . . . Out of the machine, the Commandant of the Command for Restoring Security and Order, General Soemitro, came with his suite."[134] In the speech to the prisoners at the soccer field, which Pramoedya quotes as if he could have taken notes, the general said among other things: "Men! Each time when I am called to our President in Jakarta, he always asks me: 'How are our friends on Buru doing?'"[135]

Pramoedya, in a sense, was a celebrity. Prompted by the general, there, on the spot, Pramoedya made "a kind of desiderata, an inventory, including a short list of books, documents from the National Archives to be photo-copied, some old journals and newspapers to be used for the novel I wanted to write about the Period of the National Awakening."[136] After some time, when the inspection mission had departed and Pramoedya was long back in his camp, a parcel indeed arrived from General Soemitro's headquarters in Jakarta. According to Pramoedya, "there was nothing relevant in the box . . . no newspapers whatsoever; not even the important speeches by the President that General Soemitro also had promised."[137]

General Soemitro's visit, nevertheless, made the officers and guards in the camp more forthcoming. On one other occasion, not long after the visit, Pramoedya writes: "Major Kusno . . . presented me with a golden Pilot fountain pen, a bottle of ink, and a thick notebook that he said he hoped I would use for writing. There was an inscription on the first page of the notebook: 'for your personal renown, and for the renown of Indonesia.'"[138]

It was not too difficult, clearly, for Pramoedya to keep communications with the Indonesian generals and captains on an emotionally manageable level. But Pramoedya's wife and his children, too, were and appeared to Pramoedya as being a part of the elusive and often brutal networking of the outside world:

Do my wife and children hope that I may come back? . . .
What is true? I have already received some letters from them. But how it is in reality, I do not know. . . .
I have heard, here, that my wife married again.
Lies. Those are lies. . . .
Is my wife still pretty?[139]

One visitor, on a rare occasion, and when things got a bit softer on Buru, offered to record Pramoedya's voice and take the tape with him back to Jakarta, to Pramoedya's family:

Sindhunata urged me to send a message to my family. He wanted to record my voice on a tape, so that my children may hear it.
I have never tried anything like that, I said. [Then, they tried it.] As I listened to the tape, my voice sounded so authoritative that I became afraid that my children might be scared if they heard it.
Don't worry, he pushed.
Thus, I gave him the tape. I did not even use it completely. The recording was not good technically, as I had never before worked with a cassette tape recorder. Only the prisoners who have made it somehow and enriched themselves on Buru owned things like that. I told just a little story on the tape, about a dog I knew— a story for my youngest son.[140]

Few of us, in this wonderfully interconnected world of ours, can imagine Pramoedya's isolation. In a letter that, in all probability, will never fly as far as home to Java, Pramoedya tries to explain to his daughter, fervently, why his first marriage, two decades ago, broke down.[141] At the end of another letter, in evident desperation, he attempts to remain a father on hand:

> You don't have a boyfriend yet, do you? You are still too young to go out alone. For something like that, you first have to ask permission from your father or mother. Don't do it in secret. You must be always honest with your parents.
> I kiss you.[142]

Sometimes, very exceptionally, what appears to be direct contact with the ones back at home is made. Or a short circuit, rather. A photograph from Buru reaches Pramoedya's daughter, in one instance, and the daughter's response gets back to Pramoedya: "My friends say that mamma remains young long. But how is it that you have changed so much? You have become much older. I cannot believe how fast it happened. How did it happen that you became old so fast?"[143]

Pramoedya was respected by many of the prisoners in the camp. When they could afford it, and when the guards did not seem to mind, the fellow prisoners took over some of Pramoedya's menial duties so that he could write. He even got a typewriter:

> When my typewriter breaks down, they repair it immediately so that I do not lose a moment. When I am sick, they care for me and nurse me. . . . During these past twelve years I have had so many, so great many generous persons around me. . . . They have helped me to retain in memory what otherwise I might lose completely.[144]

In a sense, Buru was warm and understanding to Pramoedya throughout the years. At the same time, however, everything on Buru, as he describes it, was turned outward, to the (elusive and often brutal) outside world.

This is the last and latest chapter on the history of the Indies, and Indonesian photography as well. One is reminded, especially, of Kartini, her enthusiasm for photography, sending photographs of herself, receiving photographs of others, arranging photographs on her bookshelves and her coffee-table. In the post-colonial 1970s, cameras naturally were still in Indonesia, and more important than ever. While Pramoedya on Buru is searching, often in vain, for his face in his half-dull sliver of a mirror or in letters from home, by camera his face is made flagrant. Pramoedya writes:

> Before he climbed into his helicopter to fly back to Namlea, the Commandant of the Command for Restoring Security and Order [General Soemitro] asked me: "Would you mind, Mr. Pram, if they take a picture of the two of us together?"
> Before the guests left, the political prisoners posed for yet another portrait with the Commandant of the Command for Restoring Security and Order and

his staff. Prof. Dr. Suprapto [fellow prisoner] and I, on either side of the General. Then, Mochtar Lubis and Rosihan Anwar [another of the guests, a journalist] let themselves be photographed with me in the middle.[145]

According to another prisoner quoted by Pramoedya, at about the time when General Soemitro visited the camp, the Unit II of the Buru camp was made into a "*show window*."[146] To that particular place on Buru, visitors in steadily increasing numbers, first from Indonesia and then from abroad, were permitted, and sometimes encouraged, to come and see the camp and its inhabitants. As the protests increased from the outside, and in the world outside Indonesia in particular, the prisoners of Buru were put on the stage. The closer to freedom, the more so: "International protests were making a stain on the image of the Indonesian authorities. Already during the first year of the existence of the forced labor camp Buru, all had been set to work out all possible sense of justification for the outside world."[147] Among the earliest visits to Buru, Pramoedya writes,

> a team came of the [Indonesian] State Film Co. to make a movie. I, Prof. Suprapto, and Dr. Syarifuddin [two fellow prisoners] were directed to make it as if working on a house that was just being built. There we were filmed. Mannequins again.
>
> In the team headed by a Lieutenant Second Class, native of Cirebon, there was also a cameraman, German, who introduced himself merely as Fritz from Wiesbaden. With the permission of the new commandant of our unit, he presented me with two shirts and a pair of corduroy trousers. . . . I will never forget the goodness of the man, whom I did not know before, whose country I have never seen, and whom I met here, in the middle of these plains.[148]

This is, also, another chapter on the Indies, and Indonesian technology and tourism—a craft to move through a landscape (be it that of oppression and suffering even) without being touched deeper than, lightly, on the skin. Since the time of Kartini and Couperus, since the early twentieth century, on Buru at least, tourism also appeared to mature. As Pramoedya writes, in another of his letters: "Those of us who will not survive the present sifting will offer their commentaries, and they will be written in white ink between the lines of tourist guidebooks on the great Indonesian culture, exquisite manners, warm hospitality, and picturesque decorum."[149]

As time passed, the prisoners aged, and the regime aged; as freedom for Buru prisoners came closer, or as it seemed so, visitors began to arrive in streams:

> One was Bur Rasuanto, whom I had known before I was arrested, a writer. . . . I recognized Alex Leo, also, a reporter for Radio Jakarta. . . . Then foreign journalists arrived. One of them, Jacob Vredenbregt, a Dutch, handed me a letter

from my wife through barbed wire with the words that it had already been read
by the prosecution. . . .

Other foreign journalists whom I knew were a Japanese, and an American
woman, Cindy Adams, the author of the biography of president Soekarno.

. . . Also among them was someone whose name, as I was later told, was
Goenawan Mohamad.[150]

Shortly afterward, Captain Sudjoso, the already mentioned dandy and fan
of *kroncong* music, ordered Pramoedya "and Prof. Dr. Suprapto to walk over
to Unit II; we did not know why. When we arrived at the place, we were told
that the two of us would be presented for interviews to journalists from
Jakarta, Australia, Hong Kong, and Holland. The interviews were conducted
in Indonesian, Dutch, and English."[151] Step by step, Pramoedya was being
connected to freedom.

The visits from Pramoedya's captors, official and semi-official missions by
representatives of the regime, appeared not too difficult to perceive in their
true dimensions or to handle: "I felt like a commodity on display for the big
shots from Java, like a monkey in a zoo."[152] Some visits by "neutrals" were not
very perplexing either. On one of these days, Pramoedya was called upon in
the Buru camp by the emissary of the International Red Cross: "a Swiss
citizen, his name was Dr. Reynard, if I am not wrong. Talking with this man
soon made me nervous because he did not speak very loudly, and I had to try
everything to keep my already imperfect hearing focused."[153] The visitor (and
hygiene enters our story at this moment again) emphasized that he was there
on a humanitarian, and not a political, mission:

> He told me what kinds of medicines, and how much of each kind, he had in
> mind [to send us]. As I do not have a slightest knowledge of pharmacology, I
> answered by just saying yes-yes. Perhaps he thought that knowledge of phar-
> macology was on the same level here as in Europe. He gave me two carbon
> copies on which the names and the amounts of the drugs were listed. "In total,
> one and a half barrels," he said. Then he examined our well and asked if we
> boiled the water before drinking, yes or no. He also checked the latrines. Then
> there was a telephone call for him from the headquarters in Mako, and he left in
> a small pickup truck.
>
> I passed the carbon copies to our friends, political prisoners who worked at the
> infirmary. Everything remained as before.[154]

This was easy. There are some moments, however, as disturbing as they are
touching: in the Buru arcades, on stage and in his prisoner's costume, Pramoedya appears almost as if meeting a friend. Particularly at these moments,
uncomfortably, it becomes clear that the glass house of Buru and our world
too are designed and built well:

Early in the morning, at half past five, David Jenkins from *Far Eastern Economic Review*, Hong Kong, is already here. . . . He notes down the measurements of my room: 2.5 × 2.75 m. Ceiling: 2.5 m. He wants to see my belongings. I open my chest: manuscripts, and some clothes.

That's all.

He is a likable man. He explains that he wants to write about me in his magazine, for the series *Behind the Lines*. He takes more than ten pictures.

It appears that my room has a great power of attraction. Mitsunori Matsumura from TV and Radio Japan asked if he might photograph the room. He moved away all my tins of food and then put them back when he was finished. Trisno Juwono, in a broadcast for [Indonesian] Radio Amara, mentioned my shoes, ashtray, tobacco box, and the earthen floor. . . .

Everything changes as it is touched by the current of change over there in the West. Indonesia has always been like that, and it still is. And political prisoners remain political prisoners. . . .

David Jenkins has already finished his job. He came to say good-bye.

"We will meet again in better times. I will send these photographs to your family back home."

"Thank you. Have a nice trip. Bye."[155]

This is an epilogue on a late-colonial culture. Acutely alone, at moments Pramoedya appears to enjoy the currents from the outside touching him, talking with visitors, the journalists especially, admittedly an easy talking. The more easy, looser, and, indeed, thinner the talk is, so it often seems, the better:

> the journalists did not give me a chance to reflect. . . . They shouted their questions, shouted, and fought among themselves to shout. And I was happy. At least to talk with free people, not slaves to their stomachs, it makes you feel as if touched by some freedom too. What a comfort. To feel oneself to be still human, not a pharaoh's subject, not a worshipper of the dust. Still human. *Kompas, Tempo, Suara Merdeka, Surabaya Post, Berita Buana, UPI, NHK, ABC, The Asian Wallstreet Journal,* Indonesian TV, *Radio, Merdeka,* and who knows what else.
>
> The questions-and-answers give me the first taste of what a liberation from here would mean. There is no tension.[156]

At first, these encounters as Pramoedya describes them may remind us of the early Indies newspapers' pages from almost a century earlier. Every triviality is thrown in, the whole wide world, every bit of imagery, is eagerly grabbed at. In the camp on Buru, however, it is the Indies, and Indonesia, that matured, or is old. Men and women are asking, it appears, who already know too much.

In lengthy paragraphs, on page after page of his letters, Pramoedya records the questions by visiting journalists and his answers. There appears to be no real order to the questions. Is it made by Pramoedya like a radio, again: a thing that is turned on, but nobody truly cares to listens? Indeed, the world is described as arriving at Buru, buzzing in.

And so the questions go. Will not Pramoedya be happy to be released? What is the first thing he would like to do in that case? Is Indonesian literature in decline? Is the writing Pramoedya does at present on a higher level than were his earlier works? Why did he agree, in the 1950s, that a fellow writer be kept in prison? Does he still believe that literature and politics are closely linked? Is politics dirty?

When you are free, will you be able to *cunin* with other people?

What do you mean by *cunin*? What language is that?

English.

Oh, *tune in*. I will, why not?[157]

Why was it, initially, in 1945 and soon after, that he used to be so gentle, and why, later, did he become so harsh? Was he disappointed by the outcome of the Indonesian revolution? If Indonesian society happens not to want him back, which country will he choose to live in instead? Has he read *The Decline of Constitutional Democracy in Indonesia* by Herbert Feith? Does he know Harry Aveling?[158]

You have a typewriter. Who gave it to you? . . .

I read in a newspaper that you got the typewriter from our President.[159]

Pramoedya, who lost more than most of us can even imagine ourselves losing, and who is writing from the soul of a nation that lost more than most of us can imagine our nations losing, is happy, physically happy, breathing a happiness of finding fast words, and lanky definitions of the "free men":

> In spite of everything, the meeting with free people—domestic and foreign journalists—gives me a fine feeling: those in the outside world still know that we live. And as long as there is some life, there must be some sense in it.[160] They dazzle everybody. The arrival of the foreign and domestic journalists, for the first time in twelve years, brings political prisoners in contact with visitors who come and do not issue new punishments and orders, do not intrude and brainwash. The prisoners get a chance to air their lungs, bile, heart, liver, yes, their kidneys, nerves, and skin. Their skin? Yes, because there is almost no prisoner whose skin is not torn and scratched into a map full of bad signs. The work of charity is a part of democracy! All of a sudden, the journalists appear as if closer to you than your own brothers and sisters. A ton of suffering they ease from our shoulders. Then the guests depart again, and, again, the political prisoners, gram after

gram, move to load up the ton. It does not matter. At least for a while, some humans visited the fortress of the Buru camp.[161]

This is how the happy end drags on. In December 1977, Pramoedya witnesses the first wave of prisoners released and sent home from Buru:

1,501 friends are leaving for Java. . . .

Happy journey to the free world. Happy departure from the continuance of the last twelve years. Happy being caught and placed into the old dictionaries with the new meanings. Happy being greeted by rotten phrases that are cloaked by palatial costumes to appear nice. You are leaving for a changing world—a world half mad by how dense its population has grown, and by how noisy its slogans have become—loud and empty—[162]

It is decades after Pramoedya left the radio school in Surabaya. Now the world is on the air again. Otherwise, nothing happens:

Catherine Randers from Oslo . . . She writes, in Indonesian, something like that: "I have heard on the radio news today that you were in the first group of people sent off to Buru." . . . She writes, besides, that her daughter is a member of Amnesty International. Another letter also came from that most northwestern region of Europe, from Father Verhaar in Haarlem. He writes in broken Malay, and his letter begins in almost the same way as the first one: "I listened to the news on the radio today, and I heard that you don't have writing paper. Would I be allowed to send you some?" Naturally, I cannot answer either of the letters, and, moreover, both return addresses were blackened out . . . more and more, I value the intellectual inner self of Europe.[163]

Television, at a certain point, also becomes available to the prisoners: "television could be watched by the political prisoners in the headquarters of our prison beginning in November 1977."[164] News of an impending freedom trickles in through wireless, first as rumors that prove to be false: "Some in the camp, and the officers, heard it from Radio Ambon."[165] Even the reality, as it followed, was a news reality. After the first wave of Buru prisoners was released, and as it reached Java, the television was already there, lights on, waiting in the innermost place in the returnees' homes, watching the faces of the relatives, and looking around the households. Pramoedya sits in front of the box on Buru: "The First Releasing, clearly, became a reality. Television already reported on the return of the prisoners to the embraces of their families. What is important: Indonesian society appears to accept them, as is clear from these warm and touching receptions."[166]

Pramoedya does not make it easy for us to learn where indeed he wishes to return on his day of freedom. His map of Indonesia is much more complex than any map we have encountered so far—IJzerman's map, Kartini's map, Mas Marco's map. In Jakarta, Pramoedya writes in one letter, "you are internationally a bit more exposed"; outside Jakarta, it is "more away from all those foreign priers."[167] And what is away? He may end his life at sea: if the ship carrying the prisoners back from Buru sinks, Pramoedya writes, we will leave a "scoop" behind us.[168] Pramoedya wishes to get away from Buru, naturally, and more than anything else. But how deeply may he wish to return home? In one letter from Buru he reminds his child: "You saw it yourself how our house was destroyed. With the big river stones that our neighbors had originally brought for building their own new house."[169]

Mochtar Lubis, who visited him on Buru, asked Pramoedya as a fellow writer: "Pram, what do you read these days?" Pramoedya answers: "I like the magazine *America* published by the Jesuits."[170]

Pramoedya appears serious. Besides, the Society of Jesus does not turn up fleetingly, either on Buru, where there is a small yet active Christian mission, or in Pramoedya's letters. To his daughter Nen, Pramoedya writes about it in a sudden, long, and enthusiastic letter:

Letter to Nen

Daddy has not gotten a single letter from you for the past two years. Perhaps they were not allowed through, or are you really writing? . . .

Recently I have by accident come upon a book by Dr. Josef Vital Kopp, "Teilhard de Chardin." . . . I myself have never read the writing of the greatest scholar of this century, Teilhard de Chardin, but the explanation by Dr. Kopp gives an interesting picture of the theories of this important scientist. . . .

. . . the Indonesian version has a title "Teilhard de Chardin, a New Synthesis concerning the Evolution," translated by Al Hastanta and Ign. Kuntara Wm., with a foreword by Prof. Dr. J. W. M. Verhaar SJ (SJ, Societatis Jesu, means that he is a member of the Jesuit Society, i.e. that he is a Jesuit). . . .

I hope that you, as an enlightened girl, have already read the book, and it would be very nice if you could get deep into it. . . .

Nen, Teilhard de Chardin is the greatest scientist of this century. . . . At a given moment in the seventeenth century, the light into the darkness of the preceding era was brought by Johann Kepler, namely in the field of astronomy; but, at present, into the darkness of the preceding era, the light is brought by Teilhard de Chardin, in the field of human evolution. . . . This is not a philosophy, this is almost a hundred percent science through which the truth is being proved.

. . . Father Teilhard de Chardin argues that mankind has not arisen from one single couple, but in polygenesis. Thus, there were all kinds of different Adams and Eves, and not just one pair: there was the Javanese man, the Neanderthal, the Peking man, and so on.

I learned about the theory of evolution for the first time in my middle school, and from the start I was really intrigued. I had read, in fact, about bits and pieces, unscientific interpretations, already in elementary school, in two Javanese books. One of them was written by a retired teacher who did not believe that it had been possible for God to create Adam on one try, because natural circumstances would not allow it. God tried at least three times before he succeeded.[171]

. . . I have always wanted everything to be logical. . . . In verse 31 of *Al Bahara* in the Koran, for instance, it was always hard for me to accept that man may not have a language until he is taught it by God. It can be scientifically proved that any language is a result of human struggle in life and society. . . . Each word, each name, each term, even acronym, is nothing but a result of human labor. . . . Thus, for instance, the theories of Teilhard de Chardin enriched the treasury of human languages by new terms: biosphere and noosphere.

. . . Nen, it makes me feel so lucky to be able to follow the liberating ideas of someone who lived in the same junction of time as me: Teilhard de Chardin!

. . . I like things that are lucid. Whenever anything remains hidden, human misery sprouts out in that place. That is always so. . . .

. . . I am in favor of making *oeuvre de raison* a measure of things. Not that it has to be the only measure, but certainly it should be most important.[172]

Rather than dense, this letter is filled with signs and signals. Everything and everybody, in a flash seems to be there, in the letter, as on a newspaper page. The young woman, Kartini, and her early-century belief in "through darkness to light." Pramoedya's old Javanese teacher with his wise little fables that go miraculously well with Johann Kepler. Evolution, *oeuvre de raison*, an almost 100 percent science, and Pramoedya's own daughter, who, we believe with him, is an "enlightened girl." Belief in logic, and belief in lucidity. Time as far back as when Pramoedya was a little boy until this very instant in the Buru camp, and perhaps a little into the future, can be safely surveyed, and it flows smoothly. A little screw, a nail, a piece of copper wire, a freedom. As long as words of the letter are being put together, the world makes sense.

Two key expressions in Pramoedya's long letter are the two Teilhard de Chardin's additions to "the treasury of human language"—biosphere and noosphere. As my laptop's electronic encyclopedia explains them:

The Humans, according to Teilhard, as they are now known are not the end of the process. . . . Everything that arises converges. Widely different human cultures around the Earth are now converging toward an omega point, identified by Teilhard as Christ, at which consciousness can find a new unity. Already humankind has covered the Earth's surface with a noosphere, a sort of collective human consciousness superimposed on the already-existing biosphere, and which Teilhard believed is evident in the present worldwide complex of transportation and communication.[173]

Pramoedya's—more than Kartini's, Mas Marco's or, for that matter, Tillema's—is an exile in the time of wirelessness. The Wall is torn down, and no longer is a wire, barbed or otherwise, needed to keep a nation, or a human being, together.

Pramoedya Ananta Toer in 1995. (Photo by Hervé Dangla)

In contrast to some other heroes, the rest of Pramoedya's story is not silence. As Pramoedya's publisher let us know, what we can no more read in the letters, "with *the very last group*, [Pramoedya] left the island of his exile on November 12, 1979."[174] This group of returnees traveled across the sea to Surabaya, East Java, and then by road, across the island, via Surakarta and Yogyakarta, to Magelang. After being locked for another month in Magelang, the prisoners were transferred to the military barracks in Banyumanik, south of Semarang. In Semarang, the first of a series of the official "ceremonies of release" took place, on December 20, 1979: "Even a number of foreign ambassadors was present."[175]

After the ceremony, the returnees who initially had come from West Java, Pramoedya among them, were transported to Jakarta, back to Salemba prison where, ten years earlier, this journey had begun. The next morning they were "handed over to the Military District Command of the region to which they belonged by their permanent or temporary residence permit." On December 21, 1979, Pramoedya was released and ordered to report every week at the local police station near his new place of residence in East Jakarta.[176]

As these lines are written, Pramoedya is free. Very many in the world outside Indonesia know about him. The trickle of eager visitors from abroad, and some more hesitant from Indonesia itself, never stops. Soeharto is gone, and the particular regime that put Pramoedya in prison is almost gone. It means that the trickle of Pramoedya's visitors becomes a stream at times. They take pictures and measure the house. There is an upgraded smoothness in the air.

What can we do? What can we decently do, indeed, except to join in the trickle, now the stream, be patient, ask Pramoedya when we get to him what he thinks about all this, and then just hope that he may switch off his new Japanese hearing aids for a moment and answer our question, with the wisest of his smiles: "I have eight chickens."

NOTES

PREFACE

1. Proust, Marcel, *Swann's Way* (New York: The Modern Library, 1992), 59–60.

2. Marx, Karl, *Capital: A Critique of Political Economy*, translated by Ben Fowkes, vol. 1 (Harmondsworth: Penguin, 1990), 546.

3. Hesitatingly, throughout the text, I have dropped framing "native" (*inlander* in Dutch) by quotation marks. This, I deliberated, was historically accurate, as thus it was also used at the time. Native is a common dictionary word, meaning "one of the original inhabitants of a place or country, especially as distinguished from strangers, foreigners, colonizers, etc." (*Webster's Dictionary*). Native, besides, had been an established legal term in the Indies: all population was divided by law into natives, foreign orientals, and Europeans. Only in the last three years of the colonial rule, after 1938, and even then far from entirely, indigenous (*inheemsche*) and Indonesian (*Indonesiër*) were substituted for native (G.F.E. Gonggrijp, *Geïllustreerde Encyclopaedie van Nederlandsch-Indië*, the 1934 edition reprinted, Wijken en Aalburg: Picture Publishers, 1992, 146; *Regeerings Almanak voor Nederlandsch-Indië 1942*, Batavia: Landsdrukkerij, 1942, 3–4). Still, native is a highly charged word and not easy to use—like, for the same reason, European, white, or even tradition, culture, and modernity. Especially with colonial nationalism maturing, these words were being politically corrected, and often they became words of abuse. Native, and the other words, are most awkward for a historian to use. Precisely this awkwardness, however, has to be left evident; quotation marks, however good they may be as a tool of evasion, cannot make the problem go away. For much more on this problem, see a wonderful book full of technology and largely on quotation marks, John Pemberton's *On the Subject of "Java"* (Ithaca: Cornell University Press, 1994).

4. Tillema, H.F., *Kromoblanda: Over 't vraagstuk van 'het Wonen' in Kromo's groote land*, vol. I, 1915–1916; vol. II, 1916, vol. III, 1920–1921; vol. V, 2, 1922; vol. VI, 1927 ('s Gravenhage, etc.: H. Uden Masihan, etc.).

CHAPTER ONE

1. Struik, Dirk J., ed., *Birth of the Communist Manifesto* (New York: International Publishers, 1971), 93.

2. Lacey, A.R., *Bergson* (London: Routledge, [1987] 1993), 191.

3. Bloembergen, M., "Exhibiting Colonial Modernity: De exposities van de Nederlandse koloniën op de wereldtentoonstellingen (1880–1931)," unpublished Ph.D. Prospectus (Amsterdam: CASA, 1995), 11.

4. *Technische Hoogeschool te Bandoeng: Overdracht der Hoogeschool aan den Lande* (Bandoeng: THS, 1924), 9. On IJzerman and the technical college in Bandung, see also Baudet, H., *De lange weg naar de Technische Universiteit Delft* (Den Haag, 1992, 1993), vol. 1, 389–390. Travel was in IJzerman's blood, clearly. In 1915 he published *Dirck Gerritsz Pomp alias Dirck Gerritsz China* ('s Gravenhage: M. Nijhoff), a biography of "the first Dutchman (1544–1604) who visited China and Japan." In 1926 he

was a moving spirit behind another publication of that sort, Olivier van Noort's (1580–1627) *De reis om de wereld* ('s Gravenhage: M. Nijhoff, 1926).

5. IJzerman, J.W., *Dwars door Sumatra: Tocht van Padang naar Siak* (Haarlem: F. Bohn, 1895), 167.

6. Ibid., 162.

7. Ibid., 110.

8. Ibid., 126.

9. Ibid., 320–321.

10. Ibid., 1.

11. Ibid., 100.

12. Ibid., 358.

13. Ibid., 358, 201.

14. Ibid., 358–359.

15. Ibid., 96, 98.

16. Ibid., 254–255; see also illustration on page 255.

17. Ibid., 162, 269; see also ibid., 507–508.

18. Ibid., 471.

19. Ibid., 93. See also ibid., 112.

20. Ibid., 339.

21. Ibid., 475–476.

22. Reitsma, S.A., *De verkeersbedrijven van den staat: spoorwegen, post-, telegraaf- en telefoondienst, havenwezen* (Weltevreden: G. Kolff, 1924), 61.

23. Ibid.

24. Doorn, J.A.A. van, *De laatste eeuw van Indië: Ontwikkeling en ondergang van een koloniaal project* (Amsterdam: Bert Bakker, 1994), 110–111.

25. Veth, P.J. , Wilken, G.A., and Klinkert, H.C., *Catalogus der afdeeling Nederlandsche koloniën van de internationale koloniale en uitvoerhandel tentoonstelling* (Leiden: Brill, 1883), vol. III, 91.

26. Brons Middel, R., *Ilmoe Boemi Hindia-Nederland* (Batavia: Albrecht, 1886), 38.

27. Reitsma, *De verkeersbedrijven van den staat*, 60.

28. Veth, Wilken, and Klinkert, *Catalogus der afdeeling Nederlandsche koloniën*, vol. III, 91.

29. "De Nieuwe Uitvindingen," *Kopiïst*, vol. 1, no. 8 (1842), 210–212.

30. Ibid., 212–215.

31. Ibid., 216.

32. Ibid., 217.

33. "Een gedenkdag op spoorweggebied," *De Indische Gids*, vol. 14 (1892), 1051.

34. Doorn, *De laatste eeuw van Indië*, 126.

35. Quoted in ibid., 129.

36. Tsuchiya, K., "Kartini's Image of the Javanese Landscape," *East Asian Cultural Studies*, vol. 25, nos. 1–4 (1986), 62. The connected towns and cities were Batavia, Bogor, Bandung, Tegal, Solo, Yogyakarta, Semarang, Cilacap, Madiun, Kediri. Blitar, Malang, Surabaya, Pasuruan, and Probolingo.

37. "Magnetische en meteorologische waarnemingen in 1899," *De Indische Gids*, vol. 22, no. 2 (1900), 1549–1550. This was more than a year earlier than in Amsterdam: "On January 1, 1900 . . . the private tram company was taken over by the

municipality and in August the first electric tram passed through the city" (Baar, Peter-Paul de, "A City on the Move: Amsterdam 1892–1912," in Baar, P.-P. de, M. Bax, et al., *Piet Mondriaan: The Amsterdam Years* [Amsterdam: Gemeentearchief Amsterdam, 1994], 15).

38. *Dertiende Jaarverslag der Batavia Electrische Tram-Maatschappij 1909* (Amsterdam: Batavia Electrische Tram-Maatschappij, 1910), 10.

39. Kartini, R.A., *Letters of a Javanese Princess* (New York: W.W. Norton, 1964), 46–47.

40. Kartini, R.A., *Brieven aan mevrouw R.M. Abendanon-Mandri en haar echtgenoot*, edited by F.G.P. Jaquet (Doordrecht: Foris, 1987), 121; compare Kartini, *Letters of a Javanese Princess*, 136.

41. Kartini, *Letters of a Javanese Princess*, 80–81.

42. Bouman, Dr. H., *Meer licht over Kartini* (Amsterdam: H.J. Paris, 1954), 10, 168.

43. Kartini, *Brieven aan mevrouw R.M. Abendanon-Mandri*, 168.

44. Ibid., 177.

45. Ibid.

46. Ibid., 195, 268, 24, 346.

47. Ibid., 32, 202.

48. Ibid., 186.

49. Ibid., 70–73.

50. Kartini, *Letters of a Javanese Princess*, 60.

51. Kartini, *Brieven aan mevrouw R.M. Abendanon-Mandri*, 170.

52. Musil, R., *The Man Without Qualities* (New York: Putnam's, 1980), 334.

53. Kartini, *Letters of a Javanese Princess*, 107. In 1911, there appeared for the first time a plane heavier than air over the Indies. M.P. Pattist, "Civiel Luchtverkeer in Nederlandsch-Indië," *Koloniale Studiën*, no. 1 (1927), 237.

54. Quoted in *De Indische Gids*, vol. 16, no. 2 (1894), 1090–1091.

55. "Maandenlijksche revue: S. Kalff's Hollandsche Boeren op Java," *De Indische Gids*, vol. 16, no. 2 (1894), 1091.

56. Sandick, R.A. van, "Indische schetsen," in *Vragen van den Dag: Populair Tijdschrift* (Amsterdam, 1891), 5.

57. De Braconier, A., "Het pauperisme onder de in Ned. Oost-Indië levende Europeanen," *Nederlandsch Indië Oud en Nieuw* (1916–1917), 291–297. It is not explicit here, but it seems that Eurasians who were legally Dutch were included in the number. According to official numbers, in 1893, 46,000 out of 58,000 Dutch men and women living in the colony had been born in the Indies. *Kol. Verslag van 1893*, quoted in *De Indische Gids*, vol. 16, no. 2 (1894), 1091.

58. Quoted in Doorn, *De laatste eeuw van Indië*, 42.

59. Vanvugt, E., *H.F. Tillema (1870–1952) en de fotografie van tempo doeloe* (Amsterdam: Mets, 1993), 12.

60. Doorn, *De laatste eeuw van Indië*, 23.

61. Ibid., 13. For more on Walraven, see Schomhardt, F., "Biographische inleidingen," in Walraven, W., *Brieven. Aan familie en vrienden, 1919–1941* (Amsterdam: van Oorschot, 1992), 6–26.

62. Moet, J.F.F., "Iets over een stoomtram op Java," *De Indische Gids*, vol. 2, no. 2 (1880), 4.

NOTES

63. *Veertiende Jaarverslag der Batavia Electrische Tram-Maatschappij 1910* (Amsterdam: Batavia Electrische Tram-Maatschappij, 1911), 7.

64. Meijer, Ir.H. met medewerking van Ir.F.A.J. Heckler, *De Deli Spoorweg Maatschappij: Driekwart eeuw koloniaal spoor* (Zutphen: De Walburg Pers, 1987), 65.

65. Ibid.

66. The main engineer de Bordes, the builder of the first rail line on Java, quoted in Doorn, *De laatste eeuw van Indië*, 154. See also Doorn, Jacques van, *The Engineers and the Colonial System: Technocratic Tendencies in the Dutch East Indies* (Rotterdam: Faculty of Social Science CASP 6, 1982), 28, quoting Bordes, J.P. de, *De spoorweg Semarang-Vorstenlanden* ('s Gravenhage, 1870), 16, 17, 24, and Homan van der Heide, J., "Landbouwtoestanden in Achter-Indië, beschouwd in verband met Java," *Handelingen Indisch Genootschap* (March 28, 1905), Doorn, 128f., 136.

67. Doorn, *De laatste eeuw van Indië*, 47; Meijer, *De Deli Spoorweg Maatschappij*, 74.

68. Meijer, *De Deli Spoorweg Maatschappij*, 74.

69. Veth, Wilken, and Klinkert, *Catalogus*, vol. III, 96.

70. *Onderzoek naar de mindere welvaart der inlandsche bevolking op Java en Madoera*, vol. IVa (Batavia: Landsdrukkerij, 1906), 60.

71. Ibid.

72. Ibid.

73. Ibid., 62.

74. Ibid.

75. Ibid., 64.

76. *Dertiende Jaarverslag der Batavia Electrische Tram-Maatschappij 1909*, 10.

77. Couperus, L., *The Hidden Force* (Amherst: University of Massachusetts Press, 1985), 218. Like the translator of the book in English, I am using "hidden force" rather than the more literal "silent force." My reason is that "hidden" implies a sensation beyond just hearing.

78. Cribb, R., "Telling the Time of Day in Indonesia," unpublished draft, 1993, 5.

79. Ibid., 5–6.

80. Kol, H.H. van, *Uit onze koloniën* (Leiden: A.W. Sijthoff, 1903), 65.

81. "Soilson" is a translation by Maier, Hendrik M.J., "Phew! Europeesche beschaving! Marco Kartodikromo's Student Hidjo," *Southeast Asian Studies* (Tonan Ajia Konkyu), vol. 34, no. 1 (June 1996), 203.

82. *Doenia Bergerak*, vol. 1, no. 20 (1915), 5–6.

83. Signed "MARCO," *Doenia Bergerak*, vol. 1, no. 19 (1915), 4–5.

84. Ibid., 3–4.

85. Ibid., vol. 1, no. 27 (1915), 5–8.

86. *De Beweging*, no. 3 (January 18, 1919), 26.

87. Shiraishi, T., *An Age in Motion: Popular Radicalism in Java, 1912–1926* (Ithaca: Cornell University Press, 1990), 244.

88. Ibid., 275.

89. Editorial in *Kritiek en Opbouw*, vol. 4, no. 22 (December 6, 1941), 335.

90. Shiraishi, *An Age in Motion*, 243.

91. Couperus, L., *Eastward* (New York: G.H. Doran, 1924), 101–102.

92. *Motorblad* (February 28, 1928), 107.

93. *De Ingenieur in N.I.*, vol. 6, no. 2 (February 1939), section I, 32. The first car

in the Indies, it seems, was "a Benz-auto" of Susuhunan, ruler in Surakarta, and a second oldest car in the Indies belonged to the Swiss consul in Batavia. According to the same source, there were, in 1912, 1,194 automobiles in the Indies, 1,095 in Java and Madura, and 99 in the rest of the colony. A.W. Naudin ten Cate, "De automobiel in Indië van 1902–1927," *De Indische Mercuur* (Gedenknummer, 1928), 47.

94. In 1935, 20,348 Europeans owned motor vehicles in the Indies, 18,455 natives, and 12,328 foreign orientals; 22,792 Europeans had driver's licenses, as compared to 51,528 natives and 16,661 foreign orientals. *Encyclopaedie van Nederlandsch Indië*, 2nd ed., vol. VIII. Supplement ('s Gravenhage: Nijhoff, 1939), 488.

95. Pedestrian error was the cause of 3.59 percent of the accidents, driver's error 88.82 percent. Most accidents happened on Saturdays between 4 and 7 in the afternoon. *Motorblad* (November 1, 1928), 1113.

96. Veth, Wilken, and Klinkert, *Catalogus*, vol. III, 101.

97. Sloos, J.M., "De invloed van de auto's op het tegenwoordige verkeerswezen," *Koloniale Studiën*, no. 2 (1923), 386–395.

98. Wynaendts van Resandt, W., "De Plaats van de Auto onder de Verkeersmiddelen, Speciaal in Nederlandsch-Indië," *Koloniale Studiën*, no. 1 (1926), 90–106. There was, *Koloniale Studiën* argued, a fundamental importance for the natives in the distances of about 30 kilometers. This was, the study wrote, how far, usually, bigger Javanese markets were from each other; it was also a distance between many significant holy sites for native rituals.

99. Ibid. As compared to 51,615 personal automobiles in the Indies in 1939, there were 9,242 buses and 12,860 lorries (*De Ingenieur in N.I.*, vol. 6, no. 2 [February 1939], I, 32.) According to the state-railways official data for 1933—the year when the great economic depression set in and the colonial system was again severely tested—the income of the state railways in Java, because of the predominantly native bus and lorry competition, fell 44 percent (Nus, J.v., "De Staatsspoorwegen in Nederlandsch-Indië gedurende de crisisjaren 1930 t/m 1934," *Koloniale Studiën*, no. 1 [1935], 2). In a statement by a spokesman of the colonial government in 1939, production of small industry in the Indies (it was native industry) survived the great depression of past years and was not completely wiped out. For this, the spokesman found "the foremost cause" in "the improvement of transport means, most importantly a more frequent use of buses and person-transporting trucks." Sitsen in *Kritiek en Opbouw*, vol. 2, no. 1 (February 16, 1939), 2. An Indonesian magazine, *Motorblad*, ten years earlier, stated the same thing only in a slightly different way: the competition of lorries and buses was good for the Indonesian traveling public, because it pushed prices down. *Motorblad* (August 25, 1928), 1081.

100. In a long article from 1938, a Dutch engineer expressed his satisfaction over the improvement of quality of the Indies asphalt over the past twenty to twenty-five years and especially after 1923, when a new kind of asphalt had been introduced. The rain water almost ceased to be a problem. But *tjikar*, *grobak*, and all the hard-wheeled vehicles still were a problem. To prove his point, the author compared the costs of upkeep of an about 3-meter-wide provincial asphalted road used mainly by *tjikar*, between Porong and Surabaya in East Java in the years 1934–1936, "bad sugar years," which meant less traffic, and an "equally well asphalted" highway used mainly for modern fast traffic, including heavy trucks, "one of the busiest roads in Java." The upkeep costs were annually 2.5 percent for a square meter of the fast highway and 20

percent for the *tjikar* road. "Only legal actions against the *tjikar*, *pedati*, or *grobak*," was the author's conclusion, might help. Ir.A. Poldervaart, "De tjikar als wegver-nieler," *Locale Techniek*, vol. 7, no. 5 (September–October 1938), 121–131.

101. Wittgenstein, Ludwig, *On Certainty* (New York: Harper Torchbooks, 1972), 6, §30, 9, §56, 82, §617.

102. *Magneet*, vol. 1, no. 3 (January 15 ,1914), 31.

103. Ibid., 32–33.

104. Ibid., 33.

105. Ibid., 31; vol. 3, no. 3 (February 1, 1916), 40.

106. Ibid., vol. 1, no. 4 (January 31, 1914), 43–44.

107. Hoogestraaten, Leo, "Rond de Gedeh," ibid., vol. 1, no. 13 (June 15, 1914), 150.

108. Ibid., vol. 1, no. 10 (January 1914), 123. The word *bingoeng* remained with the late-colonial culture until the end. In early 1941, a Dutch author (in *Java Bode* of February 11, 1941, "Indonesiërs als vliegeniers") argued that Indonesians were not to be admitted as aviators into the Dutch Indies corps in the coming war because, among other things, "the Indonesians are often 'bingoeng'; for the last word there is no equivalent in the Dutch language, thus, 'bingoeng-sensation' does not come forward among the Europeans." An Indonesian answer was to quote a "Malay-Dutch dictionary by Professor van Ronkel," where there could be found "no less than three equivalents, namely *verbijsterd* [bewildered], *versuft* [dazed], and *bedremmeld* [covered in confusion]. . . . Who, in all seriousness, may claim that this state never happens to a European?" "Indonesiërs als vliegeniers," *Nationale Commentaren*, vol. 4, no. 11 (March 15, 1941), 3258–3259.

109. *Magneet*, vol. 1, no. 11 (May 15, 1914), 136–137.

110. Ibid., vol. 1, no. 14 (June 30, 1914), 172.

111. Ibid., vol. 2, no. 8 (September 18, 1915), 117–118.

112. Ibid., 116–117.

113. Ibid., vol. 1, no. 3 (January 15, 1914), 36.

114. E.g., ibid.

115. *Magneet*, vol. 1, no. 10 (April 30, 1914), 121.

116. "Overlast van Automobielen en motorrijwielen," ibid., vol. 1, no. 4 (January 31, 1914), 41.

117. Ibid.

118. Ibid., 42–43.

119. Ibid., 43.

120. "Reglement Java-rit," ibid., vol. 2, no. 11 (October 9, 1915), 151.

121. "De Javarit," ibid., 151–153.

122. Decuop and Van der Hoeven, "De betrouwbaarheidsrit van den Gouverneur-Generaal door Midden Java," ibid., vol. 2, no. 8 (September 18, 1915), 111–115.

123. "De tocht naar Papandajan," ibid., vol. 1, no. 14 (June 30, 1914), 168.

124. Besides many other things, Tillema bought 2 milligrams of radiumbromide, shortly after the Curies announced their discovery, and he exhibited it in Semarang for the public. Vanvugt, E., *H.F. Tillema*, 25.

125. Tillema, H.F., "Een en ander over de beteekenis van weg en straat voor volks-gezondheid in Indië," *Gemeentereiniging: Officieel orgaan van de Nederlandsche Ver-eeniging van Reinigingdirecteuren. Maandblad voor reiniging en ontsmetting*, vol. 13, no. 1 (January 1920), 3.

126. Ibid.

127. Ibid., 3–4.

128. Ibid., 12.

129. Ibid., 3.

130. Ibid., 5.

131. Ibid., 6–7.

132. Ibid., 4–5.

133. Ibid., 12. There was much on castor oil in virtually every guide for the Indies at the time. "As soon as one came to the Indies, one used a little bit of castor oil every day, for stomach trouble, fever, etc. etc." Klay-Lancée, C.C., "Indië 25 jaar verleden," *Officieel orgaan van de Vereeniging van Huisvrouwen Bandoeng* (Bandoeng, October 1941), 841.

134. Tillema, H.F., "Een en ander over de beteekenis van weg en straat," 12–13.

135. Ibid., 10–11. The Netherlands Indies Encyclopaedia agrees with Tillema on this. The stone for building roads that could be found in the Indies, "with the exception of so-called Penang granite on the East Coast of Sumatra," it wrote, "is very soft and quickly turns into powder." *Encyclopaedie van Nederlandsch Indië*, vol. IV, 749.

136. Tillema, H.F., "Een en ander over de beteekenis van weg en straat," 20.

137. Ibid., 9.

138. Ibid., 17.

139. Ibid., 20–21.

140. See, e.g., Kelling, M.A.J., Secretaris van het Hoofd-Comité en van het Uitvoerend-Comité der Eerste Hygiëne Tentoonstelling, "De Eerste Hygiëne Tentoonstelling in Nederlandsch-Indië," *Koloniale Studiën* no. 2 (1927), 351.

141. Vanvugt, *H.F. Tillema*, 26.

142. "Timoer-Barat," *Indonêsia Raja*, vol. 1, no. 1 (January 1929), 24.

143. Ibid., vol. 1, no. 3 (April 1929), 62.

144. E.g., *Verslag omtrent der Post- en Telegraafdienst in Nederlandsch-Indië over het jaar 1900* (Batavia: Javaasche Bookhandel, 1901), 34.

145. Couperus, *The Hidden Force*, 224.

146. Langan, Celeste, *Romantic Vagrancy: Wordsworth and the Simulation of Freedom* (Cambridge: Cambridge University Press, 1995), 16.

147. Ibid., 15, 19.

148. Ibid., 17.

149. Ibid., 27.

150. Ibid., 17.

151. *Jaarboek TH Lustrum 1935* (Bandoeng: THS, 1935), 138.

152. Data quoted by Moelia in the debate about the 1939 budget of the colony. *Kritiek en Opbouw*, vol. 1, no. 12 (August 1, 1938), 181.

153. Wittgenstein, L., *Last Writings on the Philosophy of Psychology*. Vol. I. *Preliminary Studies for Part 2 of Philosophical Investigations* (Chicago: University of Chicago Press, 1990), 91 §712, 93 §726.

154. *Macintosh PowerBook 5300c Thesaurus*.

155. *Medan Politie Boemipoetra*, vol. 17, no. 3 (March 1941), 1–4; no. 4 (April 1940), 1–4.

156. Wittgenstein as summarized in *The Oxford Dictionary of Philosophy* by S. Blackburn (Oxford: Oxford University Press, 1994), 401; compare Wittgenstein's *Philosophical Investigations*, translated by G.E.M. Anscombe (New York: Macmillan, 1953), 19, §39.

157. There was a mildly dissident voice by one of the Chinese deputies: "The carriages of the 1st and 2nd class, even without any further improvement, are already more than good, and if there were to be some further improvement—I mean modernization—it would create a *superluxe* quality not really needed in the Indies." Ko Kwat Tiong (July 22, 1938) in *Handelingen van den Volksraad* (1938–1939), 367.

158. Van Helsdingen (July 22, 1938), in ibid. (1938–1939), 370.

159. Mevr. Razoux Schultz (July 22, 1938), in ibid. (1938–1939), 390.

160. Mevr. Razoux Schultz (July 22, 1938), in ibid. (1938–1939), 389–390.

161. Van Helsdingen (July 22, 1938), in ibid. (1938–1939), 370.

162. TOTOK, "Si Kromo itoelah toekang Automobiel jang bagoes sendiri," *Doenia Bergerak*, vol. 1, no. 10 (1915), 11–12.

163. Rietsema van Eck, S., *Koloniaal-staatkundig studiën 1912–1918* (Buitenzorg: n.p., 1919), 124f. quoted in Doorn, *De laatste eeuw van Indië*, 269.

164. Plate, A., "Het uitbreidingsplan der Indische gemeente," *Indisch Genootschap* (October 16, 1917), 12; quoted in ibid., 45.

165. Tillema, again, was reportedly the first one in the Indies, in 1901, to use reinforced concrete in one of his business buildings. Vanvugt, E., *H.F. Tillema*, 24.

166. Anderson, Benedict, "Sembah-Sumpah: The Politics of Language and Javanese Culture," in *Language and Power: Exploring Political Cultures in Indonesia* (Ithaca: Cornell University Press, 1990), 214.

167. *Motorblad* (April 14, 1928), 460.

168. Langan, *Romantic Vagrancy*, 18.

169. Ibid., 7, 21.

170. "Malay with Dutch letters," it was sometimes called. See Sandick, "Indische schetsen," 9.

171. Maier, Hendrik M.J., "Phew! Europeesche beschaving!"

172. "MARCO pro of contra Dr.RINKES!?" (*Doenia Bergerak*, vol. 1, no. 0 (January 31, 1914), 4, 7.

173. "Gantinja pakaian," *Saro Tomo*, vol. 1, no. 10 (1915), 157.

174. Marco, "Selembar soerat kepada joernaliste of redactrice," *Doenia Bergerak* vol. 1, no. 0 (January 31, 1914), 9.

175. Ibid.

176. The term is Javanese, used now by old people, Dédé Oetomo says, to describe young people's way of talking. Oetomo, Dédé, "The Chinese of Indonesia and the Development of the Indonesian Language," in *The Role of the Indonesian Chinese in Shaping Modern Indonesian Life* (Ithaca: Cornell Southeast Asia Program, 1991), 65.

177. Salmon, Claudine, "A Critical View of the Opium Farmers as Reflected in a *Syair* by Boen Sing Hoo (Semarang 1889)," in *The Role of the Modern Indonesian Chinese*, 28, 41. Oetomo, Dédé, "The Chinese of Indonesia and the Development of the Indonesian Language," in ibid., 53–66; Blussé, Leonard, "The Role of Indonesian Chinese in Shaping Modern Indonesian Life: A Conference in Retrospect," in ibid., 4.

178. Maier, "Phew! Europeesche beschaving!"

179. See, e.g., *Empatpuluhlima Tahun Sumpah Pemuda* (Jakarta: Gunung Agung, 1974).

180. Lie Kim Hok (1927), quoted in Salmon, Claudine, *Literature in Malay by the Chinese of Indonesia: A Provisional Annotated Bibliography* (Paris: Archipel, 1981), 116.

181. Tondokoesoemo, S., "Perasaan-hina-diri (Minderwaardigheidsgevoel)," *Keboedajaän dan Masjarakat*, vol. 2, no. 11 (March 1940), 269. See also the editorial "Maleisch in de raden," *Kritiek en Opbouw*, vol. 1, no. 23 (January 12, 1938), 341, and the editorial "De kloof," *Kritiek en Opbouw*, vol. 1, no. 11 (July 15, 1938), 166.

182. Soesilo, "De Maleische taal als Indonesische taal," *Kritiek en Opbouw*, vol. 2, no. 7 (May 16, 1939), 100.

183. Ibid.

184. Sarmidi Mangoensarkoro, "Koentoem bahasa Indonesia I," *Keboedajaän dan Masjarakat*, vol. 1, no. 6 (October 1939), 154–155. The best study in English on Taman Siswa is McVey, Ruth T., "Taman Siswa and the Indonesian National Awakening," *Indonesia*, no. 4 (October 1967), 128–149.

185. Ibid., 219–220.

186. Sjafroeddin Prawira Negara, "Mis- en verkenningen," *Orgaan van de USI*, vol. 7, no. 1 (November 1939), 33.

187. Ibid., 21.

188. Ibid., 24.

189. Ibid.

190. *Soeloeh Indonesia*, vol. 2, no. 2 (February 1927), 7–8.

191. S.S. 101 [Soedjojono], "Kesenian meloekis di Indonesia: sekarang dan jang akan datang," *Keboedajaän dan Masjarakat*, vol. 1, no. 6 (October 1939), 146.

192. "Nasib kaoem Sopir," *Sopir*, vol. 1, no. 1 (April 1932), 3.

193. Ibid., 7.

194. Ibid., 12.

195. "Kata pendahoeloean," ibid., vol. 1, no. 1 (April 1932), 2.

196. "Nasib kaoem Sopir(sambungan)," ibid., vol. 1, no. 2 (May 1932), 14.

197. Ibid., 15.

198. Ibid.

199. "Persatoean Chauffeur Mataram," ibid., vol. 1, no. 1 (April 1932), 5.

200. "Statuten dari P.C.M., seksi 5," ibid., vol. 1, no. 2 (May 1932), 23. For an establishing of a similar union of Indonesian and Chinese drivers in Surabaya, of about 600 members, in 1928, see "Oprichtingsvergadering B.A.T.O.J.A.," *Motorblad* (March 17, 1928), 335–336.

201. *Sopir*, vol. I, no. 1 (April 1932), 6; illustration ibid.

202. "Nasib kaoem Sopir," ibid., 3.

203. Oleh I. St. Manghoeto, "Sebabnja berdiri H.C.M.," ibid., vol. 1, no. 8 (November 1932), 96.

204. Ibid.

205. "Dokar dan kossongan moesti di keur dan koesirnja pakeh rijsbewijs. Gemeente haroes perhatikan," *Motorblad* (September 22, 1928), 1191. The usual driver's license in the Indies at the time contained name, address, photo, and the driver's fingerprint. "Nasib kaoem Sopir," *Sopir*, vol. I, no. 1 (April 1932), 3.

206. "Papreksahan dokter aken dapet rijbewijs," *Motorblad* (March 24, 1928), 367–368.

207. Holub, "Brief Thoughts on the Word Pain," in Holub, Miroslav, *Notes of a Clay Pigeon* (London: Secker and Warburg, 1985), 7. Compare Wittgenstein, *Philosophical Investigations*, 89 §244.

208. *Sinar Boeroeh Kereta Api*, vol. 2, no. 17 (May 1939), 14–16; see also the list of

the union's branches in the same issue, which does not make much sense—why that density and looseness there?—until one juxtaposes it against the list of the railway stations of the area. Ibid, vol. 2, no. 17 (May 1939), 22.

CHAPTER TWO

1. Pelt, R.J. van, and C.W. Westfall, *Architectural Principles in the Age of Historicism* (New Haven: Yale University Press, 1993), 298.

2. Barthes, Roland, *Writing Degree Zero* (New York: Farrar, Straus and Giroux, 1988), 12.

3. See chapter 1.

4. "De Nieuwe Uitvindingen," *Kopiïst*, vol. 1, no. 6 (1842), 657.

5. *Lelijke Tijd: Pronkstukken van Nederlandse interieurkunst 1835–1895* (Amsterdam: Waanders, 1995), 1.

6. Ibid., 7; portrait of family Metelerkamp ibid.

7. Ibid., 14.

8. Ibid, 20. Also: "Palms, cashmere shawls, draperied chairs, Eastern carpets, all that what was exotic, was beloved in the nineteenth-century interior." Ibid.

9. *Indische Bouwen: Architectuur en stedebouw in Indonesië* (Amsterdam: Nationale Contactcommissie Monumentenbescherming et al., 1990), 11–12.

10. Lelyveld, Th.B. van, "De indische bouwkunst en kunstnijverheid, oud en nieuw," *Nederlandsch Indië Oud en Nieuw 1916–1917*, 571.

11. Ibid., 573.

12. Catenius-van der Meijden, J.M.T., *Ons Huis in Indië. Handboek bij de keuze, de inrichting, de bewoning en de verzorging van het huis met bijgebouwen en erf, naar de eischen der hygiëne, benevens raadgevingen en wenken op huishoudelijk gebied* (Semarang: Masman en Stroink, [1908]), 229; quoting from a letter by a reader of her previous guidebook.

13. Ibid.

14. Benjamin, W., "Paris, Capital of the Nineteenth Century," in *Reflections* (New York: Harcourt Brace Jovanovich, 1978), 151.

15. "Holl. IJzeren Spoorweg-Maatschappij en Maatschappij tot Exploitatie van Staatsspoorwegen," *Uitreiking der bekroningen aan de Nederlandsche inzenders van de internationale Koloniale en Uitvoerhandel tentoonstelling* (Amsterdam: n.p., 1883), 10.

16. Ibid., 12.

17. Burkom, Frans van, and Wim de Wit, "Vormgeving als kunst, kunst als vormgeving," in Burkom, F., W. de Wit, et al., *Amsterdamse School: Nederlandse architectuur 1910–1930* (Amsterdam: Stedelijk Museum, 1975), 34.

18. *Uitreiking der Bekroningen*, 19; "Het koloniaal-geneeskundig museum te Amsterdam," *De Indische Gids*, vol. 13, no. 2 (1891), 1655–1657.

19. *Officieele wegwijzer voor Internationale Koloniale en Uitvoerhandel tentoonstelling te Amsterdam 1883* (n.p., n.p.), 9–12. (One Amsterdam ell is 0.688 m.)

20. Ibid., 41–53.

21. Ibid., 13.

22. The role of the Chinese landowner, Oei-Tiong-Ham, in helping the big Indies Semarang fairs was especially well known. See, e.g., *Nederlandsch Indië Oud en Nieuw*,

vol. 4 (1919–1920), 49–50. Fairs were sometimes celebrated, as by the anonymous Indonesian Chinese author's *Sair Pesta Jubileum 1898–1923 dan Pasar Gambir*, "Poem of the Jubilee Feast 1898–1923 and Gambir Fairs." Salmon, Claudine, *Literature in Malay by the Chinese of Indonesia: A Provisional Annotated Bibliography* (Paris: Archipel, 1981), 42.

23. "Javasche Schetsen: Batavia-De Harmonie (het Gebouw)," *Kopiïst*, vol. 1, no. 1 (1842), 29–46.

24. *Reglement voor de societeit 'De Harmonie' gevestigd te Batavia waaraan zijn toegevoegd de sedert 30 September 1935 geldende Huishoudelijke Bepalingen* (Batavia: De Unie, 1935), 1–22.

25. Ibid.

26. Maurits (P.A. Daum), *H. Van Brakel: Ing. B.O.W.* (Amsterdam, 1976), 11. This is one of the widely read novels from the turn of the century.

27. Nieuwenhuis, Dr. J.J., "Lichamelijke opvoeding en de onderwijshervorming in Ned.-Indië II," *Koloniale Studiën*, no. 1 (1920), 515.

28. Ibid.

29. Ibid.

30. Veth, P.J., Wilken, G.A., and Klinkert, H.C., eds., *Catalogus der afdeeling Nederlandsche koloniën van de internationale koloniale en uitvoerhandel tentoonstelling* (Leiden: Brill, 1883), vol. III, 100.

31. *Nederland te Parijs 1931: Gedenkboek van de Nederlansche deelneming aan de Internationale Koloniale Tentoonstelling* ('s Gravenhage: Van Stockum en Zoon, 1931?), 89.

32. Berlage, H.P., *Mijn Indische Reis: Gedachten over cultuur en kunst* (Rotterdam: W.L. en J. Brusse's, 1931), 8.

33. van Burkom and de Wit, "Vormgeving als kunst, kunst als vormgeving," 38; on Lion Cachet as a respected influence in the Amsterdam School, see also J.F. Staal in *Wendingen*, vol. 1, no. 1 (1918), quoted by van Burkom, Frans van, "Kunstvorming in Nederland," in Burkom, de Wit, et al. *Amsterdamse School: Nederlandse architectuur 1910–1930*, 88.

34. Borel, H., *Van Batavia naar Rotterdam: reisgids van de Rotterdamsche Lloyd* (Amsterdam: Rotterdamsche Lloyd, 1905), 8.

35. Berlage, *Mijn Indische Reis*, 7.

36. *Nederlandsch Indië Oud en Nieuw* (1916–1917), 33.

37. Ibid., 35.

38. Ibid., 37–38; see also an unsigned article: "Het dubbelschroefstoomschip 'Insulinde' van de Rotterdamsche Lloyd." Ibid., 145.

39. Catenius, *Ons Huis in Indië*, 102.

40. Borel, *Van Batavia naar Rotterdam*, 1.

41. Ibid., 2.

42. Ibid., 34.

43. Ibid., 11–12.

44. Ibid., 19.

45. Ibid., 15.

46. Ibid., 41–42.

47. On the occasion of the Bandung Fairs, in August 1923, a regular airplane connection was launched with military planes between Bandung and Batavia. In 1928,

the Royal Dutch Indies Airline Company *KNILM* was established. On November 1, 1928, daily Batavia to Bandung and Batavia to Semarang flights started. Drs. P.Joh. Bettink, "Enkele beschouwingen over het Luchtverkeer in Nederlandsch-Indië," *Koloniale Studiën*, no. 1 (1936), 55–57.

48. See pictures in Rusman, E., *Hollanders vliegen: van H-NACC tot Uiver: tien jaar Amsterdam-Batavia door de lucht* (Baarn: Bosch en Keuning, 193?), 41.

49. "Tropenhitte en Hollandsche koud," *KNILM Nieuws*, vol. 3, no. 1 (January 1937), 93–94.

50. "Een jaar Luchtverkeer, 1929–1930," *K.N.I.L.M. Reisgids 1930* (Batavia: K.N.I.L.M., 1930), 9–13.

51. Ir Noto Diningrat, "Grondslagen voor de bouwkunst op Java," *Nederlandsch Indië Oud en Nieuw* (1919–1929), 112.

52. Ibid.

53. Ibid., 113.

54. Ibid., 114.

55. Dewantara, "Associatie antara Timur dan Barat," *Wasita*, vol. 1, nos. 11/12 (August–September 1929), quoted in Tsuchiya, Kenji, *Democracy and Leadership: The Rise of the Taman Siswa Movement in Indonesia* (Honolulu: University of Hawaii Press, 1987), 60.

56. Nellie (Mevr. Nellie van Kol), *Brieven aan Minette* (reprint 'Soerabajasch Handelsblad) ('s Gravenhage: n.p., 1884), 143–144.

57. Dr. P.J.A. Sluijs, "Over Acclimatisatie," *De Indische Gids*, vol. 14, no. 1 (1892), 627.

58. Ibid., 629.

59. Jan Ligthart, H. Scheepstra, and A.F.Ph. Mann, *Ver van Huis* (Weltevreden: Wolters, 1918, etc.)

60. Veth, B., *Het leven in Nederlandsch-Indië*, 2nd ed. (Amsterdam: Van Kampen, 1900?), 3–4.

61. Ibid., 18.

62. Ibid., 178.

63. Ibid., 192.

64. Ibid., 199–201.

65. Irrigation in the wet-rice agriculture, of course, had already in pre-colonial time been a major issue of economy and society as well as politics. On the colonial policy in the field at the turn of the nineteenth and twentieth centuries, see, e.g., Homan van der Heide, J., *Beschouwingen aangaande de volkswelvaart en het irrigatiewezen op Java in verband met de Solovalleiwerken* (Batavia: Kolff, 1899), and Meyier, J.E. de, "Irrigatie-fanatisme," *Indische Gids*, vol. 24, no. 2 (1902), 172ff.

66. Numans, J.G., Ingenieur Waterst.N.-I., "Een en ander over irrigatiewerken op Java," *Nederlandsch-Indië Oud en Nieuw* (1918–1919), 1.

67. Akihary, Huib, *Architectuur en stedebouw in Indonesië 1870–1970* (Zeist: Rijksdienst voor de Monumentenzorg, 1988), 59. Channeling waters in urban areas was far from being simply a technical task, even in modern Europe and the Netherlands: "Even after 1883, when Koch through the discovery of the bacilli of cholera confirmed the theories of Snow, people remained holding steadfastly to the idea that water, even when polluted, could not cause any infectious disease." It was not until 1896 in Amsterdam, the most progressive city of the Netherlands, that the municipal

authorities moved to implement an overall plan for water supply and water distribution. See Roding, Juliette, *Schoonheid en net: hygiëne in woning en stad* ('s Gravenhage: Staatsuitgeverij, 1986), 49.

68. Tillema, H.F., *"Kromoblanda": over 't vraagstuk van "het wonen" in Kromo's groote land*, vols. 1–6 ('s Gravenhage: Uden Masman etc., 1915–1923).

69. Vanvugt, E., *H.F. Tillema (1870–1952) en de fotografie van tempo doeloe* (Amsterdam: Mets, 1993), 27.

70. Ibid., passim.

71. Quoted in Kelling, M.A.J., Secretaris van het Hoofd-Comité en van het Uitvoerend-Comité der Eerste Hygiëne Tentoonstelling, "De Eerste Hygiëne Tentoonstelling in Nederlandsch-Indië," *Koloniale Studiën*, no. 2 (1927), 335.

72. Dipl. Ings.B. de Vistarini (Bouwkundig Ingenieur te Soerabaia), "Luchtconditioneering en bouwkunde," *De Ingenieur in N.I.*, vol. 5, no. 11 (November 1938), VI, 46.

73. Ibid.

74. Hen, Mr.I., "Iets over het woningvraagstuk in de Indische gemeenten," *Koloniale Studiën*, no. 1 (1916–1917), 133.

75. Numans, J.G., Ingenieur Waterst.N.-I., "Een en ander over irrigatiewerken op Java," 1–9.

76. Letter by Djokobodo, *Doenia Bergerak*, vol. 1, no. 39 (1915), 2.

77. Ibid., 1.

78. Ibid., 2.

79. Ibid., 3.

80. Estépé, "Te vroeg?" *Soeloeh Indonesia*, vol. 2, no. 1 (January 1927), 6.

81. Ibid.

82. Ibid., 7.

83. Soe, "De Bevolking en de Volkshuizing," *Soeloeh Indonesia*, vol. 2, no. 3 (March 1927), 10.

84. Ibid. In 1937, an inquiry was done in a selected native city quarter, Tanah Tinggi in Batavia. Close to 5,000 inhabitants were questioned: 8.5 percent answered that they regularly bathed and defecated in the local *kali* (river); the number would be, actually, much higher, researchers noted, because most of the people using the river could not be reached by the questioners. Only 8.6 percent respondents had even a most simple outlet for feces, and the rest said they used communal latrines, "which were found in all aspects and completely unsatisfactory." Only about 10 percent of respondents had their own connection to tap water. Tesch, J.W., "Een studiewijk voor hygiëne te Batavia," *Koloniale Studiën*, no. 5 (1939), 501–506.

85. Katjoeng, "Bagaimana Wet menjerahkan nasib Kaoem Boeroeh kepada kelalimannja Madjikan," *Soeloeh Indonesia*, vol. 2, no. 6 (June 1927), 6.

86. "Bloc Notes," *Nationale Commentaren*, vol. 4, no. 13 (March 29, 1941), 3311.

87. Karsten, H.Th., "Stedebouw" in F.W.M. Kerchman, ed., *25 Jaren Decentralisatie in Nederlandsch-Indië, 1905–1930* (Weltevreden: Kolff en Vereeniging voor Locale Belangen te Semarang, 1931), 138n, 141.

88. Akihary, *Architectuur en stedebouw in Indonesië*, 55.

89. Tillema's paper "Van wonen en bewonen van bouwen, huis en erf," quoted in ibid., 61.

90. Ibid.

91. Ibid., 62.

92. van Breen, Ir.H., "Onderwijs in de techniek van assaineering," *Koloniale Studiën*, no. 1 (1925), 361–362.

93. Hygiene and quarantine, logically, always went together on these and other occasions. See, for an early appearance of the pair at the Amsterdam exhibition in 1883, *Uitreiking der bekroningen*, 20.

94. Kelling, M.A.J., Secretaris van het Hoofd-Comité en van het Uitvoerend-Comité der Eerste Hygiëne Tentoonstelling, "De Eerste Hygiëne Tentoonstelling in Nederlandsch-Indië," *Koloniale Studiën*, no. 2 (1927), 348.

95. Hen, Mr.I., "Iets over het woningvraagstuk in de Indische gemeenten," *Koloniale Studiën*, no. 1 (1916–1917), 129, 133.

96. Ibid. In a study of Indies urban planning from 1931, Dutch architect Karsten concluded that systematic building regulation and zone designing had been meaningfully attempted in only one of the Indies city, Bandung, the exceptional and to the greatest extent exclusive "hill-station," "Paris of the tropics." Overall, Karsten wrote, the urge to construct cities in the Indies as Western enclaves made handling of the cities as a whole impossible. *Kampongs* were left out; everything, indeed, that was not on the main roads was left out. If there were some efforts to achieve an aesthetic harmony, including some harmony between East and West, such efforts would be inevitably, in the Indies, "shipwrecked." Karsten, "Stedebouw," 140, 142–143.

97. Sukarno studied at the Technical College Bandung between 1921 and 1926. (He is on the school record as no. 55. *Jaarboek der Technische Hoogeschool te Bandoeng 15e cursusjaar . . . 1935*. [Bandoeng: Maks and van denKijts, 1935], 121). In 1926 he was, for a short time, employed by the Department of Public Works (*BOW*). In the same year, together with Anwari, he worked reportedly, among other things, on the design of a residence for the regent of West Java. "Ir. M. Anwari," *Ganeça*, vol. 6, no. 11 (November 1934), 4, and Akihary, *Architectuur en stedebouw in Indonesië*, 119. According to other information, Sukarno's partners in another architectural firm in Bandung a short time later were Soemono, Engineer, and Roosseno, Engineer. See *Roosseno: pakar dan perintis teknologi sipil Indonesia* (Jakarta: Pembimbing Masa, 1989), 891.

98. *Poedjangga Baroe* (1933–1938), 65; quoted in McIntyre, Angus, "Sukarno as Artist-Politician," in Angus McIntyre, ed. *Indonesian Political Biography: In Search of Cross-cultural Understanding* (Clayton: Monash Papers on Southeast Asia No. 28, 1993), 164–165.

99. Prof. J.F. Klinkhammer. b.i. en B.J. Ouëndag, Architect, "Het Administratiegebouw der Nederlandsch-Indische Spoorweg-Maatschappij te Semarang," *Nederlandsch Indië Oud en Nieuw* (1916–1917), 30.

100. Ibid.

101. Maclaine Pont, H., "Het nieuwe Hoofdbureau der Semarang-Cheribon-Stoomtram Maatschappij te Tegal," *Nederlandsch Indië Oud en Nieuw* (1916–1917), 90.

102. Akihary, *Architectuur en stedebouw in Indonesië*, picture no. 30.

103. "Op zoek naar een identiteit, Art Deco," *Indische Bouwen*, 27; there also, see for some illustrations of the buildings.

104. Akihary, *Architectuur en stedebouw in Indonesië*, 38. See also Wolff Schoemaker's speech at the Bandung Technical College on June 28, 1930: Wolff Schoe-

maker, C.P., "De Aesthetiek der Arhitectuur en de Kunst der Modernen," *Jaarboek der Technische Hoogeschool te Bandoeng 10e cursusjaar* . . . *1930* (Bandoeng: Maks and van denKijts, 1930), 92.

105. Akihary, *Architectuur en stedebouw in Indonesië*, 42.

106. Kelling, M.A.J., "Het Jaarbeurswezen in Nederlandsch-Indië," *Koloniale Studiën*, no. 2 (1925), 227.

107. Akihary, *Architectuur en stedebouw in Indonesië*, 44.

108. *Indische Bouwen*, 29; for illustration, see ibid., 30, and picture no. 33 in Akihary, *Architectuur en stedebouw in Indonesië*, 44.

109. Ibid., 46.

110. Quoted in Berlage, *Mijn Indische Reis*, 102.

111. Ibid., 31.

112. Ibid., 29.

113. Ibid., 134.

114. Akihary, *Architectuur en stedebouw in Indonesië*, 51.

115. Berlage, *Mijn Indische Reis*, 23–24.

116. Ibid., 24.

117. Wertheim, Wim, and Hetty Wertheim-Gijse Weenink, *Vier Wendingen in Ons Bestaan: Indië verloren-Indonesië geboren* (Breda: De Geus, 1991), 87.

118. Treub, M.W.F., *Nederland in de Oost: reisindrukken* (Haarlem: Willink, 1923), 332–333.

119. Berlage, *Mijn Indische Reis*, 33–34.

120. Walraven, W., "De schout als stut in het leven," *Kritiek en Opbouw*, vol. 5, no. 2 (January 20, 1942), 23.

121. Sk.[H. Samkalden], "Notitie over laster," *Kritiek en Opbouw*, vol. 2, no. 1 (February 16, 1939), 6. According to another author from the same time, the Indies city grew "mercilessly," and it was "like pieces of a jigsaw puzzle that do not fit together." Wetering, F.H. van, "Zôneering als sociaalstedebouwkundige maatregel," *Koloniale Studiën*, no. 5 (1939), 594–595.

122. Ibid.

123. Ibid.

124. Soesilo, "De maatschappelijke evolutie in de Indische stad," *Kritiek en Opbouw*, vol. 1, no. 17 (October 16, 1938), 256.

125. Ibid.

126. See, e.g., "Boomvormen," *I.B.T. Locale Techniek*, vol. 9, no. 1 (February 1940), 23–28; illustrations ibid.

127. *Nederlandsch Indië Oud en Nieuw* (1916–1917), 1.

128. Ibid., 567–568. A Dutch sociologist of the Indies pondered recently, in the mid-1990s, about some of the "memorial books" from the last decade of the late-colonial Indies, and he wrote down his impressions: "There is no need to doubt the usefulness of public hygiene, but what strikes everyone who wanders through these pages and looks at the pictures is the disappearance of the green from wherever the city quarters were built. Through the clearing, however well-meaning it might have been, the tropical sun is given a full play; even in the black and white photographs, the harsh light causes a pain in one's eyes." Doorn. *De laatste eeuw van Indië*, 156, commenting upon, e.g., F.W.M. Kerchman, ed., *25 Jaren decentralisatie in Nederlandsch-Indië 1905–1930* (Semarang, 1930), and G. Flieringa, *De zorg voor de volks-*

huisvesting in de stadsgemeenten in Nederlandsch Oost Indië in het bijzonder in Semarang ('s Gravenhage, 1930).

129. Walraven, W., "Het Indische Stadsbeeld voorheen en thans," *Kritiek en Opbouw*, vol. 3, no. 24 (February 1, 1941), 377.

130. Bonset [T. van Doesburg], I.K., "Tot een constructieve dichtkunst," *Mécano*, nos. 4/5 (1923), n.p.

131. In Magnitogorsk, J. Leonidov attempted a "linear city." "Tape urbanistics" (by N. Milyutin) and "parabola cities "(by N. Ladovsky) were other forms of the experiments in solving the problems of modern city. See *Sovjet Architectuur, 1917–1987* (Amsterdam: Art Unlimited Books, 1989), 27.

132. Akihary, *Architectuur en stedebouw in Indonesië*, 17.

133. Nieuwenhuijs, R., "Over de Europese samenleving van 'tempo doeloe' 1870–1900," *Fakkel*, vol. 1, no. 9 (July–August 1941), 782.

134. *Indische Bouwen*, 17. In the report published in 1939 by the government commission on the "improvement of the native quarters," the key phrase and aim were *orde en netheid*, "order and neatness." The report of the commission is quoted in Doorn, *De laatste eeuw van Indië*, 156.

135. P.K., "Stadskampongs," *Tecton*, vol. 2, no. 11 (November 1937), 201.

136. Ibid.

137. Ibid., 203.

138. "Bungalow-itis," *Nationale Commentaren*, vol. 4, no. 44 (November 1, 1941), title page. See also a lecture by architect de Vistarini in 1938: "This is a striking feature—numerous families that earlier had lived in reasonably large 'modern' houses in the city of Surabaia now find it desirable to give the house up for something of a dollhouse in one of the mountain areas around the city, from which the head of the family spends one hour, at least, every morning and every evening, traveling there and back; this in spite of the fact that the costs of living in these hill areas are the same as in the city, or higher." Dipl. Ing. B. de Vistarini, "Luchtconditioneering en bouwkunde," VI, 45.

139. Ibid.

140. Wetering, F.H. van de, "Sateliet-Plaatsen," *Koloniale Studiën*, no. 3 (1940), 247–269.

141. "Bungalow-itis," *Nationale Commentaren*, vol. 4, no. 44 (November 1, 1941), title page.

142. Wertheim and Wertheim, *Vier Wendingen*, 171.

143. Perron, E. du, *Brieven*, vol. 6 (November 1, 1935–June 30, 1937) (Amsterdam: Van Oorschot, 1980), 394 (Tjitjoeroeg, March 16, 1937); ibid., 360–361 (Tjitjoeroeg, February 26, 1937).

144. Perron, E. du, *Brieven*, vol. 7 (July 2, 1937–November 30, 1938) (Amsterdam: Van Oorschot, 1981), 69 (Tjitjoeroeg, August 9, 1937).

145. Ibid., 185 (Garoet, November 11, 1937).

146. Ibid., vol. 6, 314 (Bandoeng, November 17, 1936).

147. Ibid. vol. 7, 346–347 (Sitoe Goenoeng, April 24, 1938).

148. "Het bebouwingsplan voor het Koningsplein te Batavia volgens het ontwerp Fermont-Cuypers, " *De Ingenieur in N.I.*, vol. 5, no. 2 (February 1938), 26–27.

149. Ibid., 27.

150. Treub, *Nederland in de Oost*, 168.

151. Ibid., 233.

152. Ibid., 249–250, 252.

153. Ibid., 267. These, to perfection, were moveable and floating homes as well. In Tjepu *dienst park*, for instance, Treub wrote: "in the houses of the higher personnel, all the furnishings belong to the company, not merely the big furniture but also tables—and bedding, glass, etc. In this way, when needed by the company, the personnel can be transferred without a great and complicated process of leaving." Ibid., 234.

154. "The roomy houses full of air . . . in which the Indies man used to live, and we are witnesses to it, are fast disappearing." See Tillema, H.F., *"Kromoblanda, vol. IV/2*, 557.

155. See van Doesburg, quoted above.

156. Meijer, Ir.H. met medewerking van Ing. F.A.J. Heckler, *De Deli Spoorweg Maatschappij: Driekwart eeuw koloniaal spoor* (Zutphen: De Walburg Pers, 1987), 75. Another form of *dienst parks* in the Indies were the training camps for the Dutch and Eurasian colonists, getting ready to move to colonize New Guinea. See, e.g., "Nieuws uit de werk- en leerkampen: Het werk- en leerkamp in Oost Java," *Onze Toekomst* (January 2, 1936), 7. "Werkkamp en Europeesche Boerenstand in Indië," ibid. (January 15, 1936), 19–20; "Het landbouw leer- en werkkamp 'Oost-Java' te Poerwodadi bij Lawang," ibid. (February 25, 1939), 27; Tichelman, G.J. en Jhr.Mr Dr W.H. Alting van Geusau, *N.S.B.-Deportatie naar Oost en West* (Amsterdam: Uitgegeven met voorkennis van den Regeringsvoorlichtingsdienst door D.A.V.I.D, 1945). In one of the camps, Kesilir in East Java, used by the Japanese administration as a concentration camp, Willem Walraven was interned with his sons, and there, on February 13, 1943, he died. Introduction by F. Schamhardt to Walraven, W., *Eendagsvliegen. Journalistieke getuigenissen uit kranten en tijdschriften* (Amsterdam: G.A. van Oorschot 1971), 13–14; and Walraven W. Jr., "Mijn vader, de bodemloze put en de wijnkan," *Tirade*, vol. 11, no. 121 (January 1967), 18.

157. Heidegger, Martin, "Building, Dwelling, Thinking," in *Poetry, Language, Thought*, translated by Albert Hofstadter (New York: Harper, 1971), 150.

158. Ibid., 154.

159. Kierkegaard, Søren, *The Concept of Irony with Continual Reference to Socrates*, translated by Howard V. Hong and Edna Hong (Princeton: Princeton University Press, 1989), 171.

160. Overy, Paul, *De Stijl* (London: Thames and Hudson, 1991), 30–31.

161. Ibid., 24.

162. Ibid.

163. Dipl. Ing. B. de Vistarini, "Luchtconditioneering en bouwkunde," VI, 44–45.

164. Ibid., VI, 45.

165. Ibid.

166. Ibid.

167. Ibid.

168. Ibid.

169. Ed.Cuypers, arch., "De Moderne Ambtenaarswoning in Nederlandsch-Indië," *Nederlandsch Indië Oud en Nieuw* (1919–1920), 117.

170. Ibid., 119, 124. Louis Couperus, after a long absence, visited the colony in 1922, and instantly he noticed the change: "for the first time I see the new type of

Dutch-Indies houses . . . old type . . . [had] verandah, back and front . . . pillars. . . .
Modern Indies house . . . [has] front verandah . . . more like a hall . . . not quite
open . . . more private, but I wonder if it is not also more stuffy . . . shut off as much
as possible from the all-penetrating element . . . even the back gallery . . . [is] more
shut in than in the old type house." Couperus, Louis, *Eastward* (New York: G.H.
Doran, 1924), 40.

· 171. Walraven, "Sepada," *De Indische Courant* (November 17, 1933), reprinted in
Walraven, W., *Eendagsvliegen*, 223, 226, 228.

172. Walraven, W., "Het Indische Stadsbeeld voorheen en thans," 376.

173. Walraven, "Sepada," 224.

174. Walraven, W., "Het Indische Stadsbeeld voorheen en thans," 376.

175. Walraven, "Sepada," 224–225.

176. Catenius, *Ons Huis in Indië*, facing pages 16 and 48. There might be some
prehistory to this, and with a similar motivation: the houses of the Indies Chinese
reacting to the ever unpredictable and as often as not anti-Chinese Indies environment—a shop downstairs, the living quarters with a built-up (imagined) safety off the
ground, on the second floor (personal communication from Benedict Anderson).

177. Catenius, *Ons Huis in Indië: Advertentie* (n.p., n.d.), 1.

178. van Lelyveld, Th.B., "De indische bouwkunst en kunstnijverheid, oud en
nieuw," 570.

179. Van Burkom and de Wit, "Vormgeving als kunst, kunst als vormgeving," 34.

180. Van Burkom, "Kunstvorming in Nederland," 75.

181. Van Burkom and de Wit, "Vormgeving als kunst, kunst als vormgeving," 35.

182. Overy, *De Stijl*, 149.

183. Ibid., 105.

184. Le Corbusier, *Towards a New Architecture* (New York: Dover, 1986), 4ff.

185. Sluijs, Dr. P.J.A., "Over Acclimatisatie," 627.

186. *Tijdschrift van het Kon.Instituut van Ingenieurs afd N.I.*, quoted in "Verkoeling
van spoorwegrijtuigen in Indië," *De Indische Gids*, vol. 4, no. 1 (1882), 413.

187. Ibid.

188. Ibid.

189. "Over *acclimatisatie* in de tropen," *De Indische Gids*, vol. 15 (1893), 839.

190. Ibid.

191. Ibid., 840.

192. Prof. dr. ir. C.P. Mom (Uit het Laboratorium voor Technische Hygiëne
te Bandoeng), "De Ontwikkeling der Luchtconditioneering in 1937 in Noord-
Amerika," *De Ingenieur in N.I.*, vol. 5, no. 5 (May 1938), VI, 3–5; prof. dr. ir. C.P.
Mom, "De Bacteriën en de Hygiëne," ibid., VI, 16–17; prof. dr. W. Radsma et al.,
"Metingen van Lichaamstemperatuur bij Blanken in Tropische en in Koelere Lucht-
streken," ibid., no. 9 (September 1938), VI, 18; also prof. dr. J.E. Dinger et al.,
"Physiologisch reacties van gezonde personen op luchtbehandeling, "ibid., VI, 19–21.

193. Janssen, Dr. E., Arts te Bandoeng, "Luchtbehandeling en Geneeskunde,"
ibid., vol. 5, no. 9 (September 1938), VI, 23.

194. Ibid.

195. Ibid., VI, 25.

196. "Luchtcondities en behaaglijksgebied in de kuststreken en in het bergland van
Java," *De Ingenieur in N.I.*, vol. 5, no. 9 (September 1938), VI, 37.

197. Thornburh, H.A., "Air Conditioning in Hospitals," ibid., no. 11 (November 1938), VI, 30. There were, for instance, about two hundred air-cooling installations working or ready to work in Surabaya in 1938. Ir. P.Timmerman, Ingenieur bij de Algemeene Ned.-Indische Electriciteit Mij. te Soerabaia, "De bevordering van lucht-conditioneering door de electriciteitmaatschappijen en het verbruik van electrische stroom in Ned.-Indië," ibid., vol. 5, no. 11 (November 1938), VI, 48. Surabaya, in 1930, had a population of 342,000 (of it 26,000 "Europeans" and 39,000 "Chinese"). Gonggrijp, G.F.E., ed., *Geillustreerde Encyclopaedie van Nederlandsch Indië.* The 1934 edition reprinted (Wijken en Aalburg: Pictures Publishers, 1992), 1319.

198. Dipl. Ings. B. de Vistarini, "Luchtconditioneering en bouwkunde," VI, 44–46.

199. Ir. P. Timmerman, "De bevordering van lucht-conditioneering," VI, 47.

200. Dipl. Ings. B. de Vistarini, "Luchtconditioneering en bouwkunde," VI, 44–46.

201. Ibid., VI, 44.

202. Ibid.

203. Ibid.

204. Ibid., VI, 43–44.

205. Ibid.

206. Ir. P. Timmerman, "De bevordering van lucht-conditioneering," VI, 49.

207. Ibid., VI, 45.

208. Ibid.

209. Wertheim and Wertheim, *Vier Wendingen*, 204.

210. Prof. dr. ir. C.P. Mom, "Over den invloed den luchtelectriciteit op de behaaglijkheid en de gezondheid," *De Ingenieur in N.I.*, vol. 5, no. 9 (September 1938), VI, 28.

211. Ibid.

212. Treub, *Nederland in de Oost*, 42.

213. Idsinga, T. en J. Schilt, *Architect W. van Tijen 1894–1974* ('s Gravenhage, 1987), 46.

214. *Alles electrisch in huis en bedrijf,* vol. 1, no. 1 (June 1931), n.p.

215. Ibid.

216. Ibid., vol. 1, no. 2 (August 1931), n.p.; ibid., vol. 1, no. 6 (December 1931), 4.

217. Ibid., vol. 1, no. 12 (June 1932), 13; three apparatuses pictured.

218. Ibid., vol. 1, no. 5 (November 1931), title page.

219. Ibid., vol. 1, no. 2 (August 1931), n.p.

220. Musil, R., *The Man Without Qualities* (New York: Putnam's, 1980), 30.

221. Ibid.

222. Doorn, A., "Hoe men soms bouwt," *Tecton*, vol. 1, no. 8 (August 1936), 152.

223. Ibid., 154.

224. Prof. J.F. Klinkhammer. b.i. en B.J. Ouëndag, Architect, "Het Administratiegebouw der Nederlandsch-Indische Spoorweg-Maatschappij te Semarang," *Nederlandsch Indië Oud en Nieuw 1916–1917*, 25–26.

225. Ed.Cuypers, arch., "De Moderne Ambtenaarswoning in Nederlandsch-Indië," 119.

226. Ibid., 124.

227. Ibid.

CHAPTER THREE

1. Rorty, Richard, *Philosophy and the Mirror of Nature* (Princeton: Princeton University Press, 1979), 39.

2. Sandick, R.A. van, *In het Rijk van Vulcaan: de uitbarsting van Krakatau en hare gevolgen* (Zutphen: Thieme, 1890), 45–46.

3. Ibid., 84–86.

4. Ibid., 99.

5. Ibid., 48.

6. Ibid., 65.

7. Ibid., 138–139.

8. Ibid., 99.

9. Ibid., 100–102.

10. Ibid., 170.

11. Wiggers, F., *Boekoe Peringatan tjeritain dari halnja saorang prampoewan Islam Tjeng Kao bernama Fatima*, 2nd ed. (Batavia: Goan Hong, 1916), title page.

12. Ibid., 3–7.

13. Ibid., 8.

14. Ibid.

15. Ibid., 12.

16. Ibid., 16.

17. Ibid., 17.

18. His full name is given a little later as *Toewan Adolf Wilhelm-Verbond* [sic] *Hinne*. This points to an actual police officer A.W.V. Hinne. See more on him in Till, Margreet van, "In Search of Si Pitung: The History of an Indonesian Legend," *Bijdragen tot de Taal-, Land- en Volkenkunde*, vol. 152–III (1996), 462–479, and Termorshuizen, Gerard, "A Murder in Batavia or the Ritual of Power," in Peter Nas, ed. *Urban Symbolism* (Leiden: Brill, 1993), 135–152. A.W.V. Hinne is also one of the protagonists in an Indonesian feature film *Banteng Betawi*, "Bull of Batavia," released in Jakarta in 1971.

19. Wiggers, *Boekoe Peringatan*, 21–25.

20. Ibid., 21.

21. Ibid., 20–21.

22. Ibid., 24.

23. Pramoedya Ananta Toer, ed., *Tempo doeloe: antologi sastra pra-Indonesia* (Jakarta: Hasta Mitra, 1982), 17.

24. Ibid., 18–19. Possibly again with a Chinese, Lie Kim Hok, F. Wiggers authored "Count of Monte Christo" and a selection from "Arabian Nights." See Chambert-Loir, Henri, "*Sair Java-Bank di rampok:* Littérature malaise ou sine-malaise?" in Salmon, ed. *Le moment "sino-malais" de la littérature indonesiènne* (Paris: Cahier d'Archipel, 1992), 43–70, and Salmon, Claudine, *Literature in Malay by the Chinese of Indonesia: A Provisional Annotated Bibliography* (Paris: Archipel, 1981), 32, 33.

25. Wiggers, *Boekoe Peringatan*, 19.

26. Sandick, van, *In het Rijk van Vulcaan*, 185ff.

27. On the "camera" and "photography" metaphors of Dutch literary criticism at the turn of the century, see Bel, J., *Nederlandse literatuur in het fin de siècle* (Amsterdam: Amsterdam University Press, 1993), 57ff.

28. Couperus, Louis, *The Hidden Force* (Amherst: University of Massachusetts Press, 1985), 49.

29. Ibid., 151, 131, 45. See a photograph of Harmonie in Pasuruan in Buitenweg, H., *Soos en samenleving in tempo doeloe* (Den Haag: Service, 1966), 13; also Nieuwenhuijs, R., *Baren en oudgasten: tempo doeloe-een verzonken wereld* (Amsterdam: Querido, 1981), 146–166.

30. Couperus, *The Hidden Force*, 192, 145.

31. Ibid., 126.

32. Ibid., 183. Couperus wrote later, in 1911, in his recollections of his own childhood: "The child knew . . . the anguish, in the mysterious Indies . . . that tropical hour of the twilight . . . all sounds suddenly dropped silent . . . one had to go through an overgrown path, which led through the garden. . . . Take your second bath. . . . Through the darkened backyards, over the black lawns . . . towards the bathroom" (Couperus, Louis, "De Angst," *Het Vaderland* [September 16, 1911]).

33. Couperus, *The Hidden Force*, 45.

34. Kartini, R.A., *Letters of a Javanese Princess* (New York: W.W. Norton, 1964), 64. See also Kartini, R.A., *Brieven aan mevrouw R.M. Abendanon-Mandri en haar echtgenoot*, edited by F.G.P. Jaquet (Doordrecht: Foris, 1987), 135.

35. Ibid., 6.

36. Ibid., 70.

37. Ibid., 118.

38. The first edition of Kartini's letters was published by Van Dorp (Semarang, Surabaja en Den Haag), 1911, 1,000 copies; the second edition appeared in Luctor et Emergo, N.V. Electrische Drukkerij in The Hague, 1912, 3,000 copies; the third edition by the same publisher, 4,000 copies; in 1923, again, with Luctor et Emergo in The Hague, the fourth edition, 3,000 copies (Kartini, *Brieven aan mevrouw R.M. Abendanon-Mandri*, viii, ix).

39. Kartini, *Letters of a Javanese Princess*, 109.

40. Kartini quoted in Tsuchiya, Kenji, "Kartini's Image of the Javanese Landscape," *East Asian Cultural Studies*, vol. 25, nos. 1–4 (1986), 77–78; compare Kartini, *Brieven aan mevreuw R.M.A. Abendanon-Mandri*, 100–101.

41. Tsuchiya, "Kartini's Image," 70. The subjects of the drawings published are: a Priangan village; a small town (Depok) between Jakarta and Buitenzorg; the Ciliwang River in Depok; Pacet, a hill village on the road from the Puncak Pass to Bandung; the road between Depok and Bogor; Cikeumeuh, a village near Buitenzorg—all at a substantial distance from Japara.

42. "Every day, the mail brings us journals . . . heap of newspapers." Kartini, *Brieven aan mevrouw R.M. Abendanon-Mandri*, 27, 57.

43. Ibid., 309, Tsuchiya, "Kartini's Image," 77–78.

44. Kartini, *Brieven aan mevrouw R.M. Abendanon-Mandri*, 149; compare Kartini, *Letters of a Javanese Princess*, 147–148.

45. Kartini, *Brieven aan mevrouw R.M. Abendanon-Mandri*, 273; compare Kartini, *Letters of a Javanese Princess*, 215.

46. Kartini, *Brieven aan mevrouw R.M. Abendanon-Mandri*, 198, 207, 305, 232, 84, 51.

47. Ibid., 166.

48. Ibid., 19; compare Kartini, *Letters of a Javanese Princess*, 206–207.

49. Kartini, *Brieven aan mevrouw R.M. Abendanon-Mandri*, 48.

50. Kartini, *Letters of a Javanese Princess*, 169; see also Kartini, *Brieven aan mevrouw R.M. Abendanon-Mandri*, 126.

51. Ibid., 353.

52. Ibid., 290.

53. Kartini, *Letters of a Javanese Princess*, 159.

54. Kartini, *Brieven aan mevrouw R.M. Abendanon-Mandri*, 166.

55. Bouman, Dr. H., *Meer licht over Kartini* (Amsterdam: H.J. Paris, 1954), 61; Kartini, *Brieven aan mevrouw R.M. Abendanon-Mandri*, 1, 68.

56. Ibid., 273.

57. Kartini, *Letters of a Javanese Princess*, 89. Kartini actually participated in, and helped to organize, a Japara collection at the *Nationale Tentoonstelling van Vrouwenarbeid*, "National Exhibition of Women's Work," in The Hague, in 1898. Part of her and Japara's items exhibited can be found in the colonial classic, Rouffaer, G.P., and Dr. H.H. Juynboll, *De Batik-Kunst in Nederlandsch-Indië en haar Geschiedenis op grond van Materieel aanwezig in 's Rijks Ethnographisch Museum en andere Openbare Verzamelingen in Nederland* (Haarlem: Kleinmann; Utrecht: Oosthoek, 1899, 1914); see also Bouman, *Meer licht over Kartini*, 24–25.

58. Barthes, Roland, *Camera Lucida: Reflections on Photography*, translated by Richard Howard (New York: Noonday, 1989), 11. Couperus, in 1920, wrote, in a sort of Thiers-like manner, in his preface to the first American edition of Kartini's letters: "the simple language of these letters chants a poem 'From Darkness into Light.' The mist of obscurity is cleared away from her land and her people. The Javanese soul is shown as simple, gentle, and less hostile than we Westerners had ever dared to hope. . . . The mysterious 'Quiet Strength [Hidden Force]' is brought into the light, it is tender, human and full of love, and Holland may well be grateful to the hand that revealed it." Louis Couperus' foreword to *Letters of a Javanese Princess*, edited by Agnes Symmer (New York, 1920); see also Rutherford, Danilyn, "Unpacking a National Heroine: Two Kartinis and Their People," *Indonesia*, no. 55 (April 1993), 40.

59. *Gedenkboek Nederlandsch-Indische Gas-Maatschappij, 1863–1913* (Rotterdam: Nederlandsch-Indische Gas-Maatschappij, 1913), 5.

60. Ibid., 13.

61. Ibid., 12.

62. "De uitbreiding der Indische kustverlichting," *De Indische Gids*, vol. 25, no. 2 (1903), 1775.

63. Stevens, Harm, "Gietijzer voor Indische vuurtoren," in *Techniek in schoonheid: De Marinemodellenkamer van het Rijksmuseum te Amsterdam* (Wormer: Immerc, 1995), 69–70. "De uitbreiding der Indische kustverlichting," *De Indische Gids*, vol. 25, no. 2 (1903), 1774.

64. *Gedenkboek Nederlandsch-Indische Gas-Maatschappij 1913*, 15.

65. Walraven "Rembang" (from *De Indische Courant*, June 17, 1938) in Walraven, W., *Eendagsvliegen. Journalistieke getuigenissen uit kranten en tijdschriften* (Amsterdam: G.A. van Oorschot 1971), 255–256.

66. "Economische aspecten van de Electriciteit-voorziening van Nederlandsch-Indië," *Koloniale Studiën*, no. 2 (1934), 600.

67. Sparnaay (Ing.Administrateur N.V.G.E.B.E.O. te Bandoeng), Ing. D.W., "De electrificatie van 'het platteland,'" *De Ingenieur in N.I.*, vol. 5, no. 7 (July 1938), III,

33–35. See also on descriptions and estimates in *Gedenkboek Nederlandsch-Indische Gas-Maatschappij 1863–1938* (Rotterdam: Nederlandsch-Indische Gas-Maatschappij, 1938), 34–35.

68. "Indië's wereldrecords," *Ganeça*, vol. 6, no. 1 (January 1935), 7.

69. Sparnaay, Ing. D.W., "De electrificatie van 'het platteland,'" III, 33.

70. Ibid., 34.

71. Ibid., 35.

72. See, for instance, van Doorn, "Hoe men soms bouwt," *Tecton*, vol. 1, no. 8 (August 1936), 152. On *villas met kathedraalglas*, see *Nederlandsch Indië Oud en Nieuw* (1916–1917), 569.

73. Vistarini, B. de, "Luchtconditioneering en bouwkunde," *De Ingenieur in N.I.*, vol. 5, no. 11 (November 1938), VI, 46.

74. Walraven in *De Indische Courant* (August 1, 1939), in *Eendagsvliegen*, 304.

75. See, for instance, *Alles Electrisch in het Indische Huis*; see, e.g., vol. 1, no. 5 (November 1931), title page; also "De decoratief-verlichte eetkamer van Huize 'Lichtoord' op de Jaarmarkt te Soerabaja," ibid., vol. 1, vol. 3 (September 1931), 5.

76. *Gedenkboek Nederlandsch-Indische Gas-Maatschappij 1938*, 32.

77. Oudemans, A. C., *Ilmoe Alam of Wereldbeschrijving voor de inlandsche scholen* (Batavia: Landsdrukkerij, 1873), vol. 1, 1. Malay translation came later: see *Ilmoe Alam. Terkarang pada Bahasa Belanda oleh Dr. A. C. Oudemans dan dikarangkan pada Bahasa Melajoe oleh A.F. van Dewall*, 4th ed. (Betawi: Pertjetakan gouvernement, 1909).

78. Ibid., 100ff. *Kijker* from *kijken*, "to look," in a standard Dutch dictionary is translated as "spectator," and also as an instrument for looking: "(field)glasses," "binoculars," "telescope," "opera glass"; the dimunitive *kijkertje* also means "peeper." In the poem on the building of the Batavia–Krawang railway from 1890 by a Chinese-Malay author Tan Teng Kie, the Dutch surveyors are called *toewan keker*, "Mr. *Keker*," which, according to the translator, is *keker* from the Dutch *kijker*, "spectator." Salmon, Claudine, "The Batavian Eastern Railway Co. and the Making of a New 'Daerah' as Reflected in a Commemorative Syair Written by Tan Teng Kie in 1900," *Indonesia*, no. 45 (April 1988), 56. In an effort to convey the meaning of the word *kijker* as it will appear so often on the following pages and, I believe, so importantly, I turn sometimes to Robinson Crusoe, when he is "pulling out [his] *perspective glass*, which [he] had taken on purpose" (Defoe, Daniel, *Robinson Crusoe* [Danbury, Conn.: Grolier, n.d.], 207; emphasis mine).

79. Oudemans, *Ilmoe alam of wereldbeschrijving*, 101.

80. Ibid., 181.

81. Pyenson, Lewis, *Empire of Reason: Exact Sciences in Indonesia, 1840–1940* (Leiden: Brill, 1989), 21.

82. Ibid. Nonchalantly, almost naturally, the existence of a possible indigenous astronomy was ignored by all colonial and post-colonial scholars except few ethnography buffs. On the indigenous astronomy of Java before and through the nineteenth century, see Hidayat, Bambang, "Astronomy in Indonesia," unpublished paper (Bandung: Bosscha Observatory, n.d.), n.p.

83. Pyenson, *Empire of Reason*, 21.

84. Oudemans in a letter quoted in ibid., 24–25.

85. Ibid., 26.

86. Ibid., 179; see photos ibid., 28, 35.

87. Ibid., 32–33, 37.

88. Ibid., 44. See also Oudemans, Jean Abraham Chrétien, *Die Triangulation von Java ausgeführt vom Personal des geographischen Dienstes in Niederländisch Ost-Indien*, 6 vols. (Batavia: Staats-Druckerei, 1875–1900).

89. Review by M.T.H. Perelaer of *Kaart van Java met aanduiding der spoor- tram- en andere communicaties wegen* by H.H. Stemfoort and Hora Adema; and of *De nieuwe Atlas van de Nederlandsche Bezettingen in Oost-Indië naar de nieuwe bronnen samengesteld en aan de Regeering opgedragen* by J.W. Stempoort and J.J. ten Siethoff, captain of the General staff of the Netherlands Indies army, in *De Indische Gids*, vol. 7, no. 2 (1885), 1614–1615.

90. Witkamp, H.Ph.Th., "Een voorbeeld zonder voorbeeld," *De Indische Gids*, vol. 8, no. 2 (1886), 1606, 1627.

91. Ibid., 1628.

92. van Gelder, W., *Dari Tanah Hindia Berkoeliling Boemi: Kitab Pengadjaran Ilmoe Boemi bagi sekolah anak negeri di Hindia-Nederland*, 3rd ed. (Wolters Groningen, 1892), 1.

93. Ibid., 46.

94. Els Hoe, "Piet Mondrian," in Blotkamp, Carel, ed., *De Stijl: The Formative Years* (Cambridge, Mass.: The MIT Press, 1986), 59.

95. Bax, Marty, "Mondriaan and his Friends," in Baar, P.-P. de, M. Bax, et al., *Piet Mondriaan: The Amsterdam Years* (Amsterdam: Gemeentearchief Amsterdam, 1994), 22–42, 40.

96. Meijer, D.H., "Dactyloscopie voor Nederlandsch-Indië," *Koloniale Studiën*, no. 2 (1926), 912.

97. Misset, I.H. (De Chef van het Centraalkantoor voor de dactyloscopie bij de politie), *Departement van Binnenlandsch Bestuur Politieschool. Het Centraal kantoor voor de dactyloscopie bij de Politie* (Weltevreden: Visser, 1917), 20; see also Henry, E.R., Sir, *Classification and Uses of Fingerprints* (London: Darling and Son, 1913).

98. Misset, *Het Centraal kantoor* 1917, 20.

99. Ibid., 7. On anthropometry, especially, in English, see Barker, Joshua David, *The Tattoo and the Fingerprint: Crime and Security in an Indonesian City*, unpublished Ph.D. thesis (Cornell University, 1999).

100. Misset, *Het Centraal kantoor*, 23.

101. Musil, Robert, *On Mach's Theories*, translated by Kevin Mulligan (Washington, D.C.: Catholic University of America Press, 1982), 29.

102. Defoe, *Robinson Crusoe*, 82.

103. Spit, Directeur van Justitie in *Volksraad*, July 10, 1929. *Handelingen van den Volksraad* (Batavia: Volksraad van Nederlandsch Indië, 1929–1930), 775.

104. Ibid.

105. Meijer, Ir.H. met medewerking van Ing. F.A.J. Heckler, *De Deli Spoorweg Maatschappij: Driekwart eeuw koloniaal spoor* (Zutphen: De Walburg Pers, 1987), 74.

106. Misset, *Het Centraal kantoor*, 18.

107. Ibid., 20–23.

108. Ibid., 26.

109. Meijer, D.H., "Dactyloscopie voor Nederlandsch-Indië," 947.

110. Hoedt, G.H.A., "Naschrift," *Koloniale Studiën*, no. 2 (1919), 167–168.

111. "The colored fingerprints were photographed, the photographs were enlarged, and the enlargement was compared with the fingerprints photographed and enlarged before." Misset, *Het Centraal kantoor 1917*, 29.

112. van den Bovenkamp, A., "Het dactyloscopeeren als algemeene Regeerings-maatregel ter tijdelijke vervanging van een Burgerlijken stand voor de Inlandsche bevolking," *Koloniale Studiën*, no. 2 (1919), 156–164.

113. Misset, *Het Centraal kantoor 1917*, 23.

114. Ibid.

115. "De inlandsche klerken en javaansche bestellers in onzen dienst," *Indisch Tijdschrift voor Post en Telegraphie: orgaan van den Bond van Ambtenaren en Beamten*, vol. 2, no. 9 (December 1906), 340–343.

116. *Officieele wegwijzer voor Internationale Koloniale en Uitvoerhandel tentoonstelling te Amsterdam 1883* (n.p., n.p.), 14.

117. Scheltema, J.F., "Nederlandsch-Indië en de Wereldtentoonstelling te Parijs," *De Indische Gids*, vol. 22, no. 2 (1900), 967–968.

118. Bloembergen, Marieke, "Exhibiting Colonial Modernity: De exposities van de Nederlandse koloniën op de wereldtentoonstellingen (1880–1931)," unpublished Ph.D. Prospectus (Amsterdam: CASA, 1995), 3–4.

119. *Nederland te Parijs 1931: Gedenkboek van de Nederlandsche deelneming aan de Internationale Koloniale Tentoonstelling* ('s Gravenhage: Van Stockum en Zoon, 1931?), 46.

120. Bloembergen, "Exhibiting Colonial Modernity," 14, 14n.

121. *Nederland te Parijs 1931*, 31.

122. Ibid., 11–18.

123. Ibid., 57.

124. Ten years later, in connection with a demand by *Jardin d'acclimatation* in Paris and under pressure from Dutch "enlightened" circles, the Netherlands Indies government issued a circular to the local authorities not to encourage further "exhibiting of natives" in Europe. "Het tentoonstellen van inlanders in Europa," *De Indische Gids*, vol. 15, no. 1 (1893), 1022.

125. *Nederland te Parijs 1931*, 49.

126. See Siegel, James, "Georg Simmel Reappears: The Aesthetic Significance of the Face,'" *Diacritics*, vol. 29, no. 2 (1999), 100–113.

127. Scheltema, "Nederlandsch-Indië en de Wereldtentoonstelling te Parijs," 954–955.

128. Marco, "Koloniale tentoonstelling boekan boetoehnja orang Djawa," *Doenia Bergerak*, vol. 1, no. 20 (1915), 3.

129. Vanvugt, E., *H.F. Tillema (1870–1952) en de fotografie van tempo doeloe* (Amsterdam: Mets, 1993), 28, 37.

130. Ibid., 37.

131. Ibid., 160.

132. Ibid., 29, 158–159. In 1914, Wilhelmina made him the officer of the *Orde van Oranje-Nassau*. Ibid., 29.

133. Tillema, H.F., *"Kromoblanda": over 't vraagstuk van "het wonen" in Kromo's groote land*, 6 vols. ('s Gravenhage: Uden Masman etc. 1915–1923), vol. I, 7–12.

134. See especially ibid., vol. IV, part 2, 523; vol. V, part 2, 454–457, 782–783, 884.

135. See also Doorn, J.A.A. van, *De laatste eeuw van Indië: Ontwikkeling en ondergang van een koloniaal project* (Amsterdam: Bert Bakker, 1994), 156, quoting F.W.M. Kerchman, ed., *25 Jaren decentralisatie in Nederlandsch-Indië 1905–1930* (Semarang, 1930); G. Flieringa, *De zorg voor de volkshuisvesting in de stadsgemeenten in Nederlandsch Oost Indië in het bijzonder in Semarang* ('s Gravenhage, 1930).

136. Vanvugt, *H.F. Tillema*, 35.

137. Tillema, H.F., "Filmen en fotografeeren in de tropische rimboe. Met foto's van den schrijver," *Nederlandsch-Indië Oud en Nieuw* (1930–31).

138. Ibid., 109.

139. Ibid., 107.

140. Ibid., 97–98.

141. Ibid., 103.

142. Ibid., 100–101.

143. Ibid., 108.

144. Ibid., 105.

145. Ibid., 106.

146. Tillema, H.F., *Apo-Kajan, een filmreis naar en door Centraal Borneo* (Amsterdam: van Munster, 1938); see also Tillema, H.F., "The Etiquette of Dress among the Apo-Kayan Dayaks," *The Netherland Mail*, vol. 2 (n.p., n.d.), nos. 3–4.

147. "Ibid., 2–5. See also a photo by Tillema of an Apo-Kayan woman at a sewing machine in Vanvugt, *H.F. Tillema*, 163. See also a comment in Volksraad about Tillema being helped in the project by the colonial government, namely, with "military transport." Speech of Van Helsdingen in Volksraad on July 22, 1938, *Handelingen van den Volksraad* (1938–39), 371–372.

148. Marinetti's Futurist Manifesto as published in *Figaro*, quoted in McLuhan, Marshall, *The Mechanical Bride: Folklore of Industrial Man* (Boston: Beacon, 1967), 88, 90.

149. Scheltema, "Nederlandsch-Indië en de Wereldtentoonstelling te Parijs," 969.

150. *Nederland te Parijs 1931*, 54.

151. Ibid., 61.

152. *Alles Electrisch in het Indische Huis*, vol. 1, no. 1 (June 1931), title page; see also ibid., vol. 1, no. 2 (August 1931), title page, ibid., vol. 1, no. 3 (September 1931), title page.

153. E.g., ibid.

154. Kaden, Michael, "An Evening Breeze from Southeast" (Leiden: Proceedings of the European Conference on Southeast Asian Studies, EUROSEAS, 1995), 1. See also Masak, Tanete A. Pong, *Le cinéma Indonésien (1926–1967): Études d'Histoire Sociale*, unpublished Ph.D. thesis (Paris: l'École des Hautes Études en Sciences Sociales, 1989), 26.

155. Walraven on Java around 1918; *Fakkel*, vol. 1, no. 9 (July–August 1941), 704. See also, e.g., Nio Joe Lan, "De vestiging van een Indische Filmindustrie," *Koloniale Studiën*, vol. 25 no. 5 (1941), 386–400. See also: Junus Nur Arif, "Gedung bioskoop di Djakarta dahulu," *Djaja*, no. 387 (1968), 21, 31–32.

156. Dipl. Ings. B. de Vistarini, "Luchtconditioneering en bouwkunde," *De Ingenieur in N.I.*, vol. 5, no. 11 (November 1938), VI, 45. Ir. P. Timmerman, "De bevordering van lucht-conditioneering door de electriciteitmaatschappijen en het verbruik van electrische stroom in Ned.-Indië," ibid., vol. 5 , no. 11 (November 1938), VI, 47.

157. Nio Joe Lan, "De vestiging van een Indische Filmindustrie," 394–400. See also Berg, S. van den, "Notabele ingezetenen en goedwillende ambtenaren: De Nederlands-Indische filmkeuring, 1912–1942," in Berg, S. van den, R. Witte, et al., *Jaarboek Media Geschiedenis: 4 Nederlands-Indië* (Amsterdam: Stichting beheer IISG, Stichting Mediageschiedenis, 1992), 157, 160. According to the data by the Netherlands East Indies Department of Economic Affairs, in 1939, for instance, of the movies shown in the Indies (1,526,708 meters) 65 percent were American, 12 percent Chinese, 4.8 percent French, 4.5 percent German, 3.4 percent English, 3.1 percent Dutch, 2.9 percent locally made, 0.2 percent Japanese. Masak, *Le cinéma Indonésien*, 38.

158. Treub, M.W.F., *Nederland in de Oost: reisindrukken* (Haarlem: Willink, 1923), 249–250.

159. See, e.g., "Van eerbied gesproken!" *Nationale Commentaren*, vol. 4, no. 39 (September 27, 1941), 3817.

160. "VERSCHRIKKELIJK," *De Beweging*, no. 6 (February 8, 1919), 101.

161. Sir Hesketh Bell after visiting the Dutch Indies in 1926 in his *Foreign Colonial Administration in the Far East* (London: Arnold, 1928), 121–122, quoted in S. van der Berg, "Notabele ingezetenen en goedwillende ambtenaren: De Nederlands-Indische filmkeuring, 1912–1942," 157. The Dutch Indies *Filmland* journal was anxious that "the show of crime and degeneration that appear on the screen might be taken by Malays, Javanese, and even Indians and Chinese [in the Indies] as a true representation of the life of the whites in their land. The love scenes give a sad picture of the morality of the whites, and, what is more serious, of white woman. . . . From the remarks and the shouting one can hear coming from the popular sections of the theater during the love scenes, one can imagine how the blood of the young coolies is boiling." "Het Filmgevaar in de koloniën," *Filmland*, vol. 4, no. 10 (December 1926), 6, quoted in Masak, *Le cinéma Indonésien*, 41.

162. Ibid., 148.

163. Ibid., 149.

164. Ibid., 150, for more detail see ibid., 150ff.

165. Maclaine Pont, H., "Het nieuwe Hoofdbureau der Semarang-Cheribon-Stoomtram Maatschappij te Tegal," *Nederlandsch Indië Oud en Nieuw* (1916–1917), 98.

166. Vissering, C.M., "Een Pasar Gambir," ibid., 399.

167. Ibid., 399ff.

168. Ibid., 406–407.

169. Vissering, C.M., "Een Pasar Gambir," 407. There are, of course, more direct uses of late-colonial floodlighting on record. *La Lumière électrique* in 1884 reported on the British in Sudan: "When the rebels were only a few hundred meters away and had begun to attack, the electric floodlights suddenly blazed into action, bathing them in the most brilliant light. The surprise and confusion were so complete that they defy description." *La Lumière électrique*, 1884, quoted in Schivelbusch, Wolfgang, *Disenchanted Night: The Industrialization of the Light in the Nineteenth Century* (Berkeley: University of California Press, 1988), 57.

170. Buys, M., *Batavia, Buitenzorg en de Preanger-Gids voor bezoekers en toeristen* (Batavia: Kolff, 1891), quoted in *De Indische Gids*, vol. 15, no. 1 (1892), 896.

171. Treub, *Nederland in de Oost*, 268.

172. Berlage, H.P., *Mijn Indische Reis: Gedachten over cultuur en kunst* (Rotterdam: W.L. en J. Brussel's, 1931), 36.

173. Ibid., 36–37.

174. Ibid., 49.

175. Ibid., 110.

176. *Ganeça*, vol. 1, no. 6 (November 1, 1922), 83. See also, for "the Netherlands as a 'flat land,' the Indies 'the land of the third dimension,'" Sandick, van, *In het Rijk van Vulcaan*, 7.

177. Berlage, *Mijn Indische Reis*, 110. *Ueber allen Gipfeln ist Ruh* . . . , "O'er all the hilltops is quiet," is from Goethe's "*Wanderers Nachtlied*"; the English is a Longfellow translation. Bandung, beside other things, was the famous late-colonial Indies city of annual colonial fairs. "Bandoeng and the Annual Fairs are one and the same thing," S.A. Reitsma, a railway expert and also a deputy mayor of Bandung, declared in 1922. "Rede van den wd.Burgemeester van Bandoeng, den Heer S.A. Reitsma," *Orgaan van de Vereeniging 'Nederlandsch-Indische Jaarbeurs'*, vol. 1, no. 4 (February 15, 1922), 55.

178. "Batavia en Bandoeng," *Kritiek en Opbouw*, vol. 2, no. 1 (February 16, 1939), 4.

179. Picture no. 28 in Rusman, E., *Hollanders vliegen: van H-NACC tot Uiver: tien jaar Amsterdam-Batavia door de lucht* (Baarn: Bosch en Keuning, 193?), 33.

180. Rusman, *Hollanders vliegen*, 15–16.

181. Morrell, Charles M., "The 100th Flight: The Royal Dutch Airways Great Achievement," *The Java Gazette*, no. 2 (n.d.)

182. Rusman, *Hollanders vliegen*, 17–19. Also "9,000 Miles in 10–12 Days. Weekly Service (Amsterdam, Nuremberg, Budapest, Belgrade, Athens, Mersin . . .)," *KLM Air Line Amsterdam-Batavia (Holland-Java): 9000 Miles in 10/12 days weekly service* (n.p., 1931).

183. Ibid., 23.

184. *Tien (10) vervlogen jaren*. Uitgegeven ter gelegenheid van het tienjarig bestaan van de Koninklijke Nederlandsch-Indische Luchtvaart Maatschappij. 1 November 1928–1938 (Batavia: K.N.I.L.M., 1938?), 7.

185. *K.N.I.L.M. Reisgids 1930* (Batavia: K.N.I.L.M., 1930), 11. A fifty-page poem on the same explosion of Krakatoa was composed in Malay by an Indies Chinese from Tanggerang, West Java, Ong Tjong Sian, *Sair petjanja goenoeng Karakatau*, "Poem on Krakatoa explosion" (1929), "describing the terror which took hold of Chinese and Indonesian society in West Java when the volcano became active again." Salmon, *Literature in Malay by the Chinese of Indonesia*, 69.

186. Ibid., 13.

187. Illustration no. 29 in *Tien (10) vervlogen jaren*, 16b.

188. Soeroso, June 22, 1938, in *Handelingen van den Volksraad* (1938–39), 364.

189. De Hoog, June 22, 1938, in ibid., 377.

190. Soeroso, August 12, 1938, reacting in ibid., 889.

191. "Een jaar Luchtverkeer, 1929–1930," *K.N.I.L.M. Reisgids 1930*, 10; see also *Tien (10) vervlogen jaren*, 45.

192. Ibid.

193. *K.N.I.L.M. Reisgids 1930*, 57.

194. For more on Boscha and the observatory, see Hucht, K.A. van der, and C.L.M. Kerkhoven, "De Bosscha-Sterrenwacht: Van thee tot Sterrenkunde," *Zenit*, vol. 9 (1982), 292–300.

195. Couperus, *Eastward*, 166–167.

196. Pyenson, *Empire of Reason*, 56–57, 134.

197. According to Cornelis Braak, acting chief of observatory in Batavia, in June 1920, the new observatory should "'limit its work exclusively to science; the practical side of things would be left to Batavia." Ibid., 55.

198. Ibid., 49, 71, 71n.

199. Ibid., 57.

200. Prof. Dr. J. Clay on "De zonsverduistering van 21 September j.l. en te beteekenis van de expeditie ter waarneming van het Einstein-effect" *Ganeça*, vol. 1, no. 5 (October 15, 1922), 68, 96.

201. Bovene, G.A. van, *Nieuws! Een boek over pers, film en radio* (Batavia: G. Kolff, 1941?), 52.

202. Pyenson, *Empire of Reason*, 77.

203. Ibid., 81; also Bambang Hidajat, "Astronomy," 12.

204. van Bovene, *Nieuws! Een boek over pers, film en radio*, 53.

205. *Bromo-Hotel-Tosari* (Tosari: Bromo-Hotel, 1926), 20–21.

206. Kelling, M.A.J., Secretaris van het Hoofd-Comité en van het Uitvoerend-Comité der Eerste Hygiëne Tentoonstelling, "De Eerste Hygiëne Tentoonstelling in Nederlandsch-Indië," *Koloniale Studiën*, vol. 11, no. 2 (1927), 346–348.

207. Ibid., 355–356.

208. Elias, Jhr.P.J., *Dan liever de lucht in. Herinneringen van een marinevlieger* (Amsterdam: van Kampen en Zoon, 1963), 83–84.

209. Ibid., 83–86. There is a photo of *Zeven Provinciën* in the book, an air-photo, naturally (ibid., 113).

210. "The first one of the physiological principles subsumes all phenomena, as intuitions in space and time, under the concept of Quantity." Kant, Immanuel, *Prolegomena: To Any Future Metaphysics That Can Qualify as a Science*, translated by Paul Carus (La Salle: Open Court, 1989), 65.

211. Mondrian in *De Stijl*, no. 5 (1918), quoted in Els Hoe, "Piet Mondrian," in Blotkamp, ed., *De Stijl*, 48.

212. E.g., for December 31, 1892, 32,941, for December 31, 1897, 33,484. *De Indische Gids*, vol. 22, no. 2 (1900), 1443.

213. E.g., "Kosten van de Europeesche gevangenen in Ned.-Indië per hoofd en per dag," ibid., vol. 15, no. 2 (1893), 872–873. See also Stibbe, D.G., "Een mogelijke bezuiniging op de voeding de gevangenen in Indië," ibid., 1442–1444.

214. "Desertiën onder inlandsche gevangenen," *Statistische opgaven over 1889 betreffende het groote meerendeel der gevangenissen in Nederlandsch-Indië* (Weltevreden: Landsdrukkerij, 1889).

215. E.g., *De Indische Gids*, vol. 6, no. 2 (1884), 691, and Doorn, *De laatste eeuw van Indië*, 39.

216. E.g., "Hoeveel personen er gedurende 1878 in Nederlandsch-Indië omkwamen door 'ongelukken,'" *De Indische Gids*, vol. 1, no. 2 (1879), 759.

217. E.g., "Het woningvraagstuk en de Pestbestrijding," *Koloniale Studiën*, vol. 1, no. 2 (1916–1917), 448; C.D. de Langen, arts te Weltevreden, "Tuberculose in Nederlandsch-Indië en hare bestrijding," ibid., vol. 2, no. 1 (1918), 177.

218. In 1938, a journal for agriculture in the Dutch Indies published an article on fighting the rats in West Krawang, a regency east of Batavia. Tables and diagrams

showed how many "rat holes in use" there were in West Krawang, how many rats in an average hole, how many "young," "adult," "male," "female," "gravid," and "barren." Doorn, *De laatste eeuw van Indië*, 39.

219. Rusli, Mh., *Sitti Nurbaya: Kasih Tak Sampai* [1922], 24th ed. (Jakarta: Balai Pustaka, 1994), 38; "in order to keep undesirable *elements* away . . . there are night watches still common in this country, poorly armed but making loud sounds . . . on Java a little drum made of bamboo (*kentongan*) is used, sounding *tèk-tèk*." Meijer, D.H., "Internationale politioneele samenwerking," *Koloniale Studiën*, vol. 19, no. 2 (1935), 430. In English, see also Barker, *The Tattoo and the Fingerprint*, passim.

220. Rusli, *Sitti Nurbaya*, 119; see also ibid., 182.

221. Ibid., 186.

222. The phrase is borrowed from James C. Scott's *Seeing Like a State: How Certain Schemes to Improve Human Conditions Have Failed* (New Haven: Yale University Press, 1998).

223. Heidegger, M., "The Question Concerning Technology," in *The Question Concerning Technology and Other Essays*, translated by William Lovitt (New York: Harper, 1977), 16, 27.

224. Barthes, *Camera Lucida*, 41.

225. Ibid., 6.

226. *Nederlandsch Indië Oud en Nieuw* (1916–1917), 1.

227. Walraven, W., "Het Indische Stadsbeeld voorheen en thans," *Kritiek en Opbouw*, vol. 3, no. 24 (February 1, 1941), 376.

228. Ch.O.v.d.P., "Om de zuiverheid van onze taal!" *Fakkel*, vol. 1, no. 6 (April 1941), 510.

229. Plas, Ch. van der, "Opmerking over *The Tempest*," ibid., vol. 1, no. 5 (March 1941), 391–394.

230. Ibid., vol. 1, no. 1 (November 1, 1940), 16–20.

231. Jack Zeylemaker, "Where—the twain shall meet. Naar aanleiding van de tentoonstelling van 'Balische kunst van heden' te Batavia," ibid., vol. 1, no. 2 (December 1, 1940), 157–159. J.d.L., "De tentoonstelling uit particulier bezit van Hindoe-Javaansche kunst," ibid., vol. 1, no. 4 (February 1941), 327–328. Stutterheim, W.F., "Inleiding bij de opening van de tentoonstelling van Hindoe-Javaansche kunst," ibid., vol. 1, no. 5 (March 1941), 383–390. There were two Indonesians on the editorial board of twelve. Indonesians also rarely appeared as authors in the journal. The most important exceptions were two chapters of Soewarsih Djojopoespito's *Buiten het Gareel*; see *Fakkel*, vol. 1, no. 9 (July–August 1941), 715–725; and a review of the Indonesian literary journal: Armijn Pané, "De Poedjangga Baroe I,II," ibid., vol. 1, no. 9 (July–August 1941), 746–760.

232. Beb Vuyk, "Way Baroe in de Molukken," ibid., vol. 1, no. 1 (November 1, 1940), 7–15.

233. Overy, Paul, *De Stijl* (London: Thames and Hudson, 1991), 150.

234. *Ganeça*, vol. 1, no. 2 (August 15, 1922), 25.

235. Ibid., 29.

236. Ibid., vol. 1, no. 3 (September 1, 1922), 45.

237. Rolph Sarca: "Idees uit m'n dagboek," ibid., vol. 1, no. 5 (October 15, 1922), 69. And in the next issue: Red. Ganeça: *Noot van Sarca aan Soekarno*. Brave . . . ," ibid., vol. 1, no. 6 (November 1, 1922), 95.

238. Shakespeare, William, *Measure for Measure*, quoted in Rorty, *Philosophy and the Mirror of Nature*, 42.

239. "Sedikit bersoewara," *Orgaan dari Perhimpoenan Pegawai Post Telegraaf Telefoon Radiodienst Rendahan (P.T.T.R.)*, vol. 1, no. 1 (September 1937), vol. 2, no. 12 (December 1938), passim; vol. 2, no. 4 (April 1938), 6–7.

240. *Efficiency*, vol. 4, no. 11 (November 1936), last page.

241. Ibid., vol. 4, no. 3 (March 1936), 106.

242. Ibid., 108.

243. "Ketoea kita Dr. R. Soetomo kembali di tanah air," *Soeara Parindra*, no. 2 (April 1937), 14.

244. Sayid, R.M., *S.R.V. Gedenkboek* (Soerakarta: S.R.V., 1936), 45–46.

245. "Warta Hoofdbestuur," *Orgaan dari PTTR*, vol. 2, no. 6 (June 1938), 5.

246. Ibid.

247. As an example, see an advertisement page from *Insulinde*. Padang: "Toko M.Rosenberg en Co" *Insulinde*, vol. 2 (1902), 542–543.

248. See, e.g., "sarcastic" "De bril een populair mode-artikel onder de Inlanders" from 1931 (?): "wearing glasses among the natives has become truly a *mode rage* . . . not some lenses to strengthen vision . . . in large part, rather, to 'cock' [*wapen*] the eyes with an ordinary glass . . . an ordinary, uncut glass." There are photographs in the article with couplet-like captions such as: "The coachman feels himself happy for the first time / When in the most *dernier* way / He can see his beak through glass." The article was pointed out to me by Dr. M. Till of the University of Amsterdam.

249. Walraven, W., "Apologie," *Kritiek en Opbouw*, vol. 2, nos. 21–22 (December 16–January 1, 1940), 339.

CHAPTER FOUR

1. Cinderella. Adapted from C. Perrault's *Cendrillon* of 1697 (London: Jonathan Cape, 1980), 23.

2. *Catalogus van de verzameling poppen weergevende verschillende kleederdrachten van de volkeren van de Nederlandsch Oost-Indischen archipel, aangeboden aan Hare Majesteit de Koningin der Nederlandsch Oost-Indië, en welke verzameling op de tentoonstelling te Batavia van 1893 bekrond is geworden met het eerendiploma* (Batavia: Kolff, 1894), 10–12.

3. Ibid., 3.

4. Foucault, M., *Discipline and Punish: The Birth of the Prison* (New York: Vintage, 1979), 135–169, 195–230.

5. *Catalogus van de Verzameling*, 82.

6. Kleist, Heinrich von, "On the Marionette Theatre" (1810), in *Essays on Dolls*, translated by Idris Parry and Paul Keegan (London: Penguin, 1994), 6–7.

7. *Verslag omtrent den Post- en Telegraafdienst in Nederlandsch-Indië over het Jaar 1900* (Batavia: Javasche Boekhandel, 1901), 51, 53–54.

8. See above, chapter 1.

9. The legal dress code loosened, too. By the 1900s, the 1872 statute was long forgotten according to which in Java "it was forbidden to dress otherwise than in the manner customary to one's own population group." Wertheim, W.F., *Indonesian Society in Transition* (The Hague: van Hoeve, 1964), 138.

10. Mas Marco's journal *Doenia Bergerak* wrote in 1915: "In the place we live (Tjiamis), now [1915] there are not yet many Javanese who dress in the European manner, either officials or private persons. . . . Initially, as we could see by our own eyes, several people liked to wear clothes of European style: Mas Sastrawidjaja, the undercollector, Mas Sastra Atmadja, the helper clerk of second class, several candidates of the school of the first class, all of them officials, and K. Koesna Ardja, a private person. Of the people mentioned above, however, only three school candidates remained still doing so and K. Koesna Ardja." KOERANG POETIH, "Apakah ini boekan Anti Circulaire No. 2014?" *Doenia Bergerak*, vol. 1, no. 19 (1915), 7–8.

11. Tillema, H.F., *"Kromoblanda": over 't vraagstuk van "het wonen" in Kromo's groote land*, 6 vols. ('s Gravenhage: Uden Masman, etc., 1915–1923).

12. Catenius-van der Meijden, Mevrouw J.M.T., *Ons Huis in Indië. Handboek bij de keuze, de inrichting, de bewoning en de verzorging van het huis met bijgebouwen en erf, naar de eischen der hygiëne, benevens raadgevingen en wenken op huishoudelijk gebied* (Semarang: Masman en Stroink, 1908).

13. In 1905 there lived in the Indies 4,000 European women born in Europe; by 1930 their number had increased to about 26,000. The sex ratio normalized at the same time from about 471 to about 884 women for 1,000 men. As one contemporary source put it, "unbridled bachelor has slowly been dying." Nieuwenhuis, R., "Over de Europese samenleving van 'tempo doeloe' 1870–1900.," *Fakkel*, vol. 1, no. 9 (July–August 1941), 777. On *njai*, "concubine," stories in the early Indonesian and Malay-Chinese novels, see especially Salmon, Claudine, *Literature in Malay by the Chinese of Indonesia: A Provisional Annotated Bibliography* (Paris: Archipel, 1981), 33.

14. Catenius-van der Meijden, *Ons Huis in Indië*, 102–103.

15. Ibid., 75.

16. Ibid., 14–16.

17. See above, chapter 2.

18. Catenius-van der Meijden, *Ons Huis in Indië*, 280.

19. Ibid., 73. Indies *négligée* meant *sarong* and *kabaja* for women, and *indische slaapbroek*, "pajama-trousers," and *kabaja* for men; see., e.g., Borel, H., *Van Batavia naar Rotterdam: reisgids van de Rotterdamsche Lloyd* (Amsterdam: Rotterdamsche Lloyd, 1905), 30–31.

20. The increasing number of Dutch women, and thus purely Dutch families in the Indies, did not make the problem go away. At the end of the era, in 1941, a Dutch housewife in Java remembered a book similar to that of Madame Catenius (Mevr. J. Kloppenburg-Versteegh, *Het leven van de Europeesche vrouw in Indië*) that she got as a present when she arrived in the Indies in 1913. She recalled that she was strongly and repeatedly advised by the book not to put too much in the house into the hands of the servants: "Most strongly, the young mothers were admonished to take themselves care of their babies and not to take a chance on the *baboe*'s *slendeng* [a shawl that *baboe* wore over their shoulder, and in which they carried the babies]." Klay-Lancée, C.C., "Indië 25 jaar geleden," *Officieel orgaan van de Vereeniging van Huisvrouwen Bandoeng* (October 1941), 839.

21. *"Die kampongbewoners—familieleden . . . zijn de parasieten onzer huisbedienden"* (ibid., 145). At the same time, there is a special warning in Catenius' book about servants being very proficient in using *goena-goena*, "black magic," poison, or both, against their masters. In that case, it is urgent, the guide suggested, to call a doctor, by telephone if possible, as quickly as possible. Ibid., 246.

22. Lipovetsky, Gilles, *The Empire of Fashion: Dressing Modern Democracy* (Princeton: Princeton University Press, 1994), 23, 59.

23. Catenius, *Ons Huis in Indië*, 147–148.

24. Ibid., 148.

25. Ibid.

26. Couperus, Louis, "Meditatie over het mannelijk toilet," in *Kleeding en de man* ('s Gravenhage: Magazijnen 'Nederland', 1915), 7; emphasis mine.

27. Ibid., 12.

28. Couperus, Louis, *The Hidden Force* (Amherst: University of Massachusetts Press, 1985), 110–111, passim.

29. Ibid., 223.

30. Ibid., 224.

31. Ibid., 152.

32. On "construction of the dandy" in Europe one or two eras before this Indies case, see Stafford, Barbara M., *Artful Science: Enlightenment Entertainment and the Eclipse of Visual Education* (Cambridge, Mass.: The MIT Press, 1994), 170. In the Indies more than in Europe, proximity of the dandy to another modern/bourgeois concept, "savage," is clear—the savage, too, in Rousseau's classical image "n'etant attaché à aucun lieu," never suffers from rheumatism, and "of all the people of the world is least curious and least bored" (Rousseau, J.-J. *Émile, ou De l'éducation* [Paris: Granier, 1939,] 118, 132, 271).

33. On the role of "fashion dolls" in the European history of fashion since the eighteenth century, see Lipovetsky, *The Empire of Fashion*, 56, passim.

34. Kleist, von, "On the Marionette Theatre," 4.

35. *Indisch Tijdschrift voor Post en Telegraphie: orgaan van den Bond van Ambtenaren en Beamten*, vol. 7, no. 2 (March 1912), 524–525.

36. Perron, E. du, *Indies Memorandum* (Amsterdam: De Bezige Bij, 1946), 50. Compare also the Bandung technical college student journal in 1934 on "a swinging Indonesian with a pen in his breast pocket and with a cane." Corpslid, "Over de gezondheid van het Bandoengsche Studenten Corps," *Ganeça*, vol. 6, no. 7 (July 1934), 9.

37. Baudelaire, Charles, "The Painter of Modern Life," quoted and interpreted in LeGoff, Jacques, *History and Memory* (New York: Columbia University Press, 1992), 40–41.

38. Countless illustrations in the Indies journals like *Nederlandsch Indië Oud en Nieuw* are witness to this image.

39. The issue was discussed most vehemently in the 1920s in the colony advisory council. Some of the *Volksraad* deputies just could not understand "how government could be so lukewarm on that issue." Others were disappointed at those among the officers who were looking for "all possible excuses not to wear such a uniform, such as bicycle trips and hiking in the mountains." See, e.g., *Handelingen van den Volksraad* (Batavia: Volksraad van Nederlandsch Indië, 1928–29), 1344, 2047, 2213, 2223, 2229. For a summary of earlier discussions, in terms of blue color of uniforms as opposed to gray and of big as opposed to small chevrons, see, e.g., *De Indische Gids*, vol. 16, no. 1 (1894), 288.

40. Marco Kartodikromo, *Student Hidjo* (Semarang: Masman en Stroink, 1919). See also Shiraishi, Takashi, *An Age in Motion: Popular Radicalism in Java, 1912–1926* (Ithaca: Cornell University Press 1990), 31–32.

41. Maier, Hendrik M.J., "Phew! Europeesche beschaving! Marco Kartodikromo's *Student Hidjo*," *Southeast Asian Studies* (Tonan Ajia Konkyu), vol. 34, no. 1 (June 1996), 185.

42. Marco, *Student Hidjo*, 53, 14–16.

43. Ibid., 22. "Panorama" was a term fashionable in Europe, too, only a few years earlier. Young Bertold Brecht remembered the annual *Herbstplärrer*, the 'autumn barkers' fairs with their "booths, music, merry-go-round and panoramas that displayed crude pictures of historical events like 'The Shooting of the Anarchist Ferrer at Madrid' or 'Nero Surveys the Fire of Rome.'" Ewen, F., *Bertold Brecht: His Life, His Art, His Time* (New York: Citadel, 1992), 57. For more on the history of panoramas in Europe, see Oetterman, Stephen, *Das Panorama: Die Geschichte eines Massenmedium* (Frankfurt: Syndicat, 1980).

44. Anderson, Benedict, "A Time of Darkness and a Time of Light: Transposition in Early Indonesian Nationalist Thought," in *Language and Power: Exploring Political Cultures in Indonesia* (Ithaca: Cornell University Press, 1990), 243.

45. Marco, *Student Hidjo*, 14.

46. This is a Muslim-land nonalcoholic dandy, of course. James Rush pointed out a specific shift, in the Indies of the early twentieth century, from opium to soft drinks. Rush, James R., *Opium to Java: Revenue Farming and Chinese Enterprise in Colonial Indonesia, 1860–1910* (Ithaca: Cornell University Press, 1990), 222, 235, passim; and personal communication from James Rush.

47. Marco, *Student Hidjo*, 13, 16.

48. Ibid., 33.

49. Ibid., 39.

50. Ibid., 25. There is some blood, too. An illness of a Dutch woman teacher that looks very much like an abortion as it is described. Ibid., 91–93.

51. Ibid., 60–61.

52. Ibid., 94–102.

53. There is a nice photograph of Mas Marco, in all-white; see *Saro Tomo*, vol. 1, no. 17 (1915), 282.

54. Ibid., vol. 1, no. 10 (1915), 160, *Loetjon No. 35*, "Joke no. 35."

55. Sr.K., "Tidak betoel," ibid., vol. 1, no. 3 (1915), 43.

56. Decades later, Pramoedya Ananta Toer, the hero of the epilogue of this book, brought the virtually forgotten Mas Marco back to (some in) Indonesia. See, e.g., Pramoedya Ananta Toer, ed., *Tempo doeloe: antologi sastra pra-Indonesia* (Jakarta: Hasta Mitra, 1982); idem, *Sang Pemula dan karya-karya non-fiksi (journalistik), fiksi (cerpen, novel) R.M. Tirto Adhi Soerjo* (Jakarta: Hasta Mitra, 1985); idem, *Rumah Kaca* (Jakarta: Hasta Mitra, 1988).

57. Salmon, *Literature in Malay by the Chinese of Indonesia*, 40, 71–72.

58. Manusama, A.Th., *Komedie Stamboel of de Oost Indische opera* (Weltevreden: Favoriet, 1922), 22.

59. Benjamin, W., "Paris, Capital of the Nineteenth Century," in *Reflections* (New York: Harcourt Brace Jovanovich, 1978), 153–158.

60. Ibid., 156–157.

61. Sukarno, *An Autobiography as Told to Cindy Adams* (Indianapolis: Bobbs-Merrill, 1965), 156–157.

62. "The idea [of association] has already destroyed a great deal of our national

heroism, and it is cutting it off from a potential further supply of a new spirit. Perhaps, indeed, we are predestined to be overtaken by the mighty maelstrom of the European culture. Many believe so. . . . Modernism ogles us." "Indonesia en Associatie," *Soeloeh Indonesia Moeda*, no. 12 (November–December 1928), 328, 330.

63. "We understand that there is a mathematically defined dividing line between the power-holding whites and power-yielding browns. . . . We understand too that the purer and earlier the antithesis is stated, the more the character becomes involved, and the stronger the bottom line of the antagonism is drawn, the stronger the justice and aim of the struggle will be." Indonesia-Poetera, "Naar het Bruine Front," *Soeloeh Indonesia Moeda*, no. 1 (December 1927), 3.

64. "The time is not far away," nationalist *Indonésia Raja* wrote in 1929, "when most Indonesians will look down with disdain upon their brown brothers wearing those white jackets and fancy trousers of matching color. . . . These people merely try to be truly Dutch, and justly they should be condemned as mockers by both, the *sana* ["them," Dutch] side and by us, the Indonesians themselves. These people can see their ideal in the ready-made movie heroes, and in dressed-up puppets. Fortunately, these creatures are on the verge of extinction." "Nationalisme . . . ," *Indonésia Raja*, vol. 1, no. 9 (December 1929), 156. See also Katja Soengkana S.D.K., "Dr. Tjipto Oordeelt," *Soeara Parindra*, vol. 1 (October 1936), 6–7. "Dandy" was now pictured by the Indonesians themselves with hatred as "a clown with goggles [*stofbril*] and a thick fountain pen" (Soegondo Djojopoespito, "Soeroto's Indonesiër," *Kritiek en Opbouw*, vol. 2, no. 3 [March 16, 1939], 38); as a "silly . . . party animal and lemonade gulper . . . bonvivant and illicit lover . . . fop 'native' with flannel pantaloons, lacquered shoes, and jacket with an inevitable fountain pen and pencil . . . with a dirty cap on his stinky, pomaded hair! . . . [and again] with goggles." Soejitno, M., "Nogmals over Indonesiërs," *Kritiek en Opbouw*, vol. 2, no. 10 (July 1, 1939), 312–313.

65. Sukarno, *An Autobiography*, 51–52.

66. Ibid., 80.

67. Ingleson, John, *Road to Exile: The Indonesian Nationalist Movement, 1927–1934* (Singapore: Heinemann, 1979), 143.

68. In another description: "'Swadeshi' clothes were the Partindo fashion at the moment [1932], and the party's adherents appeared at the [Taman Siswa's] congress looking exceedingly beggarish, with, for example, a pitji made of bamboo matting, badju and sarung of the cheapest local stuff, and not even sandals on their feet. The atmosphere at that time was very different from the congress of 1930; then they had worn good western clothes, clean and with a tie.'" Sajoga, "Riwajat Perdjuangan Taman Siswa 1922–1952," in *Taman Siswa 30 Tahun* (Jogjakarta: Madjelis Luhur Taman Siswa, n.d.), 216, quoted in McVey, Ruth T., "Taman Siswa and the Indonesian National Awakening," *Indonesia*, no. 4 (October 1967), 147.

69. She was born on April 20, 1912. Soewarsih's father was Raden Bagoes Noersaid Djojosapoetro, of the aristocratic Cirebon family, and her mother Hatidjah, born Thio, a daughter of a Chinese merchant family that converted to Islam. Termorshuizen, Gerard, "Een leven buiten het gareel," *Engelbewaarder Winterboek 1979. Extra-uitgave van het kwartaalschrift De Engelbewaarder* (Amsterdam, 1979), 111.

70. Perhaps the only mention of clothes or fashion in the novel comes as the main woman protagonist prepares for a visit with her husband: "Soelastri improved something about her hair, brushed some powder over her face, and was ready to go, with a

cream-colored shawl over her shoulders. With disappointment, Soedarmo looked at her. 'This is too neat,' he said. 'You have to learn how to clothe yourself in a simple way. Even more so, because this is a house call.' 'But,' she was hurt by his harsh tone, 'I like to be neat. I don't harm anybody by it.' 'No,' he said, 'but it is not tactful to appear *chique* in a poor surroundings. Don't take this blouse. It is too good looking.'" Djojopoespito, Soewarsih, *Buiten het Gareel. Indonesische Roman* (Utrecht: De Haan, 1940), 33.

71. Mrázek, R., *Sjahrir: Politics and Exile in Indonesia, 1906–1966* (Ithaca: Cornell University Studies in Southeast Asia, 1994), 22.

72. Ibid., 195.

73. Ibid., 26.

74. *De Socialist* (October 5, 1929), 5.

75. Tas, Sol, "Souvenirs of Sjahrir," *Indonesia*, no. 8 (October 1969), 139–140.

76. Djojopoespito, Soewarsih, "De thuiskomst van een oud-strijder," *Tirade*, vol. 21, no. 221 (January 1977), 41–42.

77. Pringgodigdo, Soewarni, "Over du Perron en zijn invloed op de Indonesische intellectuellen (1936–39)," *Cultureel Nieuws*, no. 16 (January 1952), 145.

78. *Liederen USI* (Batavia: Unitas Studiosorum Indonesiensis, n.d.)

79. People "do not take us seriously," one of the youth complained, "they take us as kids of rich parents that are merely after fun-making and picnics . . . bashes, and hazing. . . . Indonesian people do not take us seriously." Veteraan, "Een veteraan aan het woord," *Orgaan van de U.S.I.*, vol. 8, no. 6 (October 1941), 5, 8–9.

80. Few of them take part in the radical nationalist political youth organizations, survivors of the Sukarno time, groups like *Perhimpoenan Peladjar-Peladjar* Indonesia (*PPPI*), "Union of Indonesian Students," and *Indonesia Moeda*, "Young Indonesia." Even those few, however, did it with some hesitation.

81. *Medan Politie Boemipoetra*, vol. 16, no. 11 (November 1940), 6.

82. Anderson, Benedict, "Japan 'The Light of Asia,'" in Silverstein, J., ed., *Southeast Asia in World War II: Four Essays*, Southeast Asia Studies Monograph Series no. 7 (New Haven: Yale University Press, 1966), 13–50.

83. *Imperial Japanese Army and Navy Uniforms and Equipment* (n.p., 1975), n.p.

84. Most quoted exceptions, sometimes described as the only significant ones, are Amir Sjarifoeddin and Sjahrir. For Amir, see, for instance, Leclerc, J. "La clandestinité et son double: à propos des relations d'Amir Sjarifuddin avec le communisme Indonésien," *Asian Thought and Society: An International Review*, no. 6 (1981), 36–48; Wellem, *Amir Sjarifoeddin: Pergumulai imannya dalam perjuangan kemerdekaan* (Jakarta, 1984).

85. Sukarno, *An Autobiography*, 81.

86. Mrázek, *Sjahrir: Politics and Exile in Indonesia*, 228.

87. Ibid.

88. Jansen, L.F., *In deze halve gevangenis: Dagboek van mr dr L.F. Jansen, Batavia/Djakarta 1942–1945* (Franeken: Van Wijnen, 1988), 200–201.

89. Communication to the author by Mr. Posthuma; The Hague, 1987. Also interview with M.A. Djoehana, one of the group, Prague, August 16, 1983.

90. Coast, John, *Recruit to Revolution: Adventure and Politics in Indonesia* (London: Christophers, 1952), 96.

91. On Soedjojono, see also above, chapter 1.

92. E.g., Anderson, Benedict, *Java in a Time of Revolution: Occupation and Resistance, 1944–1946* (Ithaca: Cornell University Press, 1972); *Lukisan Revolusi Indonesia, 1945–1950* (Jakarta: Kementerian Penerangan Republik Indonesia, n.d.), 23–24.

93. Musil, Robert, *The Man Without Qualities* (New York: Putnam's, 1980), 331–338.

CHAPTER FIVE

1. This I am told by Benedict Anderson; and also that this is a name of the most tasty of all Javanese mangoes.

2. Kartini, Raden Adjeng, *Letters of a Javanese Princess* (New York: Norton, 1964), 190.

3. Ibid., 48.

4. Kierkegaard explaining his and Hegel's view of metaphor in Kierkegaard, Søren, *The Concept of Irony with Continual Reference to Socrates*, translated by Howard V. Hong and Edna Hong (Princeton: Princeton University Press, 1989), 101–103.

5. For an expansion of the telegraph network in the Indies between 1870 and 1918 most graphically, see *Plaat V: Uitbreiding van het telegraafnet en toename van het telegraafverkeer sinds 1870*, in Wigman, G.M., "De Nederlandsch-Indische Gouvernements Telegraafdienst," *Koloniale Studiën*, no. 2 (1918), 22–23.

6. Reitsma, S.A., *De verkeersbedrijven van den staat: spoorwegen, post-, telegraaf- en telefoondienst, havenwezen* (Weltevreden: G. Kolff, 1924), 162, 178.

7. *Verslag omtrent den Post- en Telegraafdienst in Nederlandsch-Indië over het Jaar 1900* (Batavia: Javasche Boekhandel, 1901), 51–53. At the same time, in twelve months, 10,296 pieces of official correspondence and 24,492 private letters and postcards were sent from Japara. Japara subscribed to 26 newspapers. Ibid., *Bijlage D*, 32–35.

8. *De Indische Gids*, vol. 24 (1902), 855, quoting a reporter of *Nieuwe Rotterdamsche Courant*.

9. Kafka, F., *I Am a Memory Come Alive: Autobiographical Writings* (New York: Schocken, 1976), 196.

10. "We asked our medium, and two days later a letter came . . . repeating word for word what [the medium] had told us." Kartini quoted in Bouman, H., *Meer licht over Kartini* (Amsterdam: H.J. Paris, 1954), 49.

11. Kartini, R.A., *Brieven aan mevrouw R.M. Abendanon-Mandri en haar echtgenoot*, ed. by F.G.P. Jaquet (Doordrecht: Foris, 1987), 184–185. Van Kol, a socialist at a spiritist table, was not an exception at that time. Spiritism was denounced by the Holy See in 1898 and 1917, and Pope Pius IX spoke of "the twin evils of spiritualism and socialism." Winter, Jay, *Sites of Memory, Sites of Mourning: The Great War in European Cultural History* (Cambridge: Cambridge University Press, 1995), 55–56.

12. Kartini quoted in Bouman, *Meer licht over Kartini*, 46–48.

13. Kierkegaard, *The Concept of Irony*, 101.

14. Pyenson, Lewis, *Empire of Reason: Exact Sciences in Indonesia, 1840–1940* (Leiden: Brill, 1989).

15. Berretty, D.W., *Van 13 momenten uit een 13-jarig bestaan Aneta* (Batavia: Aneta

1931), 71; see also "Uit het Jaarverslaag 1924," *Stof over en voor Aneta Staf* (Weltevreden: Aneta), 4.

16. *Radio Bode*, vol. 14, no. 24 (June 9, 1940), 2.

17. Boon, Mr. G.A., Lid van de Tweede Kamer, *De Radio-Omroep voor Nederland en Indië: Nederland op z'n smalst* (Den Haag: Haagsche Drukkerij, 1930), 19. See also Numans, J.J. "Radiotelefonie Holland-Indië," *Radio Nieuws* (January 1928), 8.

18. Rimember en Henvo, *Hallo!—Batavia* (n.p., n.p., 1929?).

19. By 1935, Nirom's signal could be received throughout Java; see Witte, René, "Exploitatie en bevoogding: De Europese en inheemse radio-omroep in Nederlands-Indië tot 1942, " in Berg, S. van den, R. Witte, et al., *Jaarboek Media Geschiedenis: 4 Nederlands-Indië* (Amsterdam: Stichting beheer IISG, Stichting Mediageschiedenis, 1992), 29. About 3,000 listeners listened to Phohi, and about 7,000 listeners in Java listened to Nirom; see Beijering, Marjan, "Overheidscensuur op een koloniale radiozender: De Philips Omroep Holland Indië en de Indië Programma Commissie, 1933–1940," in Berg, S. van den, R. Witte, et al., *Jaarboek Media Geschiedenis*, 51.

20. *Radio-Omroep*, December 1938 and January 1939; see also the list of stations as well as times of broadcast for Batavia, Semarang, Surabaya, Magelang, and Malang— virtually all day or several hours daily at least. *De Indische Luistergids*, vol. 1, no. 2 (December 30, 1933), 11.

21. *Philips Concern, Archief NIROM* quoted by R. Witte, "Exploitatie en bevoogding," 30. Nirom declared 5,390 listeners on April 1, 1934, 54,462 listeners by December 1937. For more details see Vries, D. de, "De cultureele en sociale taak van de radio in Nederlandsch-Indië," *Koloniale Studiën*, no. 2 (1941), 167.

22. Friedrich Nietzsche in Blackburn, Simon, *The Oxford Dictionary of Philosophy* (Oxford: Oxford University Press, 1994), 240; see, e.g., Nietzsche's *The Birth of Tragedy*, section 8.

23. For instance, the November 1938 and February 1939 issues of *De Nirom Bode*; they were provided to me by Professor Wertheim.

24. Wertheim, Wim, and Hetty Wertheim-Gijse Weenink, *Vier Wendingen in Ons Bestaan: Indië verloren-Indonesië geboren* (Breda: De Geus, 1991), 228.

25. Witte, R., "Exploitatie en bevoogding," 25.

26. As an example for many cases, see advertisement by *Pianohandel Kok, Rijswijk 10*, in Batavia, selling radio models like Philips and Erres. E.g., *De Nederlandsche Leeuw*, vol. 1, no. 1 (August 1, 1938), 8.

27. *Alles Electrisch in het Indische Huis*, vol. 1, no. 1 (June 1931).

28. E.g., *Radio Bode*, vol. 14, no. 18 (April 28, 1940), 1.

29. See, e.g.: "radio is a modern substitute for the hearth-side." Contril, H., and G.W. Alport, *The Psychology of Radio* (New York: Harper, 1935).

30. At the 1883 world and colonial exhibition in Amsterdam, in "Class 18: Telegraphy, Post, Telephone, Signalization," there were *kapok* trees exhibited with a commentary that they were being used as telegraph poles, live, and that they also, bore "good fruit, potentially a useful commodity." *Officieele wegwijzer voor Internationale Koloniale en Uitvoerhandel tentoonstelling te Amsterdam 1883* (n.p., n.p.), 50. Sometimes "The wire was led over the high *kappok* trees, all of them evenly cut at their tops, and covered with iron or other metal caps to stop their growth. The branches of the trees were cut off; thus, they were smooth and bald, because otherwise the white

ants would invade them and eat them up." Nieuwenhuys, R., "Tempo Doeloe XII," *Kritiek en Opbouw*, vol. 3, no. 1 (February 16, 1940), 14.

31. Review of "Voorschrift ter vervaardiging van topographische kaarten betreffende de Nederlandsche bezittingen in Oost Indië" by H.Ph.Th. Witkamp, "Een voorbeeld zonder voorbeeld," *De Indische Gids*, vol. 8, no. 2 (1886), 1625.

32. *Officieele wegwijzer*, 50.

33. *Verslag omtrent den Post- en Telegraafdienst in Nederlandsch-Indië over het Jaar 1900*, 1861ff. One extremely rare but very flagrant exception were the so-called Tjilegon riots in 1888, during which, according to an official report, "The house of an assistant resident and of some other Europeans, as well as of many native officials, were attacked, some of the houses' inhabitants were murdered, and the household equipment was destroyed. The strongboxes of an under-collector and a state salt collector were plundered, the telegraph wire to Serang was cut, and the local prison was broken into." "Het officiële relaas van de onlusten te Tjilegon en pogingen tot oproer in Midden- en Oost-Java," quoted and commented in *De Indische Gids*, vol. 11, no. 2 (1889), 1768–1770.

34. *Verslag omtrent den Post- en Telegraafdienst in Nederlandsch-Indië over het Jaar 1914* (Batavia: Javasche Boekhandel, 1915), 39.

35. There was a hitch, in fact. Connection was lost for a few minutes. The dancers, however, went on with their graceful dancing and the *gamelan* orchestra with its playing. When the wireless connection was restored, the dancers and the orchestra met exactly on the beat. Interview with Gusti Nurul, one of the dancers; Bandung, August 1, 1999.

36. Meijer, D.H., "Dactyloscopie voor Nederlandsch-Indië," *Koloniale Studiën*, no. 2 (1926), 931. Some excitement over the technology came from the possibility of booking a hotel in the Indies at a distance, via telegraph. A special *Hotelcode* was developed. E.g. "Please reserve one single-bed room" was signaled as *Alba*, "reserve one double-bed room" as *Album*; other codes were, e.g., *Babel, Bazar, Cadeau, Daisy, Distance, Delaware, Deduct. Hotel des Indes* (Batavia: Hotel Des Indes, n.d.), n.p.

37. Letters from listeners in *Over de wereld klinkt Nederlands stem PHOHI*, typescript (Hilversum: Directie PHOHI, 1935), 10.

38. Letters from listeners in *De Indische Luistergids*, vol. 1, no. 1 (December 23, 1933), 2; ibid., vol. 1, no. 2 (December 30, 1933), 7.

39. *Over de wereld klinkt Nederlands stem PHOHI* , 3.

40. In Beijering "Overheidscensuur op een koloniale radiozender," there are some truly remarkable examples of Indies radio censorship. For instance, in 1936, a lecture by a major of the army, W.G. Bas, on the Dutch liberation struggle against the French in 1813 could not go on the air because the commission thought it simplistic, and possibly "tactless toward the native listeners, for whom, in fact, the independence has not yet come." A popular voice on the Indies radio, a certain Mr. Aletrino, offered a lecture in 1933 on Mahatma Gandhi, describing Gandhi's hunger strike as a media stunt. The talk was not allowed on the air, with a member of the Indies radio control commission commenting that Aletrino "does not appreciate the importance of even a neutral word in the tropics." Ibid., 55.

41. Ibid., 63.

42. *De Indische Luistergids* (April 28, 1934), 17. It took another four years before,

in the *Volksraad*, a deputy demanded that radio be similarly introduced into the higher classes of the Indies elementary schools and into the middle schools; this time, the model that the deputy suggested was South Africa. Razoux Schultz in *Handelingen van den Volksraad* (Batavia: Volksraad van Nederlandsch Indië, 1938–1939), 478–479.

43. E.g.: "In the first month of this year, there were 221,469 new listeners in Germany. This means an additional 7,144 listeners every day or 300 new listeners every hour." *De Indische Luistergids* (April 21, 1934), 3.) In 1938, in the *Volksraad* again, a deputy demanded "greater awareness" by the Indies authorities of the role of the radio "in this modern society"; it is a "medium of publicity and propaganda," the deputy said, "and it can work over large masses." "Of course," he added, "I do not want to sound like the German minister for propaganda, Dr. Göbbels." Van Baalen in *Handelingen van den Volksraad* (1938–1939), 234.

44. Ronell, Avital, *The Telephone Book: Technology-Schizophrenia-Electric Speech* (Lincoln: University of Nebraska Press, 1989), 30, 412.

45. "We are told that the mask used in Greek tragedy had a magical function: to give the voice a chthonic origin—to make it come from somewhere under the earth." Barthes, Roland, *A Lover's Discourse: Fragments* (New York: Hill and Wang, 1993), 115. Barthes compares that function with that of modern communication technology, namely, with the telephone.

46. Regularly published letters by listeners to radio journals, in both the colony and the Netherlands, bear witness to this. See also, e.g., "Wat luisteraars schrijven!" *Over de wereld klinkt Nederlands stem PHOHI*, 9–10.

47. "[At] the Press Pavilion at the 1928 *Pasar Gambir* [Batavia Fair], organized by Aneta . . . thanks to the cooperation by the *P.T.T.* [Post, telephone, and telegraph service], the visitors had an opportunity to speak directly from the Aneta fairgrounds pavilion, wireless, with the Netherlands." Berretty, *Van 13 momenten*, n.p.

48. *The Wireless Transmitting Station MALABAR* (Bandoeng: General Post-Tele-graph- and Telephone Service, 1929), n.p. See also "A Diorama of the Government Radio Station at Malabar as Created by Aneta in the Press Pavilion on Gambir Fair in Batavia in 1928," in Berretty, *Van 13 momenten*, n.p.

49. "Oorlog en B.R.V.," *Radio Bode* vol. 14, no. 25 (June 16, 1940), 2.

50. Ibid., 3.

51. Ibid.

52. Ibid., vol. 15, no. 49 (December 7, 1941), end page.

53. *De Indische Luistergids* (April 28, 1934), 13.

54. Beginning January 1, 1901, a spiritist journal *Orgaan de Vereeniging Lumen* appeared regularly for four years. Between 1906 and 1912, *Het Spiritistisch Tijdschrift voor Ned.-Indië*, appeared; then after April 1, 1929, *Indisch Spiritistisch Tijdschrift* appeared first irregularly and by 1930 monthly, first in Weltevreden (Batavia), then in Semarang.

55. *Indisch Spiritistisch Tijdschrift*, vol. 11, no. 3 (1940), 46. *Nederlandsch-Indische Federatie van Spiritisten* had branches in many towns of Java and other islands. Each of the branches was supposed to have a chairperson, secretary, treasurer, and control-ling, propaganda, and verifying committees.

56. In one instance, a speaker at a spiritist meeting is reported to have given a "graphic representation of electromagnetic radiation." *Indisch Spiritistisch Tijdschrift*, vol. 19, no. 12 (1938), 191; in another case a spiritist lectured about an "electric

phenomenon through a medium," namely, how, in order to connect ghosts successfully, an "electric stream of low voltage [had to be directed through] 2 silver sheets 4 mm apart [in order to achieve the best] galvanic effect" (*Indisch Spiritistisch Tijdschrift*, vol. 10, no. 11 [1939], 171).

57. *Radio Bode*, vol. 14, no. 20 (May 12, 1940), 1.

58. E.g., ibid., vol. 14, no. 27 (June 30, 1940), 8; see also for a Faust advertisement in an Indonesian journal, *Pembangoen*, no. 19 (November 4, 1939), n.p.

59. Wertheim and Wertheim, *Vier Wendingen*, 244–245.

60. Ch. van der Plas, "Opmerking over *The Tempest*," *Fakkel*, vol. 1, no. 5 (March 1941), 391.

61. Dierikx, Marc, *Bevlogen jaren: Nederlandse burgerluchtvaart tussen de wereldoorlogen* (Romen: Luchtvaart, n.d.), 135.

62. Bovene, G.A. van, *Nieuws! Een boek over pers, film en radio* (Batavia: G. Kolff, 1941?). Van Bovene himself calls the Batavia–Sydney flight and his reportage a "sensation" and a "fairy tale." Ibid., 122–123.

63. Ibid., 26.

64. Ibid., 7.

65. Ibid., 12.

66. Ibid., 160.

67. Ibid., 168.

68. Ibid., 172.

69. Ibid., 68.

70. Ibid., 222–225.

71. Dimyati, Muhammad, *Dibalik Tabir Gelombang Radio* (Atjeh: Indonesia, 1359 [1939]). On Dimyati, see, e.g., Teeuw, A., *Modern Indonesian Literature*, vol. 1 (Dordrecht: Foris, 1986), 75, 140; vol. 2 (The Hague: Nijhoff, 1979), 12.

72. "Perhaps there had been something in Amir's soul that was open to sounds; perhaps he was just a young man attracted to *Radio*." Dimyati, *Dibalik Tabir Gelombang Radio*, 17.

73. Ibid., 7.

74. Ibid., 25.

75. Ibid., 37.

76. Ibid., 40, 56.

77. Ibid., 56.

78. Ibid., 87.

79. Djojopoespito, Soewarsih, *Buiten het Gareel. Indonesische Roman Met een Inleiding van E. Du Perron* (Utrecht: De Haan, 1940), 16, 20–21.

80. Djojopoespito, Soewarsih, "De taal de Soendanese jongeren," *Kritiek en Opbouw*, vol. 1, no. 23 (January 12, 1939), 348.

81. Jacques Derrida quoted in Ronell, *Telephone Book*, 425. See also Derrida, J., "The Voice That Keeps Silence" in Derrida, J., *Speech and Phenomenon* (Evanston: Northwestern University Press, 1973).

82. *De Indische Luistergids* (January 6, 1934), 7. Compare Marinetti, Filippo Tommaso, "Destruction of Syntax—Wireless Imagination—Words in Freedom," *Lacerba* (May 11, June 15, 1913), translated in Pioli, Richard J., *Stung by Salt and War: Creative Texts of the Italian Avant-Gardist F.T. Marinetti* (New York: Lang, 1987), 45–53.

83. Ibid.

84. Benjamin, W., "Karl Kraus," in *Reflections* (New York: Harcourt Brace Jovanowich, 1978), 239, quoting Kraus.

85. From *Nirom-Bode*, quoted in Witte, R., "Exploitatie en bevoogding," 29. On the predominance of gramophone records in the Indies radio program, see also Numans, *Radiotelefonie Holland-Indië*, 6.

86. Ibid.

87. *Radio Bode*, vol. 15, no. 49 (December 7, 1941), 4, 10.

88. Reproduced, e.g., in Buitenweg, Hein, *Kind in Tempo Doeloe* (Wassenaar: Service, 1969). C. Cornelis Jetses was born in 1873 and died in 1925.

89. *Van Dale Handwoordenboek Nederlands-Engels* (Utrecht: Van Dale, 1993), 301.

90. See above, chapter 1.

91. Salmon, Claudine, *Literature in Malay by the Chinese of Indonesia: A Provisional Annotated Bibliography* (Paris: Archipel, 1981), 42–43.

92. Tsuchiya, Kenji, "Kartini's Image of the Javanese Landscape," *East Asian Cultural Studies*, vol. 25, nos. 1–4 (1986), 82 n. 8 and 9.

93. "Dixi," in *De Indische Gids*, vol. 8, no. 2 (1886), 1088–1089.

94. The "victims" of the language were described by this writer as "considering, as the songs of their own, 'krontjong' and not 'Wilhelmus' [Dutch anthem]; when thinking most often thinking in Malay, when dreaming most often dreaming in Javanese." "Kazerne-taal Atavisme," *Soeloeh Indonesia*, vol. 2, no. 9 (September 1927), 7–8.

95. *Liederen USI* (Batavia?: USI, n.d.), n.p., song no. 10.

96. Wertheim and Wertheim, *Vier Wendingen*, 201–202.

97. Witte, R., "Exploitatie en bevoogding," 34–35. On debating the follow-up of the decision, see Soetardjo July 22 and August 8, in *Handelingen van den Volksraad* (1938–1939), 369, 777–779. Dutch deputy van Helsdingen called the Eastern radio autonomy a good example of "self-activity of the indigenous." Ibid., 374. See also Ir. A. J.H. van Leuwen, voormalig Hoofd van den Indischen Radio-dienst, "De Radio-Omroep in Nederlandsch-Indië I," *Kritiek en Opbouw*, vol. 1, no. 11 (July 15, 1938), 170–172.

98. Sayid, R.M., *S.R.V. Gedenkbok* (Soerakarta: n.p., 1936), 3.

99. Ibid., 32.

100. *Philips Concern, Archief NIROM*, quoted in Witte, R., "Exploitatie en bevoogding," 30.

101. Soetardjo, August 8, 1938, *Handelingen van den Volksraad* (1938–1939), 778.

102. For information on Soetardjo, see, for instance, Sutherland, Heather, *The Making of a Bureaucratic Elite* (Kuala Lumpur: Heinemann, 1979), especially 82–85, 122–123. Another very influential agent in the development of Eastern Indies radio was the regent of Bandoeng, R.A.A. Wiranatakoesoema; equally important throughout the period had been the aristocratic circle around Mangkoenegaran palace in Surakarta.

103. *Grondslagen voor een regeling tot deelname van de Federatie PPRK aan de Oostersche Omroep in N.I. over de NIROM-zender'* are summarized in Witte, R., "Exploitatie en bevoogding," 36.

104. *Nationale Commentaren*, vol. 4, no. 39 (September 27, 1941), 3818. See also "Evacuatie van de Burgerbevolking (Radiorede van A.W. Deelman, Hoofd van het

Kantoor Bestuurszaken van het Secretariaat van den Staatsmobilisatieraad.),” ibid., vol. 4, no. 50 (December 13, 1941), 40, 47–48.

105. “Radiorede van der Plas,” in *Nationale Commentaren*, vol. 4, no. 45 (November 8, 1941), 3938–3939.

106. Ibid., vol. 4, no. 48 (November 29, 1941), 4000.

107. Ibid., vol. 5, no. 6 (February 7, 1942), 4162.

108. Soetardjo, August 8, 1938 in *Handelingen van den Volksraad* (1938–1939), 779.

109. Whatever the real life of the Indonesian listeners had been, there were causeries on Eastern Indies radio, apparently more frequent than on European radio, with themes like “city planning,” “origins of radio,” or “automobile repair.” See, e.g., *Berita-V.O.R.L.*, vol. 4, no. 3 (March 1938); vol. 4, no. 4 (April 1938); vol. 4, no. 6 (June 1938); vol. 4, no. 7 (July 1938); vol. 4, no. 8 (August 1938); vol. 4, no. 9 (September 1938); vol. 4, no. 10 (October 1938); vol. 4, no. 11 (November 1938). An Indies radio expert, in June 1938, listened to the Eastern radio, and he published his findings in *Kritiek en Opbouw*. Out of 57 hours of the broadcast, he wrote, only 4 hours and 45 minutes, 8.5 percent, could be described as “non-amusement programs.” Out of these non-amusement programs, 30 minutes were given to book reviewing, 30 minutes to Koran reading, 20 minutes to cooking lessons, 120 minutes to lectures and news, 45 minutes to Bible reading, 45 minutes to Chinese reading. If we would like to talk about a real “education broadcast,” the author wrote, “the proportion would be about 2% of the total.” Ir. A. J.H. van Leuwen, “De Radio-Omroep in Nederlandsch-Indië II,” *Kritiek en Opbouw*, vol. 1, no. 12 (August 1, 1938), 187.

110. *Berita-V.O.R.L.*, vol. 4, no. 5 (May 1938), 8.

111. “Malam Kesenian Timoer,” *Soeara Timoer*, vol. 1, no. 2 (January 12, 1941), n.p.

112. Boediardjo, “Eenige opmerkingen over en naar aanleiding van het eerste volksconcert van de ‘Perserikatan Perkoempoelan Radio Ketimoeran,’” *Kritiek en Opbouw*, vol. 4, no. 1 (February 15, 1941), 11–12.

113. Ibid., 12.

114. “Pidato Ketoea P.P.R.K. Toean M. Soetardjo pada pemboekaän malam ‘Kesenian Timoer’ di Stadsschouwburg, pada tanggal 11 Januari 1941,” *Soeara Timoer*, vol. 1, no. 3 (January 26, 1941), n.p.

115. A.S. “Dari Kaboepaten ke Microfoon,” *Soeara Timoer*, vol. 1, no. 2 (January 12, 1941), n.p.

116. J. de K. [de Kadt], “Kleine Niromiade,” *Kritiek en Opbouw*, vol. 5, no. 3 (February 5, 1942), 48.

117. *Kritiek en Opbouw* was revived after the war, for few years, in a completely different situation of the late 1940s.

118. *Soeara Timoer*, vol. 2, no. 1 (January 11, 1942), 19.

119. Ibid., vol. 2, no. 1 (January 11, 1942), 18; vol. 2, no. 2 (February 8, 1942), 22.

120. Kierkegaard, *The Concept of Irony*, 104.

121. Ibid., 103.

122. Damhoeri, A., *Zender NIROM* (Medan: Tjendrawasih, 1940), 63.

123. Ketjindoean, *Patjar Koening. Serie V. Kedjadian jang aneh diboekakan oleh detektif Raden Pandji Soebrata* (Djokjakarta: ‘Kabe’ Kolff-Buning, 1940?), 55–57.

124. Sd.Tjokronolo, Osaka, "Membangoenkan lapisan Indonesia dalam masjarakat Indonesia," *Keboedajaän dan Masjarakat*, vol. 1, no. 5 (September 1939), 118–121.

125. *Soeloeh Indonesia Moeda*, no. 9 (August 1928), 232.

EPILOGUE

1. Pramoedya Ananta Toer, *Nyanyi Sunyi Seorang Bisu*, vol. I (Jakarta: Lentera, 1995), further as *NSSB* I, 26. Text of *NSSB* was checked against a previous version of the text, typescript *Nyanyi Tunggal Seorang Bisu* (Edisi Malaysia, 1988), further quoted as *NTSB*.

2. Sjahrazad [Sutan Sjahrir], *Indonesische Overpeinzingen* (Amsterdam: De Bezige Bij, 1945); translated by H.B. Jassin as *Renungan Indonesia* (Jakarta: Poestaka Rakjat, 1947); Pramoedya Ananta Toer, "Permenungan dan Pengapungan," the introductory part of *NSSB* I, 1–13.

3. Pramoedya is skeptical about Sjahrir's role in the Indonesian resistance against Japanese fascism and of Sjahrir's role in the Indonesian revolution of 1945. See especially *NSSB* I, 119–127.

4. Pramoedya Ananta Toer, *Lied van een Stomme: Brieven van Buru* (Houten: Manus Amici, 1991), further quoted as *LSSB*, 59–60. This is a Dutch translation of the second volume of Pramoedya's letters and reflections from Buru, in original *Cacatan-cacatan pribadi Pramoedya Ananta Toer dari tahanan Pulau Buru*, which later, six years after I had read the Dutch text, has been published in a slightly modified form, as *Nyanyi Sunyi Seorang Bisu*, vol. II (Jakarta: Lentera, 1997); it is quoted here as *NSSB* II. This particular quote is from *NSSB* II, 53.

5. *NSSB* I, 121. (Here, by the way, the late-colonial and nationalist culture of college degrees is reflected. Most of the degrees recognized by Pramoedya's mother were from the local colleges in the Indies, thus of little or no value outside the colony and in the Netherlands; the "non-recognized" Sjahrir, in fact, was widely respected for having a year, however highly informal, of academic training in the Netherlands; Hatta got a *doctorandus* degree in economics from Rotterdam, a title reserved for those who finished all the study except the thesis.)

6. See Rudolf Mrázek, *Sjahrir: Exile and Politics in Indonesia* (Ithaca: Cornell Studies in Southeast Asia, 1994), chapter 5.

7. *NSSB* I, 9.

8. Ibid., 5–6.

9. Ibid., 10. Like most of Pramoedya's sensing and thinking in exile, this did not appear suddenly. See, e.g., a text by Pramoedya from the very early 1950s: "It seems, the more different things and shapes there are in this world, the more people disappear in them—rolled up by the things, like the tools are retooled." Pramoedya Ananta Toer, "Sumber tjipta dalam kesenian," *Indonesia* (Jakarta), vol. 3 (1952), 4.

10. *NSSB* I, 12.

11. Ibid., 47.

12. Ibid., 208.

13. *LSSB*, 300–301 (*NSSB* II, 277–278).

14. *LSSB*, 302 (*NSSB* II, 278–279).

15. *NSSB* I, viii.

16. Both men occasionally read their letters, before they were "sent," to small cir-

cles of close friends in the camps, as a sort of *majalah,* "journal," Pramoedya says. Ibid., viii; Mrázek, *Sjahrir,* chapter 5.

17. *NSSB* I, ix.

18. See above, chapter 5.

19. *LSSB,* 157–158 (*NSSB* II, 158).

20. *LSSB,* 65 (*NSSB* II, 58).

21. *LSSB,* 58–59 (*NSSB* II, 51).

22. Pramoedya actually caused a (partial) revival of interest in Mas Marco in the post-colonial Indonesia. See above, chapter 4.

23. *LSSB,* 173 (*NSSB* II, 174).

24. *LSSB,* 184 (*NSSB* II, 186).

25. *LSSB,* 186 (*NSSB* II, 188).

26. *LSSB,* 189 (*NSSB* II, 192).

27. Ibid.

28. *NSSB* I, 3. Memories change with time, naturally. In a text written either on the way to the Netherlands or during his stay in Amsterdam, Pramoedya describes the ship as "the old ship: Johan van Oldenbarnevelt . . . an immigration ship. . . . Not a ship, in fact! A garrison kitchen. Four thermo tanks are not strong enough to drive the heat away. . . . From the loudspeakers suspended from the ceiling in all corners, popular American songs are roaring, and, in the intervals between them, chimes announce the times to eat." Pramoedya Ananta Toer, "Kapal Gersang," *Zenith* (Jakarta) vol. 3, no. 12 (1953), 550–552.

29. *LSSB,* 193–194 (*NSSB* II, 197). Here, Pramoedya is very subtle in his irony. For his earlier, much harsher, words on the *Sticusa* role in the Indonesian culture, and on the 1953 Sticusa conference, see Pramoedya Ananta Toer, "Kemunduran kesusasteraan Indonesia modern dewasa ini," *Duta Suasana,* vol. 2, no. 25 (1953), 14–15; idem, "Tendensi kerakjatan," *Star Weekly,* vol. 11, no. 525 (1956), 32–33; idem, "Lesu; kelesuan; krisis; impasse," *Siasat,* vol. 11, nos. 515/516 (1957), 34–36.

30. *LSSB,* 194–195, 198 (*NSSB* II, 198–200). For some motifs connected with the Pramoedya's Amsterdam woman friend, see also the text that Pramoedya wrote late in 1953 in Holland. The main difference is exactly the all-pervasive sensuality and immediacy of the earlier text. "Tentang Emansipasi Buaja," *Kisah* (Jakarta), vol. 3, no. 2 (December 1953), 722–730. In a recent article in 1992, Pramoedya again mentioned his Amsterdam friend as a very important figure in his life in that period. See "De kinder chief in een rode auto: Ervaringen met Nederland als staat en volk," *Eindhoven Dagblad Boekenweekbijlage* (March, 10, 1992), 5.

31. *LSSB,* 198 (*NSSB* II, 202).

32. *LSSB,* 201 (*NSSB* II, 205).

33. *LSSB,* 199–200 (*NSSB* II, 203–204).

34. *NSSB* I, 90.

35. Ibid., 2.

36. The Dutch translator's introduction to *LSSB,* 7.

37. *LSSB,* 289–290 (*NSSB* II, 266). Pramoedya writes that it was him who gave the camp Unit II the name *Wanayasa,* which he explains, was the name of the place where the famous seventeenth-century ruler of the Javanese kingdom of Mataram, Sultan Agung, prepared his attack against the Dutch in Batavia, and which he used as the center of his *jaringan logistik,* "logistic network." *NSSB* I, 62.

38. Ibid., 16, 100.
39. *LSSB*, 75 (*NSSB* II, 69).
40. *LSSB*, 54 (*NSSB* II, 46).
41. *NSSB* I, 162.
42. *LSSB*, 150 (*NSSB* II, 151).
43. *NSSB* I, 94–95.
44. Kartini is omnipresent in Pramoedya. He wrote an extensive biography of Kartini, *Panggil aku Kartini sadja*, vols. I and II (Jakarta: Nusantara, 1962). Pramoedya's grandfather lived in Rembang, a place where Kartini spent the last period of her life, and where she died. Pramoedya's letters from Buru contain some rich memories of his Rembang grandfather's big house, where he spent some of his vacations.
45. *LSSB*, 150 (*NSSB* II, 151).
46. *LSSB*, 151, 153 (*NSSB* II, 152). In his 1964 lectures on modern Indonesian history, Pramoedya mentioned two books on the Indonesian revolution as the most serious: Adam Malik, *Riwajat dan Perdjuangan Sekitar Proklamasi Kemerdekaan Indonesia*, mimeo (1948), and Sidik Kertapati, *Sekitar Proklamasi 17 Agustus 1945*, 3rd ed. (Jakarta: Pembaruan 1956). See Pramoedya Ananta Toer, *Sedjarah Modern Indonesia*, mimeo (Jakarta: Universitas "Res Publica," 1964), IX.
47. *NSSB* I, 37.
48. Ibid., 39.
49. Ibid., 40–41.
50. *LSSB*, 171 (*NSSB* II, 171).
51. *NSSB* I, 39.
52. *LSSB*, 156–157 (*NSSB* II, 156–157).
53. *LSSB*, 147–148 (*NSSB* II, 148).
54. *LSSB*, 170 (*NSSB* II, 169).
55. *LSSB*, 55 (*NSSB* II, 47–48).
56. *LSSB*, 104 (*NSSB* II, 101). Recently, Pramoedya has returned to this. In a Dutch interview in 1992 he said of his mother: "She died of TB. My teacher told me: when somebody dies of TB, you must place a thin omelet over his or her face, and it will almost entirely consume the bacilli. But I was not concerned. I kissed her on her mouth and nose." Hans den Boef, A., and K. Snoek, *Pramoedya Ananta Toer: Essay en Interview* (Novir: De Geus, 1992), 37. Yet another description of the event, much earlier, from 1959, may be mentioned. In a typewritten and sketchy eleven-page bio-data written by Pramoedya for Professor Teeuw, there are the following lines (Pramoedya speaks about himself in the third person): "*Blora, May 1942* . . . It was clear that mother was already dead. Broken-hearted, he kissed the forehead of his mother's corpse." Pramoedya Ananta Toer, *Bio en Bibliografie* (typescript, 1959; courtesy of Prof. Teeuw), 2.
57. *LSSB*, 24–26 (*NSSB* II, 13–15). This is one of the recollections that do not seem to change. In a recent interview Pramoedya said: "I do not understand why my parents, sometimes for months, did not talk to each other. I cannot understand it at all." Kalpana, P., and E. Elburg, "Gesprekken met Pramoedya Ananta Toer." *Bzzletin*, vol. 10 (1988), 36.
58. *LSSB*, 70 (*NSSB* II, 64).
59. Pramoedya, after a four-class elementary school, entered a middle-level *MULO* school but did not get higher than to its second grade. His father also gave him for

over a year some private tutoring. See, e.g., Pramoedya Ananta Toer, *Bio en Bibliografie*, 1. See also Hans den Boef and Snoek, *Pramoedya Ananta Toer*, 7.

60. On Pramoedya's mother talking about the leaders, see *LSSB*, 76–77 (*NSSB* II, 71–72).

61. *LSSB*, 78.

62. Ibid., 79 (*NSSB* II, 74–75).

63. *LSSB*, 80 (*NSSB* II, 75).

64. Ibid. In an earlier text translated by Benedict Anderson and published in 1983, Pramoedya also refers to this particular reading: "Actually, since 1940, my mind had been liberated from metaphysics by Pak Poeh's well known Javanese book; I was more inclined to focus on reason (*rasio*) as a rider and the flesh (*daging*) as a horse it had to keep firmly reined." "Perburuan 1950 and Keluarga Gerilya 1950," *Indonesia*, no. 36 (October 1983), 25.

65. *LSSB*, 80–81, 84–85 (*NSSB* II, 75–76, 83).

66. *LSSB*, 88 (*NSSB* II, 83–84).

67. Ibid.

68. Pramoedya Ananta Toer, *Bio en Bibliografie*, 2.

69. In a later interview, Pramoedya described the moment: "At that time my father started to work again. The Japanese were tough on gamblers! I had to get out of the house. Perhaps my father thought: 'If he stays in this place, he has no future.' He told me: 'Just go west, to Jakarta.' Thus I went to Jakarta." Hans den Boef and Snoek, *Pramoedya Ananta Toer*, 36.

70. *LSSB*, 118 (*NSSB* II, 117).

71. *LSSB*, 119 (*NSSB* II, 117).

72. *LSSB*, 115–116 (*NSSB* II, 113–114).

73. Adam Malik was a prewar nationalist journalist and political activist; after 1945 he was a leader of Murba, "Proletarian," party, one-time cabinet minister and ambassador to Moscow under Sukarno, foreign minister and vice president under Soeharto. Dahler was the leader of the Eurasian group. Djawoto was a journalist, and later an Indonesian ambassador to the People's Republic of China. Mochtar Lubis is a writer and journalist.

74. *LSSB*, 115–116 (*NSSB* II, 114).

75. *LSSB*, 116–117 (*NSSB* II, 115).

76. *LSSB*, 138 (*NSSB* II, 138–139). On another occasion, Pramoedya told another story: "On the evening of my arrest in 1965, I was interrogated at the headquarters of the military command by a picket officer, a lieutenant colonel. At half past one in the morning, I saw some men coming in with the index cards I had collected on historical figures of the past. The index cards were handed over to the officer. 'A list of [Communist Party] sections, huh?' he looked at me, just making sure. All was destroyed, also the metal index boxes that had cost me much of money. I let them be made when I was working on an encyclopedia." Kalpana, P., and E. Elburg, "Gesprekken met Pramoedya Ananta Toer," 38.

77. *LSSB*, 131 (*NSSB* II, 130–131).

78. *LSSB*, 131–132 (*NSSB* II, 131–132).

79. *LSSB*, 132 (*NSSB* II, 132).

80. Ibid.

81. *LSSB*, 134 (*NSSB* II, 133).

82. *LSSB*, 134, (*NSSB* II, 133–134). Karundeng, through the 1950s and 1960s, was widely considered to be the most prominent figure of Indonesian stenography. In 1990, the 25th edition of his *Stenografia* textbook was published. By a curious coincidence the publishers in Jakarta where the textbook by Karundeng appeared most frequently bore, of all names, the name of the revolution and of Pramoedya's mother—*Pradnya Paramita*!

83. *LSSB*, 134–135, 136–137 (*NSSB* II, 134–136).

84. *LSSB*, 136 (*NSSB* II, 136–137).

85. *LSSB*, 160 (*NSSB* II, 162).

86. In Pramoedya's short story "Jang Hitam," from the collection *Tjerita dari Blora* published in 1952, Sukarno is speaking on the radio. A blind veteran of the revolution listens to one of the president's Independence Day, August 17, speeches. The blind man listens to his "12-lamp radio," one of the last things, if not the last, that is still his. The atmosphere is gloomy; the blind invalid can hear the squeaking of "the electric dial" as he searches for another station. Besides this and Sukarno's voice, some "Western music" can be heard on the radio, cheers of "Freedom, Freedom" from the celebrations in Jakarta, and "Lilly Marlene." There is not much hope, for this veteran at least. "Jang Hitam," *Tjerita dari Blora*, 2nd ed. (Jakarta: Balai Pustaka, 1963), 381–411.

87. *NSSB* I, 10.

88. *LSSB*, 160–161 (*NSSB* II, 162).

89. *NSSB* I, 73.

90. The Dutch translator's introduction to *LSSB*, 7.

91. *NSSB* I, 312.

92. Report by "W.R.," quoted by Pramoedya in ibid., 213.

93. Ibid., 73, also ibid., 281, and ibid., 299.

94. Ibid., 48.

95. Ibid., 77.

96. Ibid., 79.

97. Ibid., 185.

98. Ibid., 25. At this point Pramoedya also labels the University Team as *Team Psikologi Universitas*, "University Psychological Team."

99. Chairil Anwar is recognized as the greatest poet of the Indonesian revolution. He died in 1949 in his twenties of multiple causes, in part the consequence of his bohemian life. In Pramoedya's earlier literary work, Chairil Anwar appeared as a most dramatic figure representing the revolution. Pramoedya Ananta Toer, *Larasati (Ara): Roman Revolusi* [1960] (Jakarta: Hasta Mitra, 2000). On Parmoedya's earlier views of Chairil, see also Pramoedya Ananta Toer, "Ke arah sastra revolusioner," *Star Weekly*, vol. 11, no. 574 (1956), 6–7, and Pramoedya Ananta Toer, "Lesu; kelesuan; krisis; impasse," *Siasat*, vol. 11, nos. 515/516 (1957), 34–36.

100. Ibid., 136–137.

101. Ibid., 70.

102. *LSSB*, 63 (*NSSB* II, 56).

103. *LSSB*, 271.

104. *NSSB* I, 130.

105. Pramoedya Ananta Toer, *Rumah kaca* (Jakarta: Hasta Mitra, 1988); in English translation: Pramoedya Ananta Toer, *House of Glass* (New York: Penguin, 1992).

106. Ibid., 135.

107. Ibid., 137.

108. Ibid., 308; Again, this is how Pramoedya arranged the appendix in his (more) original 1988 edition (*NTSB*, 267–270); in *NSSB* this is placed as "Lampiran [Appendix] 3" (*NSSB* I, 308–311).

109. Ibid., 308–309.

110. Ibid., 310.

111. Ibid., 311.

112. Ibid., 54.

113. Ibid., 56.

114. Ibid.

115. Ibid., 58.

116. This is, again, how Pramoedya arranged the appendix in his 1988 edition (*NTSB*, 262f.); in *NSSB* this is placed as "Lampiran [Appendix] 1" (*NSSB* I, 290–303).

117. "Bunuh diri dengan menggantung," ibid.

118. Ibid.

119. *Roosseno: pakar dan perintis teknologi sipil Indonesia*, 2nd ed. (Jakarta: Pembimbing Masa, 1989), 897, 900.

120. *NSSB* I, 86. Pramoedya is right to say that he opposed comics already a long time before he went to prison. See, e.g., Pramoedya Ananta Toer, *Realisme-Sosialis dan Sastra Indonesia: sebuah tindjauan sosial*, mimeo (Jakarta: Seminar Fakultas Sastra, Universitas Indonesia, 1963), 28.

121. *NSSB* I, 117.

122. Ibid.

123. Ibid., 123.

124. Ibid., 137–138.

125. "W.R.," quoted in ibid., 274.

126. Ibid., 118.

127. Ibid., 92.

128. Ibid., 314; Again, this is how Pramoedya arranged the appendix in his 1988 edition (*NTSB*, 262–263); in *NSSB* this is placed as "Lampiran [Appendix] 5."

129. *NSSB* I, 26.

130. Pramoedya says to Mochtar, also, that "gods sometimes visit Buru." However Mochtar might understand this, Pramoedya often referred in his letters to the visiting military inspections as "gods."

131. Ibid., 25.

132. Ibid., 314.

133. Ibid., 6.

134. Ibid., 20.

135. Ibid., 30.

136. Ibid., 85.

137. Ibid.

138. Ibid., 63.

139. Ibid., 175.

140. Ibid., 191–192.

141. *LSSB*, 172ff. (*NSSB* II, 173ff.).

142. *LSSB*, 305 (*NSSB* II, 283).

143. *NSSB*, 212.

144. *LSSB*, 295 (*NSSB* II, 272). Buru, in a sense, and in contrast to Boven Digoel, seemed to be a compact place. First, unlike the truly Indonesian composition of Boven Digoel, virtually all the exiles on Buru appeared to be Javanese; from some of Pramoedya's remarks, also, it seems that the Buru population, compared to Boven Digoel, was politically more homogeneous, left wing. For instance, at one place Pramoedya mentions a prisoner and writes: "Maybe as a man from *PNI* [the former center to center-left Indonesian National Party] he felt isolated, and thus he longed more for some friendship." *NSSB*, 134.

145. Ibid., 29.

146. Ibid., 242, 264.

147. Ibid., 277–278.

148. Ibid., 70–71.

149. Ibid., 135.

150. Ibid., 60–61.

151. Ibid., 77.

152. Ibid., 79.

153. Ibid., 97.

154. Ibid., 98.

155. Ibid., 192–193.

156. Ibid., 177.

157. Ibid., 178–180.

158. Ibid., 180–182.

159. Ibid., 185.

160. Ibid., 194.

161. Ibid., 199.

162. Ibid., 194–195. A few pages farther, Pramoedya comments upon a declaration which each prisoner had to sign before "made free"—"contract for a new slavery," Pramoedya calls it: "Thus, I was made ever more to understand what a friend once told me: 'Truly, I had been wounded at the moment when born as an Indonesian.'" Ibid., 201.

163. Ibid., 72–73; the letter is also mentioned in a letter to Wertheim from late 1977, quoted in the Dutch translator's introduction to *LSSB*, 9.

164. *NSSB* I, 164.

165. Ibid., 167. Ambon was an island with a public radio station nearest to Buru.

166. Ibid., 201.

167. Ibid., 4.

168. Ibid., 6.

169. Ibid., 145.

170. Ibid., 27.

171. *NSSB* II, 284, 286. On Pramoedya and Pak Poeh, see above, note 64.

172. *LSSB*, 306–310 (*NSSB* II, 286–288).

173. *The Grolier 1995 Multimedia Encyclopedia* (Grolier Electronic Publishing, Inc., 1995).

174. Joesoef Isak in *NSSB* I, X.

175. Ibid., XII.

176. Ibid.

SOURCES

Only items and serial volumes directly referred to in the text are listed below.

A. NEWSPAPERS AND MAGAZINES

Alles Electrisch in het Indische Huis (1931–1932)
Alles Electrisch in Huis en Bedrijf (1931)
Berita-V.O.R.L. (Vereeniging voor Oostersche Radio Luisteraars) (1938)
De Beweging (1919)
De Indische Gids (1879–1903)
De Indische Luistergids (1933–1934)
De Indische Mercuur (1928)
De Ingenieur in Nederlandsch Indië (1938–1941)
De Nederlandsche Leeuw (1938)
De Nirom Bode (1938–1939)
De Stijl (1918)
Doenia Bergerak (1914–1915)
Duta Suasana (1953)
Efficiency (1936)
Fakkel (1940–1942)
Ganeça (1922–1942)
Handelingen van den Volksraad (Batavia: Volksraad van Nederlandsch Indië, 1918–1942)
I.B.T. Locale Techniek (1938–1940)
Indisch Spiritistisch Tijdschrift (1938–1940)
Indisch Tijdschrift voor Post en Telegraphie: orgaan van den Bond van Ambtenaren en Beamten (1906–1912)
Indonesia (Jakarta) (1952)
Indonésia Raja (1929)
Keboedajaän dan Masjarakat (1939–1941)
Kisah (1953)
KNILM Nieuws (1935–1937)
Koloniale Studiën (1916–1941)
Kopiïst (1842–1843)
Kritiek en Opbouw (1938–1942)
Magneet (1914–1916)
Mécano (1923)
Medan Politie Boemipoetra (1940–1941)
Motorblad (1928)
Nationale Commentaren (1941–1942)
Nederlandsch Indië Oud en Nieuw (1916–1931)
Officieel orgaan van de Vereeniging van Huisvrouwen Bandoeng (1940–1942)
Orgaan dari Perhimpoenan Pegawai Post Telegraaf Telefoon Radiodienst Rendahan (P.T.T.R.) (1937–1938)

Orgaan van de USI (1939–1941)
Orgaan van de Vereeniging 'Nederlandsch-Indische Jaarbeurs' (1922)
Pembangoen (1939)
Poedjangga Baroe (1933–1938)
Radio Bode (1940–1942)
Radio Nieuws (1928)
Radio-Omroep (1938–1939)
Saro Tomo (1915)
Siasat (1957)
Sinar Boeroeh Kereta Api (1939)
Soeara Parindra (1936–1937)
Soeara Timoer (1940–1942)
Soeloeh Indonesia (1927)
Soeloeh Indonesia Moeda (1927–1928)
Sopir (1932)
Star Weekly (1956)
Tecton (1936–1937)
Wendingen (1918)
Zenith (1953)

B. CATALOGS, DICTIONARIES, YEARBOOKS, GUIDES

Bromo-Hotel-Tosari. Tosari: Bromo-Hotel, 1926.
Catalogus der afdeeling Nederlandsche koloniën van de internationale koloniale en uit-voerhandel tentoonstelling. Ed. P.J. Veth, G.A. Wilken, and H.C. Klinkert. Vols. I–III. Leiden: Brill, 1883.
Catalogus van de verzameling poppen weergevende verschillende kleederdrachten van de volkeren van de Nederlandsch Oost-Indischen archipel, aangeboden aan Hare Majesteit de Koningin der Nederlandsch Oost-Indië, en welke verzameling op de tentoonstelling te Batavia van 1893 bekrond is geworden met het eerendiploma. Batavia: Kolff, 1894.
Dertiende Jaarverslag der Batavia Electrische Tram-Maatschappij 1909. Batavia: Electrische Tram-Maatschappij, 1910.
Encyclopaedie van Nederlandsch Indië. 2nd ed. Vols. IV, V, VII, VIII 's Gravenhage: Nijhoff, 1921, 1927, 1935, 1939.
Gedenkboek Nederlandsch-Indische Gas-Maatschappij 1863–1913. Rotterdam: Nederlandsch-Indische Gas-Maatschappij, 1913.
Gedenkboek Nederlandsch-Indische Gas-Maatschappij 1863–1938. Rotterdam: Nederlandsch-Indische Gas-Maatschappij, 1938.
Gedenkboek ter herinnering van het tienjarig bestaan van de Nederlandsche vereeniging voor radiotelegrafie, 1916–1926. Zutphen: Nauta, 1926.
Gedenkboek van het Korps, Marechaussee van Atjeh, 1890. Medan: Deli Courant, 1941?
Gedenkboek voor Nederlandsch-Indië ter gelegenheid van het Regeeringsjubileum van H.M. de Koningin. Batavia: G. Kolff, 1923.
Geillustreerde Encyclopaedie van Nederlandsch Indië. Ed. G.F.E. Gonggrijp. The 1934 edition reprinted. Wijken en Aalburg: Pictures Publishers, 1992.
Hotel des Indes. Batavia: Hotel Des Indes, n.d.
Imperial Japanese Army and Navy Uniforms and Equipment. N.p., n.p., 1975.

Jaarboek der Technische Hoogeschool te Bandoeng 10e cursusjaar 1930. Bandoeng: Maks and van denKijts, 1930.

Jaarboek der Technische Hoogeschool te Bandoeng 15e cursusjaar 1935. Bandoeng: Maks and van denKijts, 1935.

K.L.M . Air Line Amsterdam-Batavia (Holland-Java): 9000 Miles in 10/12 days weekly service. Amsterdam: K.L.M., 1931.

K.N.I.L.M. Reisgids 1930. Batavia: K.N.I.L.M., 1930.

Nederland te Parijs 1931: Gedenkboek van de Nederlandsche deelneming aan de Internationale Koloniale Tentoonstelling. 's Gravenhage: Van Stockum en Zoon, 1931?

Officieele wegwijzer voor Internationale Koloniale en Uitvoerhandel Tentoonstelling te Amsterdam 1883. Amsterdam: n.p., 1883?

Onderzoek naar de mindere welvaart der inlandsche bevolking op Java en Madoera. Vol. IVa. Batavia: Landsdrukkerij, 1906.

Regeerings Almanak voor Nederlandsch-Indië. Batavia: Landsdrukkerij, 1942.

Reglement voor de societeit 'De Harmonie' gevestigd te Batavia waaraan zijn toegevoegd de sedert 30 September 1935 geldende Huishoudelijke Bepalingen. Batavia: De Unie, 1935.

S.R.V. Gedenkboek. Ed. R.M. Sayid. Soerakarta: S.R.V., 1936.

Statistische opgaven over 1889 betreffende het groote meerendeel der gevangenissen in Nederlandsch-Indië. Weltevreden: Landsdrukkerij, 1889.

Stof over en voor Aneta's Staf. Weltevreden: Aneta, 1925.

Technische Hoogeschool te Bandoeng: Overdracht der Hoogeschool aan den Lande. Bandoeng: Technische Hoogeschool, 1924.

The Grolier 1995 Multimedia Encyclopedia. N.p.: Grolier Electronic Publishing, Inc., 1995.

The Oxford Dictionary of Philosophy by S. Blackburn. Oxford: Oxford University Press, 1994.

The Wireless Transmitting Station MALABAR. Bandoeng: General Post-Tel- and Telephone Service, 1929.

Tien (10) vervlogen jaren. Uitgegeven ter gelegenheid van het tienjarig bestaan van de Koninklijke Nederlandsch-Indische Luchtvaart Maatschappij. 1928–1938. Batavia: K.N.I.L.M., 1938?

Uitreiking der bekroningen aan de Nederlandsche inzenders van de internationale Koloniale en Uitvoerhandel Tentoonstelling. Amsterdam: n.p., 1883.

Van Dale Handwoordenboek Nederlands-Engels. Utrecht: Van Dale, 1993.

Veertiende Jaarverslag der Batavia Electrische Tram-Maatschappij 1910. Amsterdam: Batavia Electrische Tram-Maatschappij, 1911.

Verslag omtrent den Post- en Telegraafdienst in Nederlandsch-Indië over het Jaar 1900. Batavia: Javasche Boekhandel, 1901.

Verslag omtrent den Post- en Telegraafdienst in Nederlandsch-Indië over het Jaar 1914. Batavia: Javasche Boekhandel, 1915.

Verslag van der Dienst der Arbeidsinspectie in Nederlandsch-Indië over het jaar 1920. Weltevredeu: Visser, 1920.

C. BOOKS AND ARTICLES

Akihary, H. *Architectuur en stedebouw in Indonesië 1870–1970.* Zeist: Rijksdienst voor de Monumentenzorg, 1988.

Anderson, B. *Java in a Time of Revolution: Occupation and Resistance, 1944–1946.* Ithaca: Cornell University Press, 1972.

———. *Language and Power: Exploring Political Cultures in Indonesia.* Ithaca: Cornell University Press, 1990.

———. "A Time of Darkness and a Time of Light: Transposition in Early Indonesian Nationalist Thought," in Anderson. *Language and Power.*

———. "Japan 'The Light of Asia,'" in J. Silverstein, ed. *South-east Asia in World War II: Four Essays.* Southeast Asia Studies Monograph Series no. 7. New Haven: Yale University Press, 1966.

———. "Sembah-Sumpah: The Politics of Language and Javanese Culture," in Anderson. *Language and Power.*

Baar, P.-P. de. "A City on the Move: Amsterdam 1892–1912," in Baar, Bax, et al. *Piet Mondriaan: The Amsterdam Years.*

Baar, P.-P. de, M. Bax, et al. *Piet Mondriaan: The Amsterdam Years.* Amesterdam: Gemeentearchief Amsterdam, 1994.

Barker, J.D. *The Tattoo and the Fingerprint: Crime and Security in an Indonesian City.* Unpublished Ph.D. thesis, Cornell University, 1999.

Barthes, R. *A Lover's Discourse: Fragments.* New York: Hill and Wang, 1993.

Barthes, R. *Camera Lucida: Reflections on Photography.* Translated by Richard Howard. New York: Noonday, 1989.

———. *Writing Degree Zero.* New York: Farrar, Straus and Giroux, 1988.

"Batavia en Bandoeng." *Kritiek en Opbouw,* vol. 2, no. 1 (February 16, 1939).

Baudet, H. *De lange weg naar de Technische Universiteit Delft.* Vols. I–II. Den Haag, 1992, 1993.

Bax, M. "Mondriaan and his Friends," in Baar, Bax, et al. *Piet Mondriaan: The Amsterdam Years.*

Becker, A.L. *Beyond Translation: Essays Towards a Modern Philology.* Ann Arbor: University of Michigan Press, 1995.

Beijering, M. "Overheidscensuur op een koloniale radiozender: De Philips Omroep Holland Indië en de Indië Programma Commissie, 1933–1940," in Berg, Witte, et al. *Jaarboek Media Geschiedenis.*

Bel, J. *Nederlandse literatuur in het fin de siècle.* Amsterdam: Amsterdam University Press, 1993.

Benjamin, W. "Karl Kraus," in W. Benjamin. *Reflections.* New York: Harcourt Brace Jovanovich, 1978.

———. "Paris, Capital of the Nineteenth Century," in W. Benjamin. *Reflections.*

Berg, S. van den. "Notabele ingezetenen en goedwillende ambtenaren: De Nederlands-Indische filmkeuring, 1912–1942," in Berg, Witte, et al. *Jaarboek Media Geschiedenis.*

Berg, S. van den, R. Witte, et al. *Jaarboek Media Geschiedenis: 4 Nederlands-Indië.* Amsterdam: Stichting beheer IISG, Stichting Mediageschiedenis, 1992.

Berlage, H.P. *Mijn Indische Reis: Gedachten over cultuur en kunst.* Rotterdam: W.L. en J. Brusse's, 1931.

Berretty, D.W. *Van 13 momenten uit een 13-jarig bestaan Aneta.* Batavia: Aneta, 1931.

Bettink, P.Joh. "Enkele beschouwingen over het Luchtverkeer in Nederlandsch-Indië," *Koloniale Studiën,* no. 1 (1936).

Bloembergen, M. "Exhibiting Colonial Modernity: De exposities van de Nederlandse koloniën op de wereldtentoonstellingen (1880–1931)." Unpublished Ph.D. prospectus. Amsterdam: CASA, 1995.

Blotkamp, C., ed. *De Stijl: The Formative Years.* Cambridge, Mass.: The MIT Press, 1986.

Blussé, L. "The Role of Indonesian Chinese in Shaping Modern Indonesian Life: A Conference," in Wolff, ed. *The Role of the Modern Indonesian Chinese.*

Boediardjo. "Eenige opmerkingen over en naar aanleiding van het eerste volksconcert van de 'Perserikatan Perkoempoelan Radio Ketimoeran.'" *Kritiek en Opbouw,* vol. 4, no. 1 (February 15, 1941).

Bonset, I.K. [T. van Doesburg]. "Tot een constructieve dichtkunst," *Mécano,* nos. 4/5 (1923).

"Boomvormen." *I.B.T. Locale Techniek,* vol. 9, no. 1 (February 1940).

Boon, G.A. *De Radio-Omroep voor Nederland en Indië: Nederland op z'n smalst.* Den Haag: Haagsche Drukkerij, 1930.

Boon, J. A. *Verging on Extra-Vagance: Anthropology, History, Religion, Literature, Art . . . Showbiz.* Princeton: Princeton University Press, 1999.

Bordes, J.P. de. *De spoorweg Semarang-Vorstenlanden.* 's Gravenhage, 1870.

Borel, H. *Van Batavia naar Rotterdam: reisgids van de Rotterdamsche Lloyd.* Amsterdam: Rotterdamsche Lloyd, 1905.

Bouman, H. *Meer licht over Kartini.* Amsterdam: H.J. Paris, 1954.

Bovene, G.A. van. *Nieuws! Een boek over pers, film en radio.* Batavia: G. Kolff, 1941?

Bovenkamp, A. van den. "Het dactyloscopeeren als algemeene Regeeringsmaatregel ter tijdelijke vervanging van een Burgerlijken stand voor de Inlandsche bevolking," *Koloniale Studiën,* no. 2 (1919).

Braconier, A. De. "Het pauperisme onder de in Ned. Oost-Indië levende Europeanen," *Nederlandsch Indië Oud en Nieuw* (1916–1917).

Breen, H. van. "Onderwijs in de techniek van assaineering," *Koloniale Studiën,* no. 1 (1925).

Brons Middel, R. *Ilmoe Boemi Hindia-Nederland.* Batavia: Albrecht, 1886.

Buitenweg, H. *Kind in Tempo Doeloe.* Wassenaar: Service, 1969.

———. *Soos en samenleving in tempo doeloe.* Den Haag: Service, 1966.

"Bungalow-itis." *Nationale Commentaren,* vol. 4, no. 44 (November 1, 1941).

Burkom, F. van, and W. de Wit. "Vormgeving als kunst, kunst als vormgeving," in Burkom, de Wit, et al. *Amsterdamse School: Nederlandse architectuur 1910–1930.*

Burkom, F. van. "Kunstvorming in Nederland," in Burkom, de Wit, et al. *Amsterdamse School: Nederlandse architectuur 1910–1930.*

Burkom, F. van, W. de Wit, et al. *Amsterdamse School: Nederlandse architectuur 1910–1930.* Amsterdam: Stedelijk Museum, 1975.

Buys, M. *Batavia, Buitenzorg en de Preanger-Gids voor bezoekers en toeristen.* Batavia, Kolff, 1891.

Catenius-van der Meijden, J.M.T. *Ons Huis in Indië: Advertentie.* n.p., n.d.

———. *Ons Huis in Indië. Handboek bij de keuze, de inrichting, de bewoning en de verzorging van het huis met bijgebouwen en erf, naar de eischen der hygiëne, benevens raadgevingen en wenken op huishoudelijk gebied.* Semarang: Masman en Stroink, [1908].

Chambert-Loir, H. "Sair Java-Bank di rampok: Littérature malaise ou sino-malaise?" in Salmon, ed. *Le moment "sino-malais" de la littérature indonesiènne.* Paris: Cahier d'Archipel, 1992.

Clay, J. "De zonsverduistering van 21 September j.l. en te beteekenis van de expeditie ter waarneming van het Einstein-effect," *Ganeça*, vol. 1, no. 5 (October 15, 1922).

Coast, J. *Recruit to Revolution: Adventure and Politics in Indonesia.* London: Christophers, 1952.

Contril, H., and G.W. Alport. *The Psychology of Radio.* New York: Harper, 1935.

Corpslid. "Over de gezondheid van het Bandoengsche Studenten Corps," *Ganeça*, vol. 6, no. 7 (July 1934).

Couperus, L. *Eastward.* New York: G.H. Doran, 1924.

———. *The Hidden Force.* Amherst: University of Massachusetts Press, 1985.

———. "Meditatie over het mannelijk toilet," in *Kleeding en de man.* 's Gravenhage: Magazijnen 'Nederland,' 1915.

Cribb, R. "Telling the Time of Day in Indonesia." Unpublished draft, 1993.

Cuypers, Ed. "De Moderne Ambtenaarswoning in Nederlandsch-Indië," *Nederlandsch Indië Oud en Nieuw* (1919–1920).

Damhoeri, A. *Zender NIROM.* Medan: Tjendrawasih, 1940.

Daum, P.A., see Maurits.

"De Bevolking en de Volkshuizing." *Soeloeh Indonesia*, vol. 2, no. 3 (March 1927).

"De inlandsche klerken en javaansche bestellers in onzen dienst." *Indisch Tijdschrift voor Post en Telegraphie: orgaan van den Bond van Ambtenaren en Beamten*, vol. 2, no. 9 (December 1906).

"De Nieuwe Uitvindingen." *Kopiïst*, vol. 1, nos. 6, 8 (1842).

"De uitbreiding der Indische kustverlichting." *De Indische Gids*, vol. 25, no. 2 (1903).

Decuop, J., and W. Van der Hoeven. "De betrouwbaarheidsrit van den Gouverneur-Generaal door Midden Java," *Magneet*, vol. 2, no. 8 (September 18, 1915).

Deelman, A.W. "Radiorede van Hoofd van het Kantoor Bestuurszaken van het Secretariaat van den Staatsmobilisatieraad. Evacuatie van de Burgerbevolking," *Nationale Commentaren*, vol. 4, no. 39 (September 27, 1941).

Defoe, D. *Robinson Crusoe.* Danbury, Conn.: Grolier, n.d.

Derrida, J. "The Voice That Keeps Silence," in Derrida. *Speech and Phenomenon.* Evanston: Northwestern University Press, 1973.

Dewantara. "Associatie antara Timoer dan Barat," *Wasita*, vol. 1, nos. 11/12 (August–September 1929).

Dierikx, M. *Bevlogen jaren: Nederlandse burgerluchtvaart tussen de wereldoorlogen.* Romen: Luchtvaart, n.d.

Dimyati, M. *Dibalik Tabir Gelombang Radio.* Atjeh: Indonesia, 1359 [1939].

Dinger, J.E., et al. "Physiologisch reacties van gezonde personen op luchtbehandeling," *De Ingenieur in Nederlandsch Indië*, vol. 5, no. 9 (September 1938).

Diningrat, Noto. "Grondslagen voor de bouwkunst op Java," *Nederlandsch Indië Oud en Nieuw* (1919–1929).

Djojopoespito, Soewarsih. *Buiten het Gareel. Indonesische Roman Met een Inleiding van E. Du Perron.* Utrecht: De Haan, 1940.

———. "De taal de Soendanese jongeren," *Kritiek en Opbouw*, vol. 1, no. 23 (January 12, 1939).

———. "De thuiskomst van een oud-strijder," *Tirade*, vol. 21, no. 221 (January 1977).

Djojopoespito, Soegondo. "Soeroto's Indonesiër," *Kritiek en Opbouw*, vol. 2, no. 3 (March 16, 1939).

Doesburg, T. van, see I.K. Bonset.

"Dokar dan kossongan moesti di keur dan koesirnja pakeh rijbewijs. Gemeente haroes perhatikan." *Motorblad* (September 22, 1928).

Doorn, A. van. "Hoe men soms bouwt," *Tecton*, vol. 1, no. 8 (August 1936).

Doorn, J.A.A. van. *The Engineers and the Colonial System: Technocratic Tendencies in the Dutch East Indies.* Rotterdam: Faculty of Social Science CASP 6, 1982.

———. *De laatste eeuw van Indië: Ontwikkeling en ondergang van een koloniaal project.* Amsterdam: Bert Bakker, 1994.

"Economische aspecten van de Electriciteit-voorziening van Nederlandsch-Indië." *Koloniale Studiën*, no. 2 (1934).

"Een gedenkdag op spoorweggebied." *De Indische Gids*, vol. 14 (1892), 1051.

"Een jaar Luchtverkeer, 1929–1930." *K.N.I.L.M. Reisgids 1930.* Batavia: K.N.I.L.M., 1930.

Elias, P.J. *Dan liever de lucht in. Herinneringen van een marinevlieger.* Amsterdam: van Kampen en Zoon, 1963.

Empatpuluhlima Tahun Sumpah Pemuda. Jakarta: Gunung Agung, 1974.

Estépé. "Te vroeg?" *Soeloeh Indonesia*, vol. 2, no. 1 (January 1927).

Ewen, F. *Bertold Brecht: His Life, His Art, His Time.* New York: Citadel, 1992.

Foucault, M. *Discipline and Punish: The Birth of the Prison.* New York: Vintage, 1979.

"Gantinja pakaian." *Saro Tomo*, vol. 1, no. 10 (1915).

Gelder, W. van. *Dari Tanah Hindia Berkoeliling Boemi: Kitab Pengadjaran Ilmoe Boemi bagi sekolah anak negeri di Hindia-Nederland.* 3rd ed. Wolters Groningen, 1892.

Hans den Boef, A., and K. Snoek. *Pramoedya Ananta Toer: Essay en Interview.* Novir: De Geus, 1992.

Heckler, F.A.J. *De Deli Spoorweg Maatschappij: Driekwart eeuw koloniaal spoor.* Zutphen: De Walburg Pers, 1987.

Heide, Homan van der, J. *Beschouwingen aangaande de volkswelvaart en het irrigatiewezen op Java in verband met de Solovalleiwerken.* Batavia: Kolff, 1899.

———. "Landbouwtoestanden in Achter-Indië, beschouwd in verband met Java," *Handelingen Indisch Genootschap* (March 28, 1905).

Heidegger, M. *Poetry, Language, Thought.* Translated by Albert Hofstadler. New York: Harper 1971.

———. "The Question Concerning Technology," in Heidegger. *The Question Concerning Technology and Other Essays.* Translated by William Lovitt. New York: Harper, 1977.

———. "Building, Dwelling, Thinking," in Heidegger. *Poetry, Language, Thought.*

Hen, I. "Iets over het woningvraagstuk in de Indische gemeenten," *Koloniale Studiën*, no. 1 (1916–1917).

Henry, E.R. Sir. *Classification and Uses of Fingerprints.* London: Darling and Son, 1913.

Hesketh Bell, H. Sir. *Foreign Colonial Administration in the Far East.* London: Arnold, 1928.

"Het bebouwingsplan voor het Koningsplein te Batavia volgens het ontwerp Fermont-Cuypers." *De Ingenieur in Nederlandsch Indië*, vol. 5, no. 2 (February 1938).

"Het dubbelschroefstoomschip 'Insulinde' van de Rotterdamsche Lloyd." *Nederlandsch Indië Oud en Nieuw* (1916–1917).

"Het koloniaal-geneeskundig museum te Amsterdam." *De Indische Gids*, vol. 13, no. 2 (1891).

"Het landbouw leer- en werkkamp 'Oost-Java' te Poerwodadi bij Lawang." *Onze Toekomst* (February 25, 1939).

"Het officiële relaas van de onlusten te Tjilegon en pogingen tot oproer in Midden- en Oost-Java." *De Indische Gids*, vol. 11, no. 2 (1889).

"Het tentoonstellen van inlanders in Europa." *De Indische Gids*, vol. 15, no. 1 (1893).

"Het woningvraagstuk en de Pestbestrijding," *Koloniale Studiën*, vol. 1, no. 2 (1916–1917).

Hoe, E. "Piet Mondrian," in Blotkamp, ed. *De Stijl: The Formative Years.*

"Hoeveel personen er gedurende 1878 in Nederlandsch-Indië omkwamen door 'ongelukken.'" *De Indische Gids*, vol. 1, no. 2 (1879).

"Hollandse IJzeren Spoorweg-Maatschappij en Maatschappij tot Exploitatie van Staatsspoorwegen." *Uitreiking der bekroningen aan de Nederlandsche inzenders van de internationale Koloniale en Uitvoerhandel Tentoonstelling.* Amsterdam: n.p., 1883.

Holub, M. "Brief Thoughts on the Word Pain," in Holub, Miroslav. *Notes of a Clay Pigeon.* London: Secker and Warburg, 1985.

Hoogestraaten, L. "Rond de Gedeh," *Magneet*, vol. 1, no. 13 (June 15, 1914).

Hucht, K.A. van der, and C.L.M. Kerkhoven. "De Bosscha-Sterrenwacht: Van thee tot sterrenkunde," *Zenit*, vol. 9 (1982).

Idsinga, T. and J. Schilt. *Architect W. van Tijen 1894–1974.* 's Gravenhage: n.p., 1987.

IJzerman, J.W. *De reis om de wereld.* 's Gravenhage: M. Nijhoff, 1926.

———. *Dwars door Sumatra: Tocht van Padang naar Siak.* Haarlem: F. Bohn, 1895.

"Indië's wereldrecords." *Ganeça*, vol. 6, no. 1 (January 1935).

Indische Bouwen: Architectuur en stedebouw in Indonesië. Amsterdam: Nationale Contactcommissie Monumentenbescherming, 1990.

"Indonesia en Associatie." *Soeloeh Indonesia Moeda,* no. 12 (November–December 1928).

Indonesia-Poetera. "Naar het Bruine Front," *Soeloeh Indonesia Moeda,* no. 1 (December 1927).

"Indonesiërs als vliegeniers." *Nationale Commentaren*, vol. 4, no. 11 (March 15, 1941).

Ingleson, J. *Road to Exile: The Indonesian Nationalist Movement, 1927–1934.* Singapore: Heinemann, 1979.

"Ir.M. Anwari." *Ganeça*, vol. 6, no. 11 (November 1934).

Jansen, L.F. *In deze halve gevangenis: Dagboek van mr dr L.F. Jansen, Batavia/ Djakarta 1942–1945.* Franeken: Van Wijnen, 1988.

Janssen, E. ""Luchtbehandeling en Geneeskunde," *De Ingenieur in Nederlandsch Indië*, vol. 5, no. 9 (September 1938).

"Javasche Schetsen: Batavia-De Harmonie (het Gebouw)." *Kopiïst*, vol. 1, no. 1 (1842).

Jenkins, D. *Suharto and His Generals: Indonesian Military Politics 1975–83.* Ithaca: Cornell Modern Indonesia Project. 1984.

Junus Nur Arif. "Gedung bioskop di Djakarta dahulu," *Djaja*, no. 387 (1969).

Juynboll, H.H. *De Batik-Kunst in Nederlandsch-Indië en haar Geschiedenis op grond van Materieel aanwezig in 's Rijks Ethnographisch Museum en andere Openbare Verzamelingen in Nederland.* 2 vols. Haarlem: Kleinmann; Utrecht: Oosthoek, 1899, 1914.

Kaden, M. "An Evening Breeze from Southeast." Leiden: Proceedings of the European Conference on Southeast Asian Studies, 1995.

Kadt, J. de (J.de K.). "Kleine Niromiade," *Kritiek en Opbouw*, vol. 5, no. 3 (February 5, 1942).

Kafka, F. *I Am a Memory Come Alive. Autobiographical Writings*. New York: Schocken, 1976.

Kahin, G.McT. *Nationalism and Revolution in Indonesia*. Ithaca: Cornell University Press, 1952.

Kalff, A. "Hollandsche Boeren op Java," *De Indische Gids*, vol. 16, no. 2 (1894).

Kalpana, P., and E. Elburg, "Gesprekken met Pramoedya Ananta Toer," 36. *Bzzletin*, vol. 10 (1988).

Kant, I. *Prolegomena: To Any Future Metaphysics That Can Qualify as a Science*. Translated by Paul Carus. La Salle: Open Court, 1989.

Karsten, H.Th. "Stedebouw," in Kerchman, ed. *25 Jaren Decentralisatie in Nederlandsch-Indië*.

Kartini, R.A. *Brieven aan mevrouw R.M. Abendanon-Mandri en haar echtgenoot*. Doordrecht: Foris, 1987.

———. *Letters of a Javanese Princess*. Kuala Lumpur: Oxford University Press, 1976.

———. *Letters of a Javanese Princess*. New York: W.W. Norton, 1964.

Karundeng. *Stenografia*. 25th ed. Jakarta: Pradnya Paramita, 1990.

Katja Soengkana S.D.K. "Dr. Tjipto Oordeelt," *Soeara Parindra*, vol. 1 (October 1936).

Katjoeng. "Bagaimana Wet menjerahkan nasib Kaoem Boeroeh kepada kelalimannja Madjikan," *Soeloeh Indonesia*, vol. 2, no. 6 (June 1927).

"Kazerne-taal Atavisme." *Soeloeh Indonesia*, vol. 2, no. 9 (September 1927).

Kelling, M.A.J. "De Eerste Hygiëne Tentoonstelling in Nederlandsch-Indië," *Koloniale Studiën*, no. 2 (1927).

———. "Het Jaarbeurswezen in Nederlandsch-Indië," *Koloniale Studiën*, no. 2 (1925).

Kerchman, F.W.M. (ed.). *25 Jaren Decentralisatie in Nederlandsch-Indië, 1905–1930*. Weltevreden: Kolff en Vereeniging voor Locale Belangen te Semarang, 1931.

Ketjindoean. *Patjar Koening. Serie V. Kedjadian jang aneh diboekakan oleh detektif Raden Pandji Soebrata*. Djokjakarta: 'Kabe' Kolff-Buning, 1940?

"Ketoea kita Dr. R. Soetomo kembali di tanah air." *Soeara Parindra*, no. 2 (April 1937).

Kierkegaard, S. *The Concept of Irony with Continual Reference to Socrates*. Translated by Howard V. Hong and Edna Hong. Princeton: Princeton University Press, 1989.

Klay-Lancée, C.C. " Indië 25 jaar geleden," *Officieel orgaan van de Vereeniging van Huisvrouwen Bandoeng* (October 1941).

Kleist, H. von. "On the Marionette Theatre" (1810), in *Essays on Dolls*. Translated by I. Parry and P. Keegan. London: Penguin, 1994.

Klinkhammer, J.F., and B.J. Ouëndag. "Het Administratiegebouw der Nederlandsch-Indische Spoorweg-Maatschappij te Semarang," *Nederlandsch Indië Oud en Nieuw* (1916–1917).

Koesna Ardja, K. "KOERANG POETIH, Apakah ini boekan Anti Circulaire No. 2014?" *Doenia Bergerak*, vol. 1, no. 19 (1915).

Kol, H.H. van. *Uit onze koloniën*. Leiden: A.W. Sijthoff, 1903.

Kol, Nellie, see Nellie.

"Kosten van de Europeesche gevangenen in Ned.-Indië per hoofd en per dag." *De Indische Gids*, vol. 15, no. 2 (1893).

Lacey, A.R. *Bergson*. London: Routledge, 1993.

Langan, C. *Romantic Vagrancy: Wordsworth and the Simulation of Freedom*. Cambridge: Cambridge University Press, 1995.

Langen, C.D. de. "Tuberculose in Nederlandsch-Indië en hare bestrijding," *Koloniale Studiën*, vol. 2, no. 1 (1918).

Le Corbusier. *Towards a New Architecture*. New York: Dover, 1986.

Leclerc, J. "La clandestinité et son double: à propos des relations d'Amir Sjarifuddin avec le communisme Indonésien," *Asian Thought and Society: An International Review*, no. 6 (1981).

LeGoff, J. *History and Memory*. New York: Columbia University Press, 1992.

Lelijke Tijd: Pronkstukken van Nederlandse interieurkunst 1835–1895. Amsterdam: Waanders, 1995.

Lelyveld, Th.B. van. "De indische bouwkunst en kunstnijverheid, oud en nieuw," *Nederlandsch Indië Oud en Nieuw* (1916–1917).

Leuwen, A.J.H. van. "De Radio-Omroep in Nederlandsch-Indië I," *Kritiek en Opbouw*, vol. 1, no. 11 (July 15, 1938).

Liederen USI. Batavia: Unitas Studiosorum Indonesiensis, n.d.

Ligthart, J., H. Scheepstra, and A.F.Ph. Mann. *Ver van Huis*. Weltevreden: Wolters, 1918, etc.

Lipovetsky, G. *The Empire of Fashion: Dressing Modern Democracy*. Princeton: Princeton University Press, 1994.

Lukisan Revolusi Indonesia, 1945–1950. Jakarta: Kementerian Penerangan Republik Indonesia, n.d.

Maclaine Pont, H. "Het nieuwe Hoofdbureau der Semarang-Cheribon-Stoomtram Maatschappij te Tegal," *Nederlandsch Indië Oud en Nieuw* (1916–1917).

"Magnetische en meteorologische waarnemingen in 1899." *De Indische Gids*, vol. 22, no. 2 (1900).

Maier, H.M.J. "Phew! Europeesche beschaving! Marco Kartodikromo's Student Hidjo," *Southeast Asian Studies* (Tonan Ajia Konkyu), vol. 34, no. 1 (June 1996).

"Malam Kesenian Timoer." *Soeara Timoer*, vol. 1, no. 2 (January 12, 1941).

Malik, A. *Riwajat dan Perdjuangan Sekitar Proklamasi Kemerdekaan Indonesia*. N.p., 1948 mimeo.

Mangoensarkoro, S. "Koentoem bahasa Indonesia I," *Keboedajaän dan Masjarakat*, vol. 1, no. 6 (October 1939).

Manusama, A.Th. *Komedie Stamboel of de Oost Indische opera*. Weltevreden: Favoriet, 1922.

Marco Kartodikromo. *Student Hidjo*. Semarang: Masman en Stroink, 1919.

———. "Koloniale tentoonstelling boekan boetoehnja orang Djawa," *Doenia Bergerak*, vol. 1, no. 20 (1915).

———. "Pro of contra Dr.RINKES!?" *Doenia Bergerak*, vol. 1, no. 0 (January 31, 1914).

———. "Selembar soerat kepada joernaliste of redactrice," *Doenia Bergerak* vol. 1, no. 0 (January 31, 1914).

Masak, Tanete A. Pong. *Le cinéma Indonésien (1926–1967): Études d'Histoire Sociale*. Thèse de Doctorat. Paris: l'École des Hautes Études en Sciences Sociales, 1989.

Maulina, D., et al. *Roosseno: pakar dan perintis teknologi sipil Indonesia*. Jakarta: Pembimbing Masa, 1989.

Maurits (P.A. Daum). *H. Van Brakel: Ing. B.O.W.* Amsterdam, 1976.

McIntyre, A. "Sukarno as Artist-Politician," in Angus McIntyre, ed. *Indonesian Political Biography: In Search of Cross-Cultural Understanding*. Clayton: Monash Papers on Southeast Asia No. 28, 1993.

McLuhan, M. *The Mechanical Bride: Folklore of Industrial Man*. London: Routledge and Kegan Paul, n.d.

McVey, R. T. "Taman Siswa and the Indonesian National Awakening," *Indonesia*, no. 4 (October 1967).

Meijer, D.H. "Dactyloscopie voor Nederlandsch-Indië," *Koloniale Studiën*, no. 2 (1926).

———. "Internationale politioneele samenwerking," *Koloniale Studiën*, vol. 19, no. 2 (1935).

Meijer, Ir.H. *De Deli Spoorweg Maatschappij: Driekwart eeuw koloniaal spoor*. Zutphen: De Walburg Pers, 1987.

Meyier, J.E. de. "Irrigatie-fanatisme," *Indische Gids*, vol. 24, no. 2 (1902).

Misset, I.H. *Departement van Binnenlandsch Bestuur. Politieschool. Het Centraal kantoor voor de dactyloscopie bij de Politie*. Weltevreden: Visser, 1917.

Moet, J.F.F. "Iets over een stoomtram op Java," *De Indische Gids*, vol. 2, no. 2 (1880).

Mom, C.P. "De Bacteriën en de Hygiëne," *De Ingenieur in Nederlandsch Indië*, vol. 5, no. 5 (May 1938).

———. "De Ontwikkeling der Luchtconditioneering in 1937 in Noord-Amerika," *De Ingenieur in Nederlandsch Indië*, vol. 5, no. 5 (May 1938).

———. "Over den invloed den luchtelectriciteit op de behaaglijkheid en de gezondheid," *De Ingenieur in Nederlandsch Indië*, vol. 5, no. 9 (September 1938).

Morrell, Ch.M. "The 100th Flight: The Royal Dutch Airways Great Achievement," *The Java Gazette*, vol. 2 (n.d.).

Mrázek, R. *Sjahrir: Politics and Exile in Indonesia, 1906–1966*. Ithaca: Cornell University Studies in Southeast Asia, 1994.

Musil, R. *On Mach's Theories*. Translated by Kevin Mulligan. Washington, D.C.: Catholic University of America Press, 1982.

———. *The Man Without Qualities*. New York: Putnam's, 1980.

"Nasib kaoem Sopir." *Sopir*, vol. 1, nos. 1–8 (April 1932).

"Nationalisme." *Indonésia Raja*, vol. 1, no. 9 (December 1929).

Naudin ten Cate, A.W. "De automobiel in Indië van 1902–1927," *De Indische Mercuur* (Gedenknummer, 1928).

Nellie (Nellie van Kol). *Brieven aan Minette* (reprint 'Soerabajasch Handelsblad*). 's Gravenhage: n.p., 1884.

Nieuwenhuijs, R. "Over de Europese samenleving van 'tempo doeloe' 1870–1900," *Fakkel*, vol. 1, no. 9 (July–August 1941).

———. "Tempo Doeloe XII," *Kritiek en Opbouw*, vol. 3, no. 1 (February 16, 1940).

———. *Baren en oudgasten: tempo doeloe-een verzonken wereld*. Amsterdam: Querido, 1981.

Nieuwenhuis, Dr. J.J. "Lichamelijke opvoeding en de onderwijshervorming in Ned.-Indië II," *Koloniale Studiën*, no. 1 (1920).

"Nieuws uit de werk- en leerkampen: Het werk- en leerkamp in Oost Java." *Onze Toekomst* (January 2, 1936).

Nio Joe Lan. "De vestiging van een Indische Filmindustrie," *Koloniale Studiën*, vol. 25, no. 5 (1941).

Numans, J.G. "Een en ander over irrigatiewerken op Java," *Nederlandsch-Indië Oud en Nieuw* (1918–1919).

Numans, J.J. "Radiotelefonie Holland-Indië," *Radio Nieuws* (January 1928).

Nus, J. van. "De Staatsspoorwegen in Nederlandsch-Indië gedurende de crisisjaren 1930 t/m 1934," *Koloniale Studiën*, no. 1 (1935).

Oetomo, D. "The Chinese of Indonesia and the Development of the Indonesian Language," in Wolff, ed. *The Role of the Modern Indonesian Chinese.*

Oetterman, S. *Das Panorama: Die Geschichte eines Massenmedium.* Frankfurt: Syndicat, 1980.

"Oorlog en B.R.V." *Radio Bode*, vol. 14, no. 25 (June 16, 1940).

Oudemans, A.C. *Ilmoe Alam of Wereldbeschrijving voor de inlandsche scholen.* Batavia: Landsdrukkerij, 1873.

———. *Ilmoe Alam. Terkarang pada Bahasa Belanda oleh Dr. A.C. Oudemans dan dikarangkan pada Bahasa Melajoe oleh A.F. van Dewall.* 4th ed. Betawi: Pertjetakan gouvernement, 1909.

Oudemans, J.A.C. *Die Triangulation von Java ausgeführt vom Personal des geographischen Dienstes in Niederländisch Ost-Indien.* 6 vols. Batavia: Staats-Druckerei, 1875–1900.

"Over *acclimatisatie* in de tropen." *De Indische Gids*, vol. 15 (1893).

Over de wereld klinkt Nederlands stem PHOHI. Typescript. Hilversum: Directie PHOHI, 1935.

Overy, P. *De Stijl.* London: Thames and Hudson, 1991.

P.K. "Stadskampongs," *Tecton*, vol. 2, no. 11 (November 1937).

Pané, A. "De Poedjangga Baroe," *Fakkel*, vol. 1, no. 9 (July–August 1941).

Pattist, M.P. "Civiel Luchtverkeer in Nederlandsch-Indië," *Koloniale Studiën*, no. 1 (1927).

Pelt, R.J. van, and C.W. Westfall. *Architectural Principles in the Age of Historicism.* New Haven: Yale University Press, 1993.

Pemberton, J. *On the Subject of "Java."* Ithaca: Cornell University Press, 1994.

Perrault, C. *Cendrillon* [1697]. London: Jonathan Cape, 1980.

Perron, E. du. *Brieven.* Vols. VI, VII. Amsterdam: Van Oorschot, 1980, 1981.

———. *Indies Memorandum.* Amsterdam: De Bezige Bij, 1946.

"Persatoean Chauffeur Mataram." *Sopir*, vol. 1, no. 1 (April 1932).

Pioli, R.J. *Stung by Salt and War: Creative Texts of the Italian Avant-Gardist F.T. Marinetti.* New York: Lang, 1987.

Plas, Ch. van der (Ch.O.v.d.P.). "Om de zuiverheid van onze taal!" *Fakkel*, vol. 1, no. 6 (April 1941).

———. "Opmerking over The Tempest," *Fakkel*, vol. 1, no. 5 (March 1941).

———. "Radiorede," *Nationale Commentaren*, vol. 4, no. 45 (November 8, 1941).

Plate, A. "Het uitbreidingsplan der Indische gemeente," *Indisch Genootschap* (October 16, 1917).

Poldervaart, A. "De tjikar als wegvernieler," *Locale Techniek*, vol. 7, no. 5 (September–October 1938).

Pramoedya Ananta Toer. "De kinder chief in een rode auto: Ervaringen met Nederland als staat en volk," *Eindhoven Dagblad Boekenweekbijlage* (March 10, 1992).

——. "Jang Hitam," in *Tjerita dari Blora*. 2nd ed. Jakarta: Balai Pustaka, 1963.

——. "Kapal Gersang," *Zenith* (Jakarta), vol. 3, no. 12 (1953).

——. "Ke arah sastra revolusioner," *Star Weekly*, vol. 11, no. 574 (1956).

——. "Kemunduran kesusasteraan Indonesia modern dewasa ini," *Duta Suasana*, vol. 2, no. 25 (1953).

——. "Lesu; kelesuan; krisis; impasse," *Siasat*, vol. 11, nos. 515/516 (1957).

——. "Perburuan 1950 and Keluarga Gerilya 1950," *Indonesia*, no. 36 (October 1983).

——. "Sumber tjipta dalam kesenian," *Indonesia* (Jakarta), vol. 3 (1952).

——. "Tendensi kerakjatan," *Star Weekly*, vol. 11, no. 525 (1956).

——. "Tentang Emansipasi Buaja," *Kisah* (Jakarta), vol. 3, no. 2 (December 1953).

——, ed. *Tempo doeloe: antologi sastra pra-Indonesia*. Jakarta: Hasta Mitra, 1982.

——. *Bio en Bibliografie*. Typescript, 1959.

——. *House of Glass*. New York: Penguin, 1992.

——. *Larasati (Ara): Roman Revolusi* [1960]. Jakarta: Hasta Mitra, 2000.

——. *Lied van een Stomme: Brieven van Buru*. Houten: Manus Amici, 1991.

——. *Nyanyi Sunyi Seorang Bisu*. Vols I, II. Jakarta: Lentera, 1995, 1997.

——. *Nyanyi Tunggal Seorang Bisu*. Edisi Malaysia, 1988.

——. *Panggil aku Kartini sadja*. Vols. I, II. Jakarta: Nusantara, 1962.

——. *Realisme-Sosialis dan Sastra Indonesia: sebuah tindjauan sosial*. Jakarta: Seminar Fakultas Sastra, Universitas Indonesia, 1963, mimeo.

——. *Rumah kaca*. Jakarta: Hasta Mitra, 1988.

——. *Sang Pemula dan karya-karya non-fiksi (journalistik), fiksi (cerpen, novel) R.M. Tirto Adhi Soerjo*. Jakarta: Hasta Mitra, 1985.

——. *Sedjarah Modern Indonesia*. Jakarta: Universitas "Res Publica," 1964, mimeo.

Pringgodigdo, S. "Over du Perron en zijn invloed op de Indonesische intellectuellen (1936–39)," *Cultureel Nieuws*, no. 16 (January 1952).

Proust, M. *Swann's Way*. New York: The Modern Library, 1992.

Pyenson, L. *Empire of Reason: Exact Sciences in Indonesia, 1840–1940*. Leiden: Brill, 1989.

Radsma, W., et al. "Metingen van Lichaamstemperatuur bij Blanken in Tropische en in Koelere Luchtstreken," *De Ingenieur in Nederlandsch Indië*, vol. 5, no. 9 (September 1938).

Reitsma, S.A. *De verkeersbedrijven van den staat: spoorwegen, post-, telegraaf- en telefoondienst, havenwezen*. Weltevreden: G. Kolff, 1924.

——. "Rede van den wd.Burgemeester van Bandoeng, den Heer S.A.Reitsma," *Orgaan van de Vereeniging 'Nederlandsch-Indische Jaarbeurs'*, vol. 1, no. 4 (February 15, 1922).

Rietsema van Eck, S. *Koloniaal-staatkundig studiën 1912–1918*. Buitenzorg: n.p., 1919.

Rimember and Henvo. *Hallo!—Batavia*. N.p., n.p., 1929.

Roding, J. *Schoonheid en net: hygiëne in woning en stad*. 's Gravenhage: Staatsuitgeverij, 1986.

Ronell, A. *The Telephone Book: Technology-Schizophrenia-Electric Speech*. Lincoln: University of Nebraska Press, 1989.

Rorty, R. *Philosophy and the Mirror of Nature*. Princeton: Princeton University Press, 1979.

Rush, J.R. *Opium to Java: Revenue Farming and Chinese Enterprise in Colonial Indonesia, 1860–1910*. Ithaca: Cornell University Press, 1990.

Rusli, Mh. *Sitti Nurbaya: Kasih Tak Sampai* [1922]. 24th ed. Jakarta: Balai Pustaka, 1994.

Rusman, E. *Hollanders vliegen: van H-NACC tot Uiver: tien jaar Amsterdam-Batavia door de lucht*. Baarn: Bosch en Keuning, 193?.

Rutherford, D. "Unpacking a National Heroine: Two Kartinis and Their People," *Indonesia*, no. 55 (April 1993).

Salmon, C. *Literature in Malay by the Chinese of Indonesia: A Provisional Annotated Bibliography*. Paris: Archipel, 1981.

———. "A Critical View of the Opium Farmers as Reflected in a *Syair* by Boen Sing Hoo (Semarang 1889)," in Wolff, ed. *The Role of the Modern Indonesian Chinese*.

———. "The Batavian Eastern Railway Co. and the Making of a New 'Daerah' as Reflected in a Commemorative Syair Written by Tan Teng Kie in 1900," *Indonesia*, no. 45 (April 1988).

Samkalden, H. (Sk.). "Notitie over laster," *Kritiek en Opbouw*, vol. 2, no. 1 (February 16, 1939).

Sandick, R.A. van. *In het Rijk van Vulcaan: de uitbarsting van Krakatau en hare gevolgen*. Zutphen: Thieme, 1890.

———. "Indische schetsen," *Vragen van den Dag: Populair Tijdschrift*. Amsterdam, 1891.

Sarca, R. "Idees uit m'n dagboek," *Ganeça*, vol. 1, no. 5 (October 15, 1922).

———. "Noot van Sarca aan Soekarno Brave." *Ganeça*, vol. 1, no. 6 (November 1, 1922).

Scheltema, J.F. "Nederlandsch-Indië en de Wereldtentoonstelling te Parijs," *De Indische Gids*, vol. 22, no. 2 (1900).

Schivelbusch, W. *Disenchanted Night: The Industrialization of the Light in the Nineteenth Century*. Berkeley: University of California Press, 1988.

———. *The Railway Journey: The Industrialization of Time and Space in the Nineteenth Century*. Berkeley: University of California Press, 1986.

Schoemaker, C.P. Wolff. "De Aesthetiek der Architectuur en de Kunst der Modernen," in *Jaarboek der Technische Hoogeschool te Bandoeng 1930*.

Scott, J.C. *Seeing Like a State: How Certain Schemes to Improve Human Conditions Have Failed*. New Haven: Yale University Press, 1998.

"Sedikit bersoewara." *Orgaan dari Perhimpoenan Pegawai Post Telegraaf Telefoon Radiodienst Rendahan (P.T.T.R.)*, vol. 1, no. 1 (September 1937); vol. 2, no. 4 (April 1938); vol. 2, no. 12 (December 1938).

Shiraishi, T. *An Age in Motion: Popular Radicalism in Java, 1912–1926*. Ithaca: Cornell University Press, 1990.

Sidik Kertapati. *Sekitar Proklamasi 17 Agustus 1954*. 3rd ed. Jakarta: Pembaruan, 1956.

Siegel, J. *Fetish Recognition Revolution*. Princeton: Princeton University Press, 1997.

———. "Georg Simmel Reappears: 'The Aesthetic Significance of the Face." *Diacritics*, vol. 29, no. 2 (1999).

Sjafroeddin, P.N. "Mis- en verkenningen," *Orgaan van de USI*, vol. 7, no. 1 (November 1939).

Sjahrir, S. (Sjahrazad). *Indonesische Overpeinzingen*. Amsterdam: De Bezige Bij, 1945.

————. *Renungan Indonesia*. Translated by H.B. Jassin. Djakarta: Poestaka Rakjat, 1947.

Sjarifoeddin, A. (A.S.). "Dari Kaboepaten ke Microfoon," *Soeara Timoer*, vol. 1, no. 2 (January 12, 1941).

Sloos, J.M. "De invloed van de auto's op het tegenwoordige verkeerswezen," *Koloniale Studiën*, no. 2 (1923).

Sluijs, P.J.A. "Over Acclimatisatie," *De Indische Gids*, vol. 14, no. 1 (1892).

Soedjojono (S.S.101). "Kesenian meloekis di Indonesia: sekarang dan jang akan datang," *Keboedajaän dan Masjarakat*, vol. 1, no. 6 (October 1939).

Soejitno, Mangoenkoesoemo. "Nogmals over Indonesiërs," *Kritiek en Opbouw*, vol. 2, no. 10 (July 1, 1939).

Soekarno. "TAO Rede," *Ganeça*, vol. 1, no. 2 (August 15, 1922); no. 3 (September 1, 1922).

Soesilo. "De maatschappelijke evolutie in de Indische stad," *Kritiek en Opbouw*, vol. 1, no. 17 (October 16, 1938).

————. "De Maleische taal als Indonesische taal," *Kritiek en Opbouw*, vol. 2, no. 7 (May 16, 1939).

Soetardjo, Kartohadikoesoemo. "Pidato Ketoea P.P.R.K. pada pemboekaän malam 'Kesenian Timoer' di Stadsschouwburg, pada tanggal 11 Januari 1941," *Soeara Timoer*, vol. 1, no. 3 (January 26, 1941).

Sovjet Architectuur, 1917–1987. Amsterdam: Art Unlimited Books, 1989.

Sparnaay, D.W. "De electrificatie van 'het platteland,'" *De Ingenieur in Nederlandsch Indië*, vol. 5, no. 7 (July 1938).

Stafford, B.M. *Artful Science: Enlightenment Entertainment and the Eclipse of Visual Education*. Cambridge, Mass.: The MIT Press, 1994.

Stevens, H. "Gietijzer voor Indische vuurtoren," in *Techniek in schoonheid: De Marinemodellenkamer van het Rijksmuseum te Amsterdam*. Wormer: Immerc, 1995.

Stibbe, D.G. "Een mogelijke Bezuiniging op de voeding de gevangenen in Indië," *De Indische Gids*, vol. 15, no. 2 (1893).

Struik, D.J. (ed.). *Birth of the Communist Manifesto*. New York: International Publishers, 1971.

Stutterheim, W.F. "Inleiding bij de opening van de tentoonstelling van Hindoe-Javaansche kunst," *Fakkel*, vol. 1, no. 5 (March 1941).

Sukarno, see also Soekarno

————. *An Autobiography as Told to Cindy Adams*. Indiannapolis: Bobbs-Merrill, 1965.

Sutherland, H. *The Making of a Bureaucratic Elite*. Kuala Lumpur: Heinemann, 1979.

Sutton, J. *Philosophy and Memory Traces: Descartes to Connectionism*. Cambridge: Cambridge University Press, 1999.

Tas, S. "Souvenirs of Sjahrir," *Indonesia*, no. 8 (1969).

Teeuw, A. *Modern Indonesian Literature*, vol. 1. Dordrecht: Foris, 1986; vol. 2. The Hague: Nijhoff, 1979.

Termorshuizen, G. "Een leven buiten het gareel," in *Engelbewaarder Winterboek 1979. Extra-uitgave van het kwartaalschrift De Engelbewaarder*. Amsterdam, 1979.

————. "A Murder in Batavia or the Ritual of Power," in Peter Nas, ed. *Urban Symbolism*. Leiden: Brill, 1993.

Tesch, J.W. "Een studiewijk voor hygiëne te Batavia," *Koloniale Studiën*, no. 5 (1939).

Thornburh, H.A. "Air Conditioning in Hospitals," *De Ingenieur in Nederlandsch In-dië*, vol. 5, no. 11 (November 1938).

Tichelman, G.J., and W.H. Alting van Geusau. *N.S.B.-Deportatie naar Oost en West.* Amsterdam: DAVID, 1945.

Till, M. van. "In Search of Si Pitung: The History of an Indonesian Legend," *Bij-dragen tot de Taal-, Land- en Volkenkunde*, vol. 152–III (1996).

Tillema, H.F. "Een en ander over de beteekenis van weg en straat voor volks-gezondheid in Indië," *Gemeentereiniging: Officieel orgaan van de Nederlandsche Ver-eeniging van Reinigingdirecteuren. Maandblad voor reiniging en ontsmetting*, vol. 13, no. 1 (January 1920).

———. *"Kromoblanda": over 't vraagstuk van "het wonen" in Kromo's groote land.* Vols. 1–6. 's Gravenhage: Uden Masman etc., 1915–1923.

———. "The Etiquette of Dress among the Apo-Kayan Dayaks," *The Netherland Mail*, vol. 2. N.p., n.d.

———. *Apo-Kajan, een filmreis naar en door Centraal Borneo.* Amsterdam: van Muns-ter, 1938.

———. "Filmen en fotografeeren in de tropische rimboe. Met foto's van den schrij-ver," *Nederlandsch-Indië Oud en Nieuw* (1930–1931).

"Timoer-Barat." *Indonêsia Raja*, vol. 1, no. 1 (January 1929).

Tjokronolo, O. "Membangoenkan lapisan Indonesia dalam masjarakat Indonesia," *Keboedajaän dan Masjarakat*, vol. 1, no. 5 (September 1939).

Tondokoesoemo, S. "Perasaan-hina-diri (Minderwaardigheidsgevoel)," *Keboedajaän dan Masjarakat*, vol. 1, no. 11 (March 1940).

Totok. "Si Kromo itoelah toekang Automobiel jang bagoes sendiri," *Doenia Bergerak*, vol. 1, no. 10 (1915).

Treub, M.W.F. *Nederland in de Oost: reisindrukken.* Haarlem: Willink, 1923.

"Tropenhitte en Hollandsche koud." *KNILM Nieuws*, vol. 3, no. 1 (January 1937).

Tsuchiya, K. *Democracy and Leadership: The Rise of the Taman Siswa Movement in Indonesia.* Honolulu: University of Hawaii Press, 1987.

———. "Kartini's Image of the Javanese Landscape," *East Asian Cultural Studies*, vol. 25, nos. 1–4 (1986).

Vanvugt, E. *H.F. Tillema (1870–1952) en de fotografie van tempo doeloe.* Amsterdam: Mets, 1993.

"Verkoeling van spoorwegrijtuigen in Indië." *De Indische Gids*, vol. 4, no. 1 (1882).

"VERSCHRIKKELIJK." *De Beweging*, no. 6 (February 8, 1919).

Veteraan. "Een veteraan aan het woord," *Orgaan van de U.S.I.*, vol. 8, no. 6 (October 1941).

Veth, B. *Het leven in Nederlandsch-Indië.* Tweede druk. Amsterdam: Van Kampen 1900?

Vissering, C.M., "Een Pasar Gambir," *Nederlandsch-Indië Oud en Nieuw* (1916–1917).

Vistarini, B. de. "Luchtconditioneering en bouwkunde," *De Ingenieur in Nederlandsch Indië*, vol. 5, no. 11 (November 1938).

Vries, D. de. "De cultureele en sociale taak van de radio in Nederlandsch-Indië," *Koloniale Studiën*, no. 2 (1941).

Vuyk, B. "Way Baroe in de Molukken," *Fakkel*, vol. 1, no. 1, (November 1, 1940), 7–15.

Walraven W. Jr. "Mijn vader, de bodemloze put en de wijnkan," *Tirade*, vol. 11, no. 121 (January 1967).

Walraven, W. *Brieven. Aan familie en vrienden, 1919–1941*. Amsterdam: van Oorschot, 1992.

———. *Eendagsvliegen. Journalistieke getuigenissen uit kranten en tijdschriften*. Amsterdam: G.A. van Oorschot, 1971.

———. "Apologie," *Kritiek en Opbouw*, vol. 2, nos. 21/22 (December 16–January 1, 1940).

———. "De schout als stut in het leven," *Kritiek en Opbouw*, vol. 5, no. 2 (January 20, 1942).

———. "Het Indische Stadsbeeld voorheen en thans," *Kritiek en Opbouw*, vol. 3, no. 24 (February 1, 1941).

———. "Sepada," *De Indische Courant* (November 17, 1933).

Wellem. *Amir Sjarifoeddin: Pergumulai imannya dalam perjuangan kemerdekaan*. Jakarta, 1984.

"Werkkamp en Europeesche Boerenstand in Indië." *Onze Toekomst* (January 15, 1936).

Wertheim, W.F., and H. Wertheim-Gijse Weenink. *Vier Wendingen in Ons Bestaan: Indië verloren-Indonesië geboren*. Breda: De Geus, 1991.

Wertheim, W.F. *Indonesian Society in Transition*. The Hague: van Hoeve, 1964.

Wetering, F.H. van de. "Sateliet-Plaatsen," *Koloniale Studiën*, no. 3 (1940).

———. "Zôneering als sociaalstedebouwkundige maatregel," *Koloniale Studiën*, no. 5 (1939).

Wiggers, F. *Boekoe Peringatan tjeritain dari halnja saorang prampoewan Islam Tjeng Kao bernama Fatima*. 2nd ed. Batavia: Goan Hong, 1916.

Wigman, G.M. "De Nederlandsch-Indische Gouvernements Telegraafdienst," *Koloniale Studiën*, no. 2 (1918).

Winter, J. *Sites of Memory, Sites of Mourning: The Great War in European Cultural History*. Cambridge: Cambridge University Press, 1995.

Witkamp, H.Ph.Th. "Een voorbeeld zonder voorbeeld," *De Indische Gids*, vol. 8, no. 2 (1886).

Witte, R. "Exploitatie en bevoogding: De Europese en inheemse radio-omroep in Nederlands-Indië tot 1942," in Berg, Witte, et al. *Jaarboek Media Geschiedenis*.

Wittgenstein, L. *Last Writings on the Philosophy of Psychology*. Vol. I. *Preliminary Studies for Part 2 of Philosophical Investigations*. Chicago: University of Chicago Press, 1990.

———. *On Certainty*. New York: Harper Torchbooks 1972.

———. *Philosophical Investigations*. New York: Macmillan, 1953.

Wolff, J.U., ed. *The Role of the Modern Indonesian Chinese in Shaping Modern Indonesian Life*. Ithaca: Cornell Southeast Asia Program, 1991.

Wynaendts van Resandt, W. "De Plaats van de Auto onder de Verkeersmiddelen, Speciaal in Nederlandsch-Indië," *Koloniale Studiën*, no. 1 (1926).

INDEX

High Religion: A Cultural and Political History of Sherpa Buddhism
by Sherry B. Ortner

A Place in History: Social and Monumental Time in a Cretan Town
by Michael Herzfeld

The Textual Condition *by Jerome J. McGann*

Regulating the Social: The Welfare State and Local Politics in Imperial Germany
by George Steinmetz

Hanging without a Rope: Narrative Experience in Colonial and Postcolonial
Karoland *by Mary Margaret Steedly*

Modern Greek Lessons: A Primer in Historical Constructivism *by James Faubion*

The Nation and Its Fragments: Colonial and Postcolonial Histories
by Partha Chatterjee

Culture/Power/History: A Reader in Contemporary Social Theory
edited by Nicholas B. Dirks, Geoff Eley, and Sherry B. Ortner

After Colonialism: Imperial Histories and Postcolonial Displacements
edited by Gyan Prakash

Encountering Development: The Making and Unmaking of the Third World
by Arturo Escobar

Social Bodies: Science, Reproduction, and Italian Modernity *by David G. Horn*

Revisioning History: Film and the Construction of a New Past
edited by Robert A. Rosenstone

The History of Everyday Life: Reconstructing Historical Experiences and
Ways of Life *edited by Alf Lüdtke*

The Savage Freud and Other Essays on Possible and Retrievable Selves
by Ashis Nandy

Children and the Politics of Culture *edited by Sharon Stephens*

Intimacy and Exclusion: Religious Politics in Pre-Revolutionary Baden
by Dagmar Herzog

What Was Socialism, and What Comes Next? *by Katherine Verdery*

Citizen and Subject: Contemporary Africa and the Legacy of Late Colonialism
by Mahmood Mamdani

Colonialism and Its Forms of Knowledge: The British in India *by Bernard S. Cohn*

Charred Lullabies: Chapters in an Anthropography of Violence
by E. Valentine Daniel